Biological Roles of Sialic Acid

Edited by

Abraham Rosenberg

and

Cara-Lynne Schengrund

Pennsylvania State University, Hershey

PLENUM PRESS • NEW YORK AND LONDON

Library of Congress Cataloging in Publication Data

Main entry under title:

Biological roles of sialic acid.

Includes bibliographies and index.
1. Sialic acid. I. Rosenberg, Abraham, 1924-   II. Schengrund, Cara-Lynne.
QP801.S47B56   574.1'9245   76-16502
ISBN 0-306-30903-3

© 1976 Plenum Press, New York
A Division of Plenum Publishing Corporation
227 West 17th Street, New York, N.Y. 10011

All rights reserved

No part of this book may be reproduced, stored in a retrieval system, or transmitted,
in any form or by any means, electronic, mechanical, photocopying, microfilming,
recording, or otherwise, without written permission from the Publisher

Printed in the United States of America

Contributors

Roscoe O. Brady, Developmental and Metabolic Neurology Branch, National Institute of Neurological Diseases and Stroke, National Institutes of Health, Bethesda, Maryland

John F. Codington, Laboratory for Carbohydrate Research, Departments of Biological Chemistry and Medicine, Harvard Medical School, and Massachusetts General Hospital, Boston, Massachusetts

Joel A. Dain, Department of Biochemistry, University of Rhode Island, Kingston, Rhode Island

Peter H. Fishman, Developmental and Metabolic Neurology Branch, National Institute of Neurological Diseases and Stroke, National Institutes of Health, Bethesda, Maryland

Roger W. Jeanloz, Laboratory for Carbohydrate Research, Departments of Biological Chemistry and Medicine, Harvard Medical School, and Massachusetts General Hospital, Boston, Massachusetts

R. W. Ledeen, Departments of Biochemistry and Neurology, Albert Einstein College of Medicine of Yeshiva University, Bronx, New York

Edward John McGuire, National Jewish Hospital and Research Center, Denver, Colorado

Sai-Sun Ng, Department of Biochemistry, University of Rhode Island, Kingston, Rhode Island

Abraham Rosenberg, Department of Biological Chemistry, The Milton S. Hershey Medical Center, Pennsylvania State University, Hershey, Pennsylvania

Cara-Lynne Schengrund, Department of Biological Chemistry, The Milton S. Hershey Medical Center, Pennsylvania State University, Hershey, Pennsylvania

Kunihiko Suzuki, The Saul R. Korey Department of Neurology, Department of Neuroscience, and the Rose F. Kennedy Center for Research in Mental Retardation and Human Development, Albert Einstein College of Medicine, Bronx, New York

John F. Tallman, Developmental and Metabolic Neurology Branch, National Institute of Neurological Diseases and Stroke, National Institutes of Health, Bethesda, Maryland

Leonard Warren, Wistar Institute of Anatomy and Biology, Philadelphia, Pennsylvania

R. K. Yu, Department of Neurology, Yale University School of Medicine, New Haven, Connecticut

Preface

There is a startling amount of research activity concerning the role of sialic acid in mammalian cells and in the mammalian organism. One may discern in the early literature premonitions of compounds containing sialic acid, traceable by descriptions of color reactions, as far back as the turn of the century. Work spanning the 1930s to the 1950s culminated in the crystallization of sialic acid from a wide variety of biological materials. The ubiquitous nature of the sialic acids, and the biological importance of the substances in which they occur, then became generally manifest. Since then, the chemistry and metabolism of sialic acid and its occurrence, notably, but not exclusively, in the outer cell surfaces of mammalian cells and in key extracellular glycoproteins, have received great attention. The involvement of sialic acid-containing substances in tumorigenicity and in numerous metabolic and infectious pathological conditions, and in the growth, development, and integrity of mammalian cells has achieved widespread recognition. Intensive inquiry into the biological roles of sialic acid continues in a large number of research laboratories throughout the world. This book is intended to represent for the uninitiated as well as for the expert a wide and detailed overview of the current state of knowledge. Major efforts and pioneering breakthroughs have emerged from several laboratories, located on both sides of the Atlantic, of which we make no special individual mention here since they will to some extent appear in the pages that follow. In bringing together the contents of this book on the biological roles of sialic acid, with the able assistance of Mary Horst and Marlene Bowser, we dedicate our own efforts to Ruth Ann and Jonathan, to Kevin and Karin-Ann, to Estelle and to David, and we wish to pay homage to the memory of Ernest Klenk and to Gunnar Blix whose former students in Sweden and in Germany are among the outstanding leaders in the field.

A. Rosenberg
C.-L. Schengrund

Contents

Abbreviations xvii

Chapter 1
Chemistry and Analysis of Sialic Acid
R. W. Ledeen and R. K. Yu

    I. Historical Background 1
   II. Natural Occurrence of Sialic Acids 5
  III. Isolation and Purification 8
  IV. Chemistry of Sialic Acids 10
       A. Basic Structures 10
       B. Stereochemistry 16
       C. Chemical Reactions and Derivatives 22
   V. Synthesis 36
  VI. Quantification of Sialic Acids 39
       A. Colorimetric and Fluorometric Assays 39
       B. Enzymatic Assay 45
       C. Gas–Liquid Chromatography 45
 VII. References 48

Chapter 2
The Natural Occurrence of Sialic Acids
Sai-Sun Ng and Joel A. Dain

    I. Introduction 59
   II. The Natural Occurrence of Sialic Acids 59
       A. Viruses 59
       B. Bacteria 60
       C. Plants 61
       D. Invertebrates 62

	E. Primitive Chordates	64
	F. Vertebrates	65
III.	Evolution of Sialic Acids	84
IV.	References	86

Chapter 3
The Distribution of Sialic Acids Within the Eukaryotic Cell
Leonard Warren

I.	Introduction	103
II.	Extracellular Sialic Acids	104
III.	Distribution within the Cell	105
	A. The Plasma Membrane	105
	B. Endoplasmic Reticulum	109
	C. Mitochondria	114
	D. Nuclei	114
	E. Other Fractions	115
IV.	Conclusions	116
V.	References	117

Chapter 4
Anabolic Reactions Involving Sialic Acids
Edward John McGuire

I.	Introduction: Perspective and Directions	123
II.	Biosynthesis of the Sialic Acids	125
	A. Glucose to Sialic Acid	125
	B. Activation	135
	C. Regulatory Problems	136
	D. Other Derivatizations	137
III.	Biosynthesis of Polymers, Glycoproteins, Mucins, and Glycolipids Containing Sialic Acid	138
	A. Colominic Acid Synthesis	141
	B. CMP-Sialic Acid: Lactose (β-Galactosyl) Sialyltransferase	142
	C. CMP-Sialic Acid: Glycoprotein (β-Galactosyl) Sialyltransferases	143
	D. CMP-Sialic Acid: Mucin (α-N-Acetylgalactosaminyl) Sialyltransferase	146

  E. CMP-Sialic Acid: Ganglioside (Glycolipid) Sialyltransferases 148
IV. Thoughts on Physiological Function of Sialic Acids . . 150
V. References 152

Chapter 5
Catabolism of Sialyl Compounds in Nature
Kunihiko Suzuki

  I. Introduction 159
 II. Pathways of Degradation 160
   A. Degradation of Gangliosides 160
   B. Degradation of Glycoproteins 165
 III. Cellular Mechanism of Degradation 169
   A. Lysosomes 169
   B. Uptake and Disposition of Substrates 169
 IV. Functional Implications 172
 V. Concluding Remarks 174
 VI. References 175

Chapter 6
Disorders of Ganglioside Catabolism
John F. Tallman and Roscoe O. Brady

  I. Introduction—The Catabolism of Gangliosides . . . 183
 II. Tay–Sachs Disease (Type I G_{M2}-Gangliosidosis) . . . 184
   A. Clinical Aspects 184
   B. Pathology 184
   C. Chemistry of the Storage Material 185
   D. Nature of the Metabolic Defect 185
   E. Enzymology of Type I G_{M2}-Gangliosidosis . . . 189
   F. Prenatal Diagnosis and Treatment 190
 III. Type II G_{M2}-Gangliosidosis 190
   A. Clinical and Pathological Aspects 190
   B. Chemistry of the Storage Material 191
   C. Metabolic Defect—Diagnosis and Treatment . . . 191
 IV. Other Variant Forms 192
   A. Type III G_{M2}-Gangliosidosis 192
   B. Hexosaminidase-A-Deficient Adults 192

V.	Generalized Gangliosidosis (G_{M1}-Gangliosidosis)	192
	A. Clinical Aspects	192
	B. Pathology	193
	C. Chemistry of the Stored Material	193
	D. Metabolic Defect	193
VI.	Potentially Related Disorders	195
	A. Hematoside (G_{M3})-Gangliosidosis	195
	B. Animal Model Gangliosidoses	195
	C. *In Vitro* Model Studies	196
VII.	References	197

Chapter 7
The Biological Role of Sialic Acid at the Surface of the Cell
Roger W. Jeanloz and John F. Codington

I.	Introduction	201
II.	Occurrence, Forms, and Amounts of Sialic Acid Residues at the Surface of the Cell	202
III.	The Masking of Cell-Surface Antigens by Sialic Acid	203
IV.	Sialic Acid as a Receptor at Cell Surfaces	207
	A. Receptor for Lectins	207
	B. Receptor for Viruses	208
	C. Receptor for Mycoplasma	210
	D. Receptor for Hormones	211
	E. Receptor for Antibodies	211
	F. Receptor for Circulating Glycoproteins	212
	G. Receptor for Tetanus Toxin	213
V.	Sialic Acid in Normal and Malignant or Transformed Cells	213
VI.	Role of Sialic Acid in Cell-to-Cell Interaction	215
	A. Cellular Adhesion	217
	B. Intercellular Aggregation	217
	C. Agglutination	220
VII.	Physiological Role of Sialic Acid Residues	223
	A. Transport of Ions, Amino Acids, and Proteins	223
	B. Phagocytosis	223
	C. Anaphylactic Shock, Hypercapnia, and Brain Excitability	224
	D. Lymphocyte Stimulation	224
	E. Sperm Capacitation	225

VIII.	Conclusion	225
IX.	References	227

Chapter 8
The Altered Metabolism of Sialic-Acid-Containing Compounds in Tumorigenic-Virus-Transformed Cells
Peter H. Fishman and Roscoe O. Brady

I.	Introduction	239
II.	Experimental Procedures	240
	A. Cells and Cell Culture	240
	B. Isolation, Identification, and Quantification of Gangliosides	241
	C. Assay of Enzymes Involved in Glycolipid Metabolism	242
III.	Ganglioside Metabolism in Cultured Mouse Cell Lines	244
	A. Distribution of Gangliosides in Normal and Virally Transformed Cells	244
	B. Enzymatic Studies	246
	C. Effect of Growth and Culture Conditions on Ganglioside Metabolism	248
	D. Sialic-Acid-Containing Glycolipids in Transformed Cells Obtained from Other Species	250
VI.	Sialic Acid and Glycoproteins in Transformed Cells	253
	A. Sialic Acid and Sialyltransferase Activity in Transformed Cells	253
	B. Membrane Glycoproteins	254
	C. Glycopeptides of Transformed Cells	255
	D. Role of Sialic Acid and Sialyltransferase	255
	E. Comments	256
V.	Relationship between Viral Transformation and Altered Ganglioside Metabolism	257
	A. Productive Infection of Mouse Cells	257
	B. Ganglioside Metabolism in Flat Revertant Cell Lines	257
	C. Specificity of the Altered Ganglioside Metabolism	259
	D. Generality of the Phenomenon	260
	E. Transformation of Mouse Cells by RNA Tumor Viruses and Other Agents	261

VI. Discussion 264
    A. Molecular Basis of Altered Ganglioside
       Metabolism 264
    B. Significance 267
VII. Concluding Remarks 269
VIII. References 270

Chapter 9
Circulating Sialyl Compounds
Abraham Rosenberg and Cara-Lynne Schengrund

  I. Introduction 275
 II. Normal Plasma Constituents 278
    A. Circulating Sialoenzymes 278
    B. Serum Sialoglobulins 279
    C. Sialoglycoprotein Hormones 279
III. Circulating Sialoglycoproteins in Abnormal
    Physiological States 282
    A. Diabetes 283
    B. Inflammatory Reactions 283
    C. Infectious Psychoses 284
    D. The Effect of Steroid Hormones 284
    E. Liver Disease 284
    F. Virus Inhibition of Hemagglutination 285
    G. Cancer 285
    H. Diet 286
IV. Role of Sialic Acid in Circulating Sialoglycocompounds . 286
 V. References 288

Chapter 10
Sialidases
Abraham Rosenberg and Cara-Lynne Schengrund

  I. Background and Nomenclature 295
 II. Bacterial Sialidases 296
    A. Occurrence of Microbial Sialidases 296
    B. Organismic Characterization and Induction of
       Bacterial Sialidases 297
    C. Purification of Bacterial Sialidases 298
    D. Size and Properties of Bacterial Sialidases . . . 300

	E. Mode of Action of Bacterial Sialidases	302
	F. Biological Roles for Bacterial Sialidases	305
III.	Viral Sialidases	313
	A. Morphology and Genetics of Viral Sialidases	313
	B. Purification of Viral Sialidases	316
	C. Size of Viral Sialidases	318
	D. Properties of Viral Sialidases	319
	E. Possible Biological Roles for Viral Sialidases	320
IV.	Experimental Use of Microbial Sialidases	322
V.	Mammalian Sialidases	322
	A. Organ Distribution of Mammalian Sialidases	323
	B. Subcellular Distribution of Mammalian Sialidases	324
	C. Purification of Mammalian Sialidases	326
	D. Assay of Mammalian Sialidases	327
	E. Physical Properties of Mammalian Sialidases	328
	F. Developmental Studies of Mammalian Sialidases	336
	G. Possible Biological Roles of Mammalian Sialidases	337
	H. Sialidase Activity in Cells in Tissue Culture	341
VI.	References	342

Index 361

Abbreviations

ACTH	adrenocorticotropic hormone
ADP	adenosine diphosphate
AMP	adenosine monophosphate
ATP	adenosine triphosphate
BHK	baby hamster kidney
CD	circular dichroism
CDP	cytidine diphosphate
CEF	chicken embryo fibroblast
Cer	ceramide
CM	carboxymethyl
CMP	cytidine monophosphate
CoA	coenzyme A
C. perfringens	*Clostridium perfringens*
CTP	cytidine triphosphate
DEAE	diethylaminoethyl
DF	decapitation factor
DNA	deoxyribonucleic acid
E. coli	*Escherichia coli*
FAD	flavine adenine dinucleotide
FMN	flavine mononucleotide
Fuc	fucose
Gal	galactose
GalNAc	N-acetylgalactosamine
GDP	guanosine diphosphate
Glc or Glu	glucose
GlcNAc	N-acetylglucosamine
Gm	glucosamine
HCG	human chorionic gonadotropin
Ki-MSV	Kirsten isolate of murine sarcoma virus
LSD	lysergic acid diamide
MCB	membranous cytoplasmic bodies

MSV	murine sarcoma virus
NAD	nicotinamide adenine dinucleotide
NANA or NAN	N-acetylneuraminic acid
NGNA or NGN	N-glycolylneuraminic acid
NMN	nicotinamide mononucleotide
NRK	newborn rat kidney
PEP	phosphoenolypyruvate
Py	polyoma virus
RDE	receptor destroying enzyme
RNA	ribonucleic acid
RSV	Rous sarcoma virus
SDS	sodium dodecyl sulfate
SV	simian virus
TDP	thymidine diphosphate
TS	temperature-sensitive
UDP	uridine diphosphate

Chapter 1

Chemistry and Analysis of Sialic Acid

R. W. Ledeen and R. K. Yu

I. HISTORICAL BACKGROUND

The term "sialic acid" first appeared in the literature in 1952 to describe an unusual acidic aminosugar present in gangliosides and submaxillary mucin (Blix et al., 1952). Although the name was new, the substance had been isolated in crystalline form many years before by Blix (1936), and its general properties were already well known by the early 1950s. Other names employed over the years included lactaminic acid (Kuhn and Brossmer, 1956b), hemataminic acid (Yamakawa and Suzuki, 1952), gynaminic acid (Zilliken et al., 1955), and O-sialic acid, but eventually the nomenclature proposal of Blix et al. (1957) was adopted by consensus. This nomenclature designated "neuraminic acid" as the unsubstituted parent structure (Figure 1) and "sialic acid" as the generic term for the family of related derivatives having an acyl group on the amino nitrogen and frequently other substituents elsewhere.

Neuraminic acid itself has not been isolated due to its intrinsic instability. The naturally occurring N-acylated forms (i.e., sialic acids) are stabilized through conversion of the amino group to amide. The most common of the sialic acids, N-acetylneuraminic acid (NANA), has the systematic name* 5-acetamido-3,5-dideoxy-D-*glycero*-D-*galacto*-nonulosonic acid when named as an open-chain compound. N-Glycolylneura-

---

* Rules of Carbohydrate Nomenclature: J. Organic Chemistry (1963) **28**:281.

R. W. LEDEEN • Departments of Biochemistry and Neurology, Albert Einstein College of Medicine of Yeshiva University, 1300 Morris Park Avenue, Bronx, New York 10461. R. K. YU • Department of Neurology, Yale University School of Medicine, 333 Cedar Street, New Haven, Connecticut 06510.

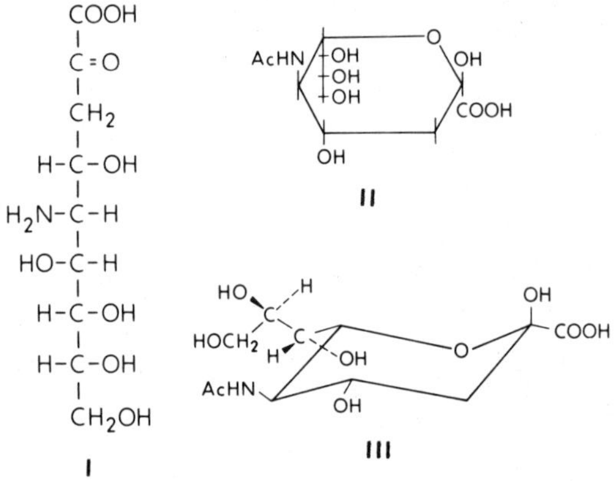

FIGURE 1. Structure of neuraminic acid in the open-chain (keto) form (I) and N-acetylneuraminic acid in the pyranose form as represented by Haworth (II) and Reeves (III). A Fischer projection formula is shown in Figure 3.

minic acid (NGNA) is another widely distributed species, and these two compounds comprise the structural backbone for most of the known sialic acids. They occur in the pyranose ring form (Figure 1) when glycosidically linked to other sugars, and this is also the preferred conformation in the free state. Figure 2 shows a space-filling model of NANA and a lipid containing this molecule, the major monosialoganglioside of brain.

The first sialic acid isolated by Blix (1936) analyzed for two acetyl groups, showed reducing power, and produced a purple color on direct reaction with Ehrlich's reagent. The latter reaction differed from that of most hexosamines, which require prior heating in alkali for color production. Humin formation in hot mineral acids was another unusual feature. The substance was itself a relatively strong acid, and this property was utilized to effect autohydrolytic cleavage from the other sugars. The crystallized compound gave analytical data indicating $C_{14}H_{25}O_{11}N$ as the empirical formula. This departed from the correct composition of N,O-diacetylneuraminic acid by one addition CH_2 unit and thus suggested that Blix's substance might have been a mixture. Blix at that time incorrectly perceived sialic acid as a disaccharide composed of acetylated hexosamine linked to a polyhydroxy acid, but he later corrected this.

Soon afterward, Blix (1938) demonstrated that this carbohydrate from submaxillary mucin generated a purple chromophore with Bial's orcinol reagent that had the same absorption maximum as that reported by Klenk (1935) for a new lipid he isolated from the brain of a patient with Niemann–Pick's disease. In his study of this lipid, which he subsequently called "ganglioside," Klenk succeeded in releasing the color-producing carbohydrate by heating with 5% methanolic hydrochloric acid and crystallized the product. This "neuraminsäure" of Klenk's

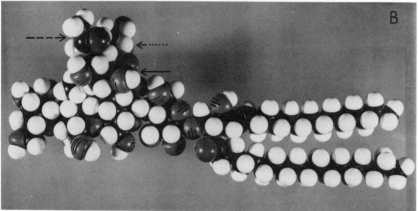

FIGURE 2. Space-filling models of (A) N-acetylneuraminic acid, β-configuration at C_2; (B) monosialoganglioside (G_{M1}) of brain; NANA linked by α-ketoside to internal galactose. Makers of sialic acid unit: → carboxyl; → C_2-hydroxyl [not visible in (B)]; ---→ CH_3CONH-; ·····→ C_7-hydroxyl. In B the plane of the sialic pyranose ring is perpendicular to that of the remaining oligosaccharide chain.

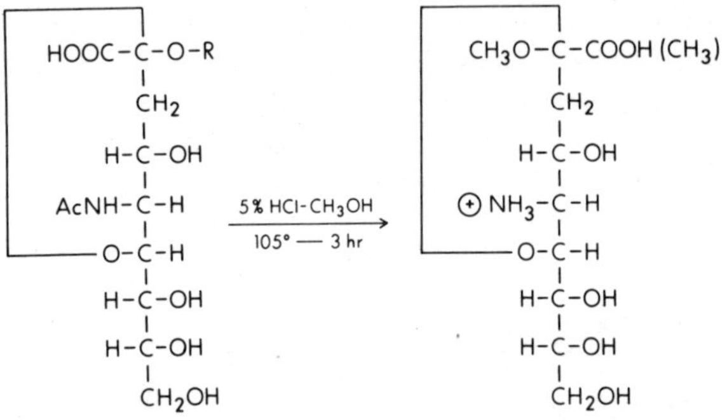

```
        HOOC-C-O-R                CH₃O-C-COOH (CH₃)
           |                             |
          CH₂                           CH₂
           |                             |
         H-C-OH                        H-C-OH
           |         5% HCl-CH₃OH        |
      AcNH-C-H       ─────────────→  ⊕NH₃-C-H
           |           105° — 3 hr       |
        ─O-C-H                        ─O-C-H
           |                             |
         H-C-OH                        H-C-OH
           |                             |
         H-C-OH                        H-C-OH
           |                             |
         CH₂OH                         CH₂OH

           IV                             V
```

FIGURE 3. Formation of methoxyneuraminic acid (V) from ketosidically bound NANA (IV), shown as Fischer projection formulas. Configuration of C_2-ketoside changes from α (IV) to β (V).

was eventually shown to bear a close resemblance to Blix's carbohydrate, despite some obvious structural differences. Klenk's substance, for example, had a free amino group, lacked reducing power, and contained a methoxyl unit. When it was determined that the latter had been introduced during methanolysis, Klenk (1942) renamed his compound "methoxyneuraminic acid." The conditions he employed are now recognized as too drastic for sialic acid isolation due to concomitant deacylation and O-methylketoside formation (Figure 3). The same modified substance was later obtained by Yamakawa and Suzuki (1952) from the "hematoside" preparation which they had extracted from horse erythrocyte stroma. Elemental analysis in their laboratory gave the correct composition $C_{10}H_{19}NO_8$.

The use of mild hydrolytic conditions to release unaltered sialic acids from glycoproteins and glycolipids proved a key factor in subsequent isolation of several more sialic acids. Another important development was the introduction of improved separation techniques such as paper and cellulose column chromatography. NANA was soon obtained in crystalline form from a number of sources including bovine submaxillary mucin (Klenk and Faillard, 1954), cow colostrum (Kuhn et al., 1954), and human milk (Zilliken et al., 1955). Four additional sialic acids isolated in the 1950s by Blix et al. (1955; 1956) from submaxillary mucins provided more evidence for structural diversity among sialic acids. Much of the early history of these discoveries has been recounted by Blix (1959), Zilliken and Whitehouse (1958) and Gottschalk (1960), while more recent reviews by Blix and Jeanloz (1969), Tuppy and Gottschalk

(1972), and Schauer (1973) describe the later developments in sialic acid chemistry.

II. NATURAL OCCURRENCE OF SIALIC ACIDS

Many different species of sialic acid have now been isolated and identified. Following the comprehensive monograph by Gottschalk (1960), which described the five naturally occurring forms known at that time, improved methods of detection and analysis resulted in a surge of new discoveries. Current literature (through 1975) reveals a total of 17 (Table I). Although the majority of these are known structures, the list includes some that are not yet fully characterized. For the most part these sialic acids occur in nature as components of glycoproteins, glycolipids, oligosaccharides, and polysaccharides, and only very small quantities are found in the free state.

Sialic acids were reported to occur in all species of vertebrates and in certain invertebrates as well (Warren, 1963a). NANA is present in some strains of *E. coli* and other bacteria as a homopolymer, of which colominic acid is one example (Barry and Goebel, 1957; Barry *et al.*, 1962; Watson *et al.*, 1958; Liu *et al.*, 1971). However, it is lacking in other bacteria (Barry, 1959; Barry *et al.*, 1963; Jann, 1969). Earlier claims for the presence of sialic acid in plant extracts have been disputed (Cabezas and Feo, 1969) since the positive test with thiobarbituric acid could not be duplicated by other colorimetric assays. Gielen (1968) demonstrated that the color-producing substances in certain plant hydrolysates were 2-keto-3-deoxyaldonic acids rather than sialic acid. To date there has been no convincing evidence for the presence of sialic acid in plants.

Most mammals contain more than one type of sialic acid in their tissues. Man has been described as an exception in having only NANA (Gottschalk, 1960). Detection of trace amounts of NGNA in HeLa cells (Carubelli and Griffin, 1968) was possibly the result of absorption of NGNA-containing substances from the fetal calf serum in which the cells were grown (Hof and Faillard, 1973). Another report (Walkowiak *et al.*, 1968) has claimed the presence of trace amounts of NGNA in human serum. A gas-liquid chromatography study of human plasma gangliosides (Yu and Ledeen, 1970) reported the absence of NGNA, but glycoproteins were not analyzed. A glycopeptide from normal human urine gave primarily NANA on mild acid hydrolysis, but a second very minor species was detected having the same migratory rate on paper as *N*-acetyl-4-*O*-acetylneuraminic acid (Carrion *et al.*, 1969). Another brief

TABLE I. Naturally Occurring Sialic Acids: Nomenclature

Common name	Systematic name	Molecular composition
1 N-Acetylneuraminic acid[a]	5-Acetamido-3,5-dideoxy-D-glycero-D-galacto-nonulosonic acid	$C_{11}H_{19}NO_9$
2 N-Acetyl-4-O-acetylneuraminic acid[b,k]	5-Acetamido-4-O-acetyl-3,5-dideoxy-D-glycero-D-galacto-nonulosonic acid	$C_{13}H_{21}NO_{10}$
3 N-Acetyl-7-O-acetylneuraminic acid[b]	5-Acetamido-7-O-acetyl-3,5-dideoxy-D-glycero-D-galacto-nonulosonic acid	$C_{13}H_{21}NO_{10}$
*4 N-Acetyl-8-O-acetylneuraminic acid[c] or N-Acetyl-9-O-acetylneuraminic acid[k]	5-Acetamido-8(or 9)-acetyl-3,5-dideoxy-D-glycero-D-galacto-nonulosonic acid	$C_{13}H_{21}NO_{10}$
*5 N-Acetyl-7,8-di-O-acetylneuraminic acid[c,d] or N-Acetyl-7,9-di-O-acetylneuraminic acid[k]	5-Acetamido-7,8(or 7,9)-di-O-acetyl-3,5-dideoxy-D-glycero-D-galacto-nonulosonic acid	$C_{15}H_{23}NO_{11}$
6 N-Acetyl-4,9-di-O-acetylneuraminic acid[d,k]	5-Acetamido-4,9-di-O-acetyl-3,5-dideoxy-D-glycero-D-galacto-nonulosonic acid	$C_{15}H_{23}NO_{11}$
*7 N-Acetyl-x,y,z-tri-O-acetylneuraminic acid[d]	5-Acetamido-3,5-dideoxy-x,y,z-tri-O-acetyl-D-glycero-D-galacto-nonulosonic acid	$C_{17}H_{25}NO_{12}$
8 N-Acetyl-4-O-glycolylneuraminic acid[e]	5-Acetamido-3,5-dideoxy-4-O-glycolyl-D-glycero-D-galacto-nonulosonic acid	$C_{13}H_{21}NO_{11}$

*9	N-Acetyl-7-O-glycolylneuraminic acid[e] or N-Acetyl-8-O-glycolylneuraminic acid[e]	$C_{13}H_{21}NO_{11}$
10	N-Acetyl-8-O-methylneuraminic acid[f]	$C_{12}H_{21}NO_9$
11	N-Acetyl-9-O-L-lactylneuraminic acid[l]	$C_{14}H_{23}NO_{11}$
12	N-Glycolylneuraminic acid[b]	$C_{11}H_{19}NO_{10}$
13	N-Glycolyl-4-O-acetylneuraminic acid[e,g,k]	$C_{13}H_{21}NO_{11}$
*14	N-Glycolyl-8-O-acetylneuraminic acid[e] or N-Glycolyl-9-O-acetylneuraminic acid[k]	$C_{13}H_{21}NO_{11}$
15	N-Glycolyl-8-O-methylneuraminic acid[h]	$C_{12}H_{21}NO_{10}$
*16	N-Acetoglycolyl-4-methyl-4,9-dideoxy-neuraminic acid[i]	$C_{14}H_{23}NO_9$
*17	Hf-neuraminic acid[j]	Unknown

*9	5-Acetamido-3,5-dideoxy-7(or 8)-O-glycolyl-D-glycero-D-galacto-nonulosonic acid		
10	5-Acetamido-3,5-dideoxy-8-O-methyl-D-glycero-D-galacto-nonulosonic acid		
11	5-Acetamido-3,5-dideoxy-9-O-L-lactyl-D-glycero-D-galacto-nonulosonic acid		
12	3,5-Dideoxy-5-glycolamido-D-glycero-D-galacto-nonulosonic acid		
13	4-O-Acetyl-3,5-dideoxy-5-glycolamido-D-glycero-D-galacto-nonulosonic acid		
*14	8(or 9)-O-Acetyl-3,5-dideoxy-5-glycolamido-D-glycero-D-galacto-nonulosonic acid		
15	3,5-Dideoxy-5-glycolamido-8-O-methyl-D-glycero-D-galacto-nonulosonic acid		
*16	N-Acetoglycolyl-3,4,5,9-tetradeoxy-4-C-methyl-D-glycero-D-lyxo-nonulosonic acid		

* Structural determinations incomplete.
[a] Klenk and Faillard (1954).
[b] Blix et al. (1955,1956).
[c] Blix and Lindberg (1960).
[d] Schauer and Faillard (1968).
[e] Buscher et al. (1974).
[f] Kochetkov et al. (1973).
[g] Hakomori and Saito (1969).
[h] Warren (1964).
[i] Hotta et al. (1970).
[j] Isemura et al. (1973).
[k] Kamerling et al. (1975a).
[l] Schauer et al. (1976).

report from the same group (Cabezas and Ramos, 1972) claimed the presence of both NANA and N,O-diacetylneuraminic acid in human bile mucins, so the question of human sialic acid types remains an open one.

Many animals contain two or more sialic acids in a given tissue or fluid, the nature and proportions of these depending on the animal and tissue in question. Glycoproteins from hog submaxillary mucin, for example, contain predominantly NGNA, whereas the related proteins from beef, sheep, and horse submaxillary mucin have mainly NANA and its O-acetyl derivatives. The same general considerations apply to gangliosides, although here the variety of sialic acid types appears to be more restricted than for glycoproteins. The mixture of gangliosides from bovine adrenal medulla, for example, contains approximately equal amounts of NANA and NGNA, and in at least one case the two types are present together in the same molecule (Ledeen et al., 1968). It was noted that bovine plasma gangliosides had approximately the same ratio of these two sialic acids (Yu and Ledeen, 1972). Mammalian brain is an exceptional organ in that virtually all ganglioside sialic acid is NANA. A few species such as cow, sheep, and pig were shown to have small quantities of NGNA in brain, amounting to 2% or less of total sialic acid (Tettamanti et al., 1964, Yu and Ledeen, 1970); however it is not clear whether this small component arises from vascular elements or the nervous tissue proper. Brain gangliosides of certain other species (Human, rat, rabbit, chicken, etc.) showed no trace of NGNA. The gangliosides of fish brain, on the other hand, have both NANA and its 8-O-acetyl derivative (Ishizuka et al., 1970). Further information on the natural occurrence of sialic acids is given in Chapter 2.

III. ISOLATION AND PURIFICATION

Isolation is usually the first step in identifying a given sialic acid. Since these compounds almost always occur glycosidically linked to other sugars in complex carbohydrate-containing materials, they must first be cleaved from the parent substance. Frequently this is accomplished by chemical means in the form of acid-catalyzed hydrolysis or methanolysis. Dilute acid and relatively short heating periods are necessary to minimize destruction of the sensitive molecule and prevent cleavage of O-acyl groups. Relative lability of the ketosidic bond was indicated by the observed autohydrolysis of sialic acid from glycoproteins in water, even at 37° (Blix and Lindberg, 1960). N-acyl groups are more stable and require vigorous acid hydrolysis for complete cleavage.

Mild methanolysis gives rise to the methylketoside methyl ester with the N-acyl group still intact (Yu and Ledeen, 1970).

Enzymatic hydrolysis is an alternative method of cleavage which has the important advantage of mildness. A number of neuraminidase preparations from bacterial and viral sources are commercially available for this purpose (cf. Chapter 10). It has also been recognized, however, that while the acid-sensitive O-acetyl groups are preserved with this method, these same groups may in some instances render the unit resistant to cleavage by bacterial neuraminidase. Thus, sialic acids which are O-acetylated on C_7 or C_8 are released at a slower rate than unacetylated forms (Schauer and Faillard, 1968) and sialic acids which are O-acetylacted at C_4 or di-O-acetylated are strongly resistant to the enzyme (Pepper, 1968; Neuberger and Ratcliff, 1972). Such inhibition is removed by prior treatment with mild alkali. It is interesting to note that O-acetyl substituents can also reduce the rate of acid hydrolysis by a considerable factor (Neuberger and Ratcliffe, 1972; 1973). Another limitation of bacterial neuraminidase is the intrinsic resistance of certain ganglioside species, such as Tay–Sachs and the major monosialoganglioside of normal brain (reviewed by Ledeen and Yu, 1973a).

Ion exchange resins have been used to purify such liberated sialic acids from biological samples. Absorption onto the anion-binding resin is followed by elution with dilute formic acid (Martensson et al., 1958) or acetate buffer (Svennerholm, 1963b). This operation may be preceeded by passage through a cation exchange resin if ionic contaminants are abundant. The O-acetyl groups of di- and triacetylneuraminic acids have been reported to be somewhat labile when absorbed to ion exchange resins (Blix and Jeanloz, 1969). Dialysis or gel filtration is often used to separate the liberated low molecular weight sialic acids from the remaining macromolecule or micelle-forming lipid. A charcoal absorption column was used on at least one occasion to purify sialic acid (Saito, 1956). Detailed procedures have been given for preparative isolation of NANA from human milk and meconium (Zilliken and O'Brien, 1960), cow colostrum (Clark et al., 1962), human serum (Svennerholm, 1963a), and gangliosides (Blix and Odin, 1955; Svennerholm, 1956). Methods have also been described for isolating NGNA from hog submaxillary mucin and O,N-diacetylneuraminic acid and O,N-triacetylneuraminic acids from bovine submaxillary mucin (Svennerholm, 1963a).

Hydrolysates containing more than one species of sialic acid require additional resolution procedures. Crystallization has not proved successful in this regard due to the tendency of sialic acid species to form solid solutions (Gottschalk, 1960). Paper chromatography has been one of the more useful tools (Svennerholm, 1963a; Schauer and Faillard, 1968). Blix and Lindberg (1960), for example, were able to isolate five different sialic acids from bovine and equine submaxillary mucins and to detect a sixth. A pressurized paper column system has also been described (Blix, 1962; Svennerholm, 1963a). Thin-layer chromatography (TLC), which has found use in analytical identification of sialic acids (Granzer, 1962; Ledeen et al., 1968; Schauer and Faillard, 1968) has obvious potential for preparative separation and was used recently to isolate a new species (Hotta et al., 1970). Two-dimensional TLC has been used to good advantage to separate and characterize complex mixtures of sialic acids (Buscher et al., 1974). Crystallization, if desired as a final stage of purification, is generally carried out with mixed solvents such as methanol–ether, water–ethanol–ether, water–acetic acid, etc. The advantage of including water to reduce esterification has been cited (Svennerholm, 1963a).

IV. CHEMISTRY OF SIALIC ACIDS

A. Basic Structures

Following the first isolation of sialic acid in 1936, many laboratories contributed over 30-odd years to revealing the full three-dimensional structure. Discrepancies in the early analysis were finally resolved by the analytical data of Blix and coworkers which established the elementary composition of four sialic acids obtained from the submaxillary glands of four different animals (Blix et al., 1955; 1956). An important observation in determining the position of functional groups was identification of pyrrole-2-carboxylic acid (VII) as a product of alkali treatment of NANA (Figure 4) (Klenk and Faillard, 1954; Gottschalk, 1955). The fact that this product formed with great ease under mild alkaline conditions indicated the keto and amino groups to be situated at α- and δ-carbons, respectively, relative to carboxyl. The presence of a methylene (deoxy) group next to the reducing unit accounted for the high vulnerability to acid while facile decarboxylation with heat and strong acid suggested an α-keto acid. These facts led Gottschalk (1955) to propose a basic structure for sialic acid which viewed it (correctly) as an aldol condensation product of N-acetylhexosamine and pyruvic acid.

Evidence for this basic structure came from the observation (Kuhn and Brossmer, 1956a) that heating NANA in pyridine with nickelous acetate resulted in cleavage to N-acetyl-D-glucosamine, carbon dioxide and a 2-carbon fragment. A similar reaction involving cleavage to N-

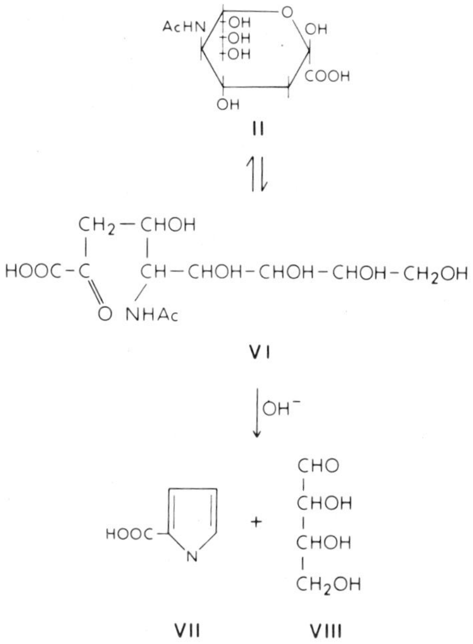

FIGURE 4. Base-cleavage of NANA to pyrrole-2-carboxylic acid (VII) and tetraose (VIII).

acetyl-D-glucosamine and pyruvic acid was effected by alkali treatment (Zilliken and Glick, 1956) (Figure 5). Reverse aldolization was also accomplished enzymatically, the first such report (Heimer and Meyer, 1956) describing N-acetylglucosamine and pyruvate as the products from reaction of NANA with an enzyme from *Vibrio cholerae*. Comb and Roseman (1958), using a purified aldolase from *Clostridium perfringens*, identified N-acetylmannosamine and pyruvate as the true cleavage products (Figure 5) and thereby established a portion of the stereochemistry (see below). The fact that N-acetylmannosamine and N-acetylglucosamine are intercoverted by base (Comb and Roseman, 1958; Kuhn and Brossmer, 1958a) would account for isolation of the latter sugar after reverse aldolization.

Structure determination of NGNA unfolded along similar lines (Blix *et al.*, 1956). Glycolic acid was cleaved with base and identified as the calcium salt by X-ray diffraction. Quantitative acetylation revealed six free hydroxyl groups, one more than for NANA. The absence of C-methyl groups, determined by the Kuhn–Roth method, implied that the extra hydroxyl is located on the N-acetyl group. Blix found a close similarity in the chemical reactions of NGNA as compared to NANA, and included the observation that heating with methanolic-HCl gave

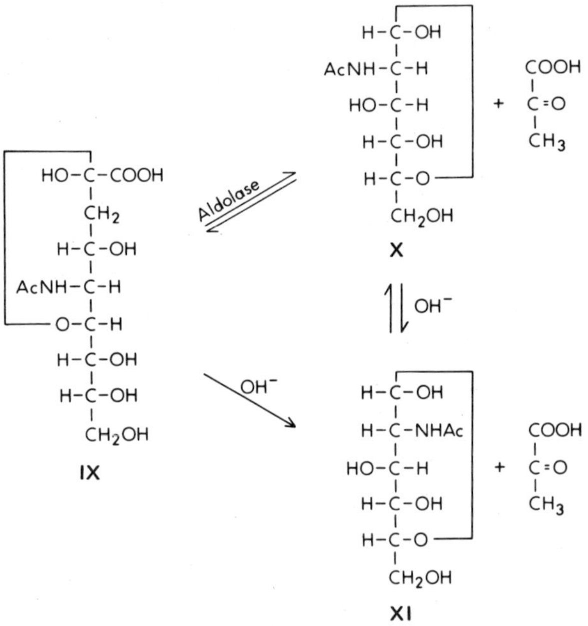

FIGURE 5. Reverse aldolization cleavage of NANA to N-acetylmannosamine (X) and pyruvate with aldolase enzyme, and to N-acetylmannosamine, N-acetylglucosamine (XI) and pyruvate with base.

the same methoxyneuraminic acid that Klenk (1942) had earlier prepared from NANA (Figure 3).

Most of the species listed in Table I are O-acylated derivatives of NANA and NGNA, which are readily converted to one or the other of these basic structures by mild base treatment. Location of the ester-linked acetates was accomplished originally by periodate oxidation and more recently by mass spectrometry. N-acetyl-4-O-acetylneuraminic acid (*2*),* the principal species in equine submaxillary mucin, was shown by Blix and Lindberg (1960) to react with 2 moles of periodate, liberating 1 mole each of formaldehyde and formic acid. The N,O-diacetylneuraminic acid isolated from bovine submaxillary mucin (Blix *et al.*, 1956) was concluded to be the 7-O-acetyl derivative (*3*) from the fact that only 1 mole of periodate was consumed with formation of 1 mole of formaldehyde. The N-acetyl-O-diacetylneuraminic acid from the same source was originally thought to have O-acetyl groups at positions 4 and

---

* Italic numbers in parentheses refer to entries in Table I.

7, but when the carefully isolated compound failed to consume periodate the acetates were placed at C_7,C_8 or C_7,C_9 (Blix and Lindberg, 1960). Periodate studies (Schauer and Faillard, 1968; Schauer, 1970) indicated substitution at C_7 and C_8 for this species *(5)*, but a more recent investigation with mass spectrometry of the trimethylsilyl ether derivative (Kamerling *et al.*, 1975a) placed the *O*-acetyl groups at C_7 and C_9. The latter study indicated that other species originally suggested to have *O*-acetyl groups at C_8 on the basis of slow periodate oxidation had these substituents on C_9 instead. This applied to the substances formerly identified as *N*-acetyl-8-*O*-acetylneuraminic acid *(4)* and *N*-glycolyl-8-*O*-acetylneuraminic acid *(14)* (Schauer and Faillard, 1968; Buscher *et al.*, 1974). A sialic acid from equine submaxillary mucin was shown by this means to be *N*-acetyl-4,9-di-*O*-acetylneuraminic acid *(6)*.

The sialic acids whose structures were recently questioned by the above mass spectral data are indicated in Table I according to both the original and new assignments. In support of the revised structures, synthetic 9-*O*-acetyl derivatives were shown to react slowly with periodate comparable to the naturally occurring sialic acids (Haverkamp *et al.*, 1975). Thus far no reports have appeared concerning the location of acetyl groups on the *N*-acetyl-tri-*O*-acetylneuraminic acid *(7)* obtaind from bovine submaxillary mucin. Very recently a new sialic acid was discovered (Schauer *et al.*, 1976) in bovine submandibular gland: *N*-aceytl-9-*O*-L-lactylneuraminic acid *(11)*.

Four naturally occurring derivatives of NGNA have been described to date. The first to be reported was *N*-glycolyl-8-*O*-methylneuraminic acid *(15)*, isolated by Warren (1964) from the starfish *Asterias forbesi*. The *O*-methyl group was detected by the Zeisel method and its location established by failure of the molecule to consume periodate. An *O*-acetyl derivative of NGNA was discovered by Hakomori and Saito (1969) in the hematoside they isolated from equine erythrocytes. The fact that periodate oxidation gave 1 mole each of formaldehyde and formic acid ruled out substitution at C_7,C_8, or C_9. The glycolipid was fully acetalyzed by reaction with methylvinyl ether, then treated with base and methylated. Vigorous methanolysis of the resulting product gave only neuraminic acid methylketoside methyl ester, indicating the *O*-acetyl to have been on the hydroxyl of the *N*-glycolyl group. However, these investigators did not consider this assignment unequivocal due to the possibility of acetyl migration between the glycolyl and C_4-hydroxyl positions. A more recent study of this compound (Buscher *et al.*, 1974) supported *O*-acylation at the C_4-hydroxyl, based on a retarded rate of cleavage with acylneuraminate pyruvate-lyase. Substituents at that position slow the enzymatic reaction to one-fifth or less that of sialic acids with a free C_4-hydroxyl. Neuraminidase from *Vibrio cholerae* is also inhibited by such substitution, but the authors (Buscher *et al.*, 1974) considered the former enzyme a more reliable indicator; neuraminidase is also blocked by bulky *N*-acyl substituents (such as *N*-acetoglycolyl). The mass spectrometry study of Kamerling *et al.* (1975a) also supported the 4-*O*-acetyl structure (*13*).

An unusual sialic acid that appears to be an NGNA derivative was isolated by Hotta *et al.* (1970) from the jelly coat of sea urchin eggs. Elemental analysis indicated an empirical formula of $C_{14},H_{23}NO_9$, while Kuhn–Roth analysis revealed three C-methyl groups. The presence of one of these at the end of the three-carbon side chain (in place of the C_9-hydroxymethyl group) was indicated by the formation of acetyldehyde rather than formaldehyde on periodate oxidation. An *O*-acetyl group was detected analytically and postulated to be present on the *N*-glycolyl hydroxyl. The full structure proposed by these workers was *N*-acetoglycolyl-4-methyl-4,9-dideoxy neuraminic acid (*16*), but this must be regarded as tentative.

A sialic acid of ambiguous identity was recently detected in a high-molecular-weight glycopeptide fraction from the Cuvierian tubules of *Holothuria forskali* (Isemura *et al.*, 1973). This substance (termed "Hf-neuraminic acid") was readily cleaved from the oligosaccharide chain by mild acid hydrolysis but could not be liberated with neuraminidase, even after mild base treatment. It migrated differently from other known sialic acids on TLC and paper chromatography but appeared to form methoxyneuraminic acid (methylester) on methanolysis. The nature of the *N*-acyl group was not determined.

An unsaturated derivative of NANA, 2-deoxy-2,3-dehydro-*N*-acetylneuraminic acid, was recently isolated from the urine of a patient with sialuria (Kamerling *et al.*, 1975b). Evidence was presented that it was not an artificial degradation product of NANA. Absence of a 2-hydroxy ketoside group would prevent the molecule from forming a covalent linkage with the oligosaccharide chain of a ganglioside or glycoprotein, and hence its status as a true sialic acid is in doubt. Its capacity to inhibit neuraminidase is discussed below.

The recent report of Buscher *et al.* (1974), in which three new sialic acids were described, presents a combination of methods that together comprise a powerful tool for separating and characterizing the components of complex mixtures. TLC in two dimensions with intermediate ammonia treatment to remove *O*-acyl groups led to identification of the basic *N*-acylneuraminic acids, while substitution of hydroxylamine for ammonia permitted identification of the *O*-acyl groups as hydroxamates. By this means two sialic acids containing an *O*-glycolyl group were characterized for the first time. The positions of O-acyl substituents were determined by periodate oxidation and the enzymatic methods referred to which measure retardation of neuraminidase and acylneuraminate pyruvate-lyase. Gas–liquid chromatography was also used to good advantage: seven sialic acids from bovine submandibular glands were resolved as trimethylsilyl derivatives on OV-17.

As indicated previously, mass spectrometry has considerable utility as a tool for microanaylsis. This was demonstrated in the structure determination of 8-O-methyl-N-acetylneuraminic acid (10) (Kochetkov et al., 1973). This sialic acid was isolated from *Distolasterias nipon* and is thus the second O-methylated species to be found in starfish. The peracetylated derivative of the methylketoside methyl ester gave an intense band at m/e 117 which was absent from the mass spectra of acetylated NANA and NGNA. This corresponded to the fragment $[CH_2OAcCHOCH_3]^+$ arising from cleavage between C_7 and C_8, and established location of the O-methyl group on C_8. In the same study Kochetkov and coworkers also found characteristic peaks from NANA and NGNA derivatized in the same manner. Dawson and Sweeley (1971) had previously differentiated these two sialic acids as trimethylsilyl ether derivatives by detection of fragment ions at m/e 186 and 274:

$$\begin{bmatrix} CH-C=CH_2 \\ | \quad | \\ NH \quad OTMSi \\ | \\ COCH_3 \end{bmatrix}^+ \qquad \begin{bmatrix} CH-C=CH_2 \\ | \quad | \\ NH \quad OTMSi \\ | \\ COCH_2OTMSi \end{bmatrix}^+$$

m/e 186 $\qquad\qquad\qquad$ m/e 274

Further study of trimethylsilyl derivatives by Kamerling et al. (1974) demonstrated characteristic fragmentation patterns that could be used to identify certain of the basic derivatives.

Mass spectrometry has also been useful in revealing sialic acid as a bound component of glycolipids (gangliosides). Permethylated gangliosides have proved more useful for this purpose than the trimethylsilyl derivatives (Karlsson, 1973; Ledeen et al., 1974; Karlsson, 1974; Karlsson et al., 1974; Price et al., 1975). Abundant primary ions are produced

m/e 376 $\qquad\qquad\qquad$ m/e 406

from terminal sialic acid at m/e 376 and 406 for NANA and NGNA, respectively. Intense peaks are also seen at m/e 344 and 374 resulting from loss of methanol from NANA and NGNA, respectively. The best results have been obtained with monosialogangliosides, but the NANA ions at m/e 376 and 344 were recently observed by Karlsson (1974) for permethylated disialoganglioside of brain. An alternative procedure for studying lipid bound sialic acid was described (Karlsson, 1973; 1974) in which the amide and carboxyl groups of permethylated ganglioside are reduced with lithium aluminum hydride, and the resulting primary hydroxyl is converted to the trimethylsilyl ether. The sialic acid parent ion, now appearing at m/e 406, seemed to be stabilized by this reaction sequence.

B. Stereochemistry

Sialic acid in its cyclic form has six centers of asymmetry and two possible chair conformations for the pyranose ring. Determining the full stereochemistry proved an elusive undertaking due to two main problems: (a) isomerization of hexosamine by the base used in its cleavage from sialic acid, (b) breakdown of Hudson's empirical rotation rules. It was appropriately stated in a recent review on carbohydrate stereochemistry (Bentley, 1972) that "many textbook authors have come to grief over the stereochemical problems of this molecule." The work of Comb and Roseman (1958; 1960) dealing with the enzymatic cleavage to pyruvate and hexosamine established the configurations of carbons 5–8 of NANA as identical to carbons 2–5, respectively, of *N*-acetyl-D-mannosamine. The observation that the latter sugar could be used in the reverse synthetic reaction, as opposed to *N*-acetyl-D-glucosamine and *N*-acetyl-D-galactosamine which could not, provided additional strong evidence (Comb and Roseman, 1958; Warren and Felsenfeld, 1962). As mentioned previously, the *N*-acetyl-D-glucosamine structure was incorrectly assigned in the beginning due to base-catalyzed interconversion between this sugar and *N*-acetyl-D-mannosamine.

The steric configuration of C_4 was first studied by comparing optical rotations of the parent compound with lactone derivatives. Two lactones were prepared from NANA by Kuhn and Brossmer (1957; 1959), the diethyl dithioacetal γ-lactone (XII) and the desulfurized γ-lactone (XIII) (Figure 6). The former was found to be significantly more levorotatory than NANA, while the latter also showed a negative optical rotation. Application of Hudson's lactone rule thus led to an assignment of the L-*glycero* configuration to C_4 of NANA but this was revised to the D-

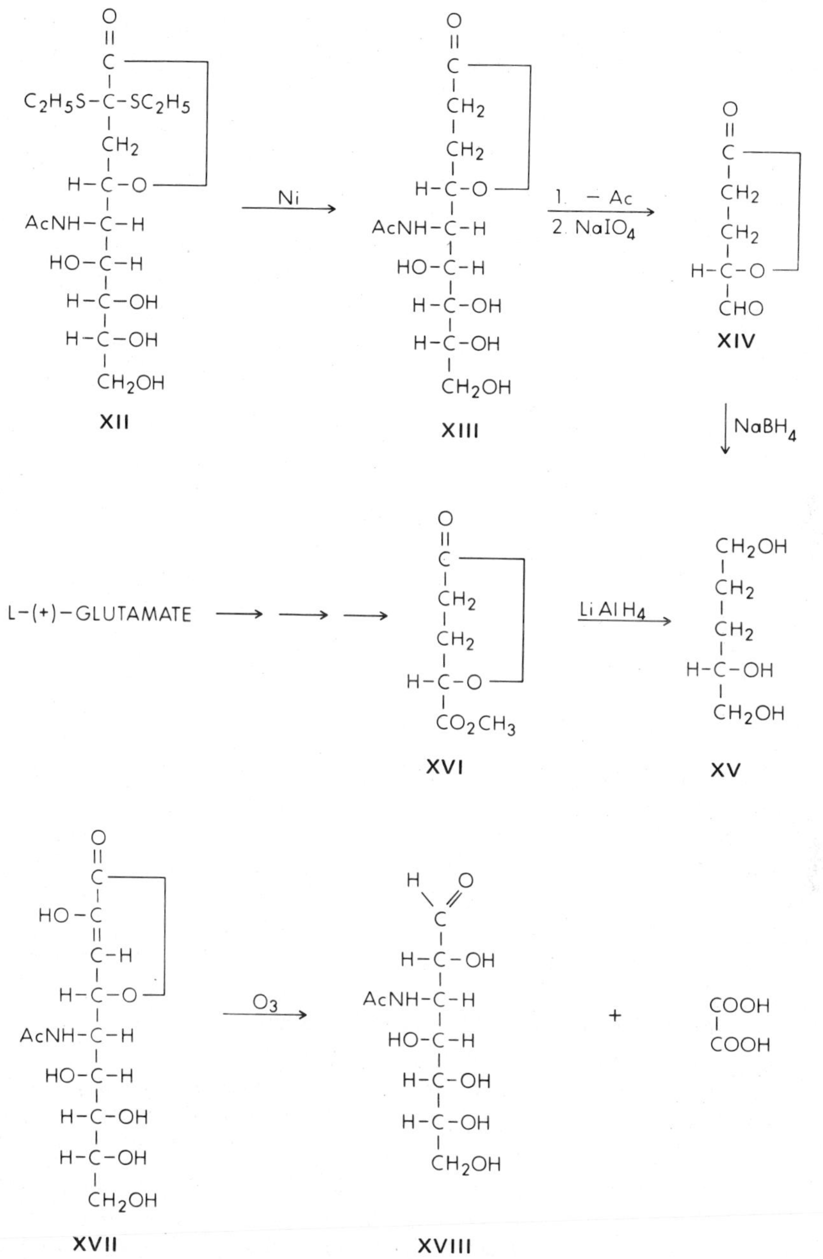

Chemistry and Analysis of Sialic Acid

FIGURE 6. Reactions used to establish D-*glycero* configuration at C_4 of sialic acid (Kuhn and Brossmer, 1962; Kuhn and Baschang, 1962).

configuration when Kuhn and Brossmer (1962) converted lactone XIII to 1,4,5-*R*(—)-pentanetriol (XV). The latter enantiomer was synthesized from L-glutamic acid (Figure 6) and also from D-glyceraldehyde. Further evidence for the D-configuration came from conversion of NANA-γ-lactone XVII (enol form) to 3-acetamido-3-deoxy-D-*glycero*-D-*galacto*-heptose (XVIII) by oxidation with ozone (Figure 6) (Kuhn and Baschang, 1962*a*).

These studies revealed the stereochemistry of all asymmetric centers except the cyclic hemiketal generated by ring formation between the C_6-hydroxyl and C_2-carbonyl. Existence of the pyranose ring in solution was indicated by the rapid consumption of 2 moles of periodate by NANA, further reaction being much slower (Blix *et al.*, 1956; Schauer and Faillard, 1968). Infrared spectroscopy revealed that the pyranose form also predominates in the crystalline state (Fischmeister, 1958). Early attempts to establish the C_2 configuration of free NANA were based on classical optical rotation methods. Kuhn and Brossmer (1957) were able to detect appreciable mutarotation on dissolving crystalline NANA in dimethyl sulfoxide, the specific rotation changing from $-115°$ after 7 min to $-24°$ at equilibrium. Application of Hudson's isorotation rules led these workers to assign the β-configuration to crystalline NANA. Reservations were subsequently expressed by Jeanloz (1963) concerning the suitability of Hudson's rules to compounds to this type. The fact that the pyranose ring of NANA has the general configuration of an L rather than D sugar (although it is formally named as a D sugar) and the possible influence of the carboxyl group on rotation were sources of doubt. It was recently pointed out (Blix and Jeanloz, 1969) that according to the concepts developed by Bose and Chatterjee (1958) the carboxyl groups could exert an influence that might conceivably reverse Hudson's rules.

Despite these reservations, the assignment turned out to be correct on the basis of nuclear magnetic resonance studies in which long-range coupling constants were used to establish an axial orientation for the C_2-hydroxyl (Jochims *et al.*, 1967). NMR also demonstrated that the methyl ester of NANA exists in the 1C conformation (Lutz *et al.*, 1968). Recent X-ray diffraction studies of NANA dihydrate (Flippen, 1973) and NANA methyl ester monohydrate (O'Connell, 1973) have verified the β-configuration in the solid state with the C_2-hydroxyl axial and all other substituents equatorial. A more limited application of X-ray diffraction was the use of powder diagrams to identify individual sialic acids in an empirical manner (Gottschalk, 1960; Abrahamson *et al.*, 1962). This technique was used to establish identity of natural and synthetic NANA (Cornforth *et al.*, 1958) and also to compare NANA

Chemistry and Analysis of Sialic Acid

from human brain ganglioside and ovine submaxillary mucin (Blix and Odin, 1955).

The remaining problem in stereochemistry was to establish the configuration of the C_2 group in ketosidic linkages, such as occur naturally in biological substances, and relate this configuration to neuraminidase specificity. The early attempts to solve this problem utilized the change in optical rotation accompanying neuraminidase cleavage of 3-*O*-*N*-acetylneuraminosyl lactose (Kuhn and Brossmer, 1958*b*; Gottschalk, 1958). Since the change was negative, the ketoside

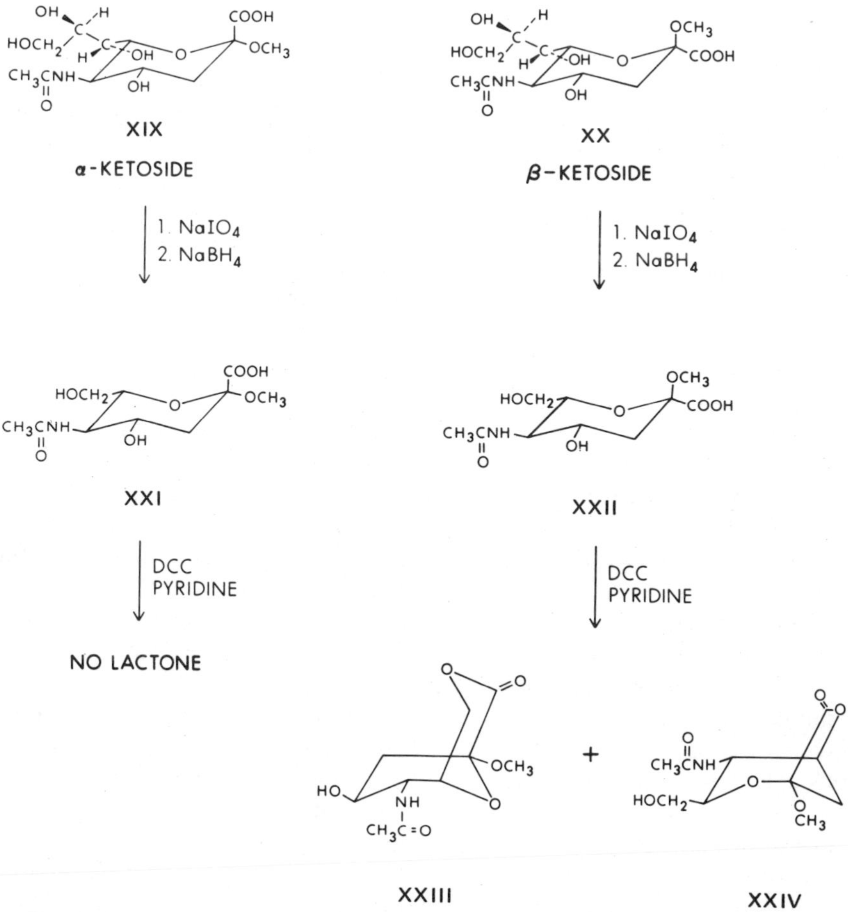

FIGURE 7. Reactions used to assign C_2-ketoside configuration of NANA (Yu and Ledeen, 1969).

bond was assigned the α-configuration in accord with Hudson's isorotation rules. However, the reservations discussed above in using such rotational changes to assign the C_2 configuration of free sialic acid applied equally to the ketosides. This was underscored by the observation, noted previously, that application of Hudson's lactone rule had led to the wrong configurational assignment at C_4 (Kuhn and Brossmer, 1957).

Assignment of the C_2-ketoside-configuration eventually became possible through a study (Yu and Ledeen, 1969) of the methylketosides of NANA prepared synthetically. One isomer was synthesized by the Koenigs–Knorr reaction and the other by direct methylation with methanol and an acid catalyst. Each was treated in sequence with sodium periodate and sodium borohydride (Figure 7) to form the N-acetylheptulosaminic acid methyl ketosides (XXI and XXII), and these were reacted with dicyclohexylcarbodiimide (DDC) in hot pyridine to force lactone formation. The neuraminidase-resistant isomer gave rise to two lactones (XXIII and XXIV) while the enzyme-susceptible anomer gave no lactone. Inspection of molecular models revealed that structure (XXI) is intrinsically unable to lactonize due to *trans*-orientation of the carboxyl and hydroxyl groups. These results led to the conclusion that the configuration of the neuraminidase-labile ketoside corresponds to structure XIX, with the carboxyl group axial and the ketosidic bond equatorial to the pyranose ring. This was assigned the α-D-configuration in accord with the rules of nomenclature.

Once it became possible to assign the α-D-configuration to sialic acid ketosides which are cleaved by neuraminidase, a question arose regarding those substances whose sialic acid is not susceptible to the enzyme. The suggestion was made several years ago that such resistance could be the result of inverted (β) ketosidic configuration, but this has not turned out to be the case. Tay–Sachs ganglioside, for example, although resistant to bacterial neuraminidase (in the absence of detergents), becomes an active substrate after selective removal of the terminal N-acetylgalactosamine (Ledeen and Salsman, 1965). The major monosialoganglioside of brain, another species unaffected by the enzyme, was also shown to have an α-ketosidic linkage for NANA through degradation to a simpler oligosaccharide (Huang and Klenk, 1972). Two sequences of periodate-oxidation–borohydride-reduction gave a product (XXV) that was identical to the one obtained by similar treatment of 3-O-N-acetylneuraminosyl lactose. Since the latter is readily cleaved by neuraminidase, it was apparent that the ganglioside sialic acid has the same (α) configuration. The resistance of the original

```
                    CH₂OH
                    |
              H-C ─────────┐
                |         |
                CH₂OH  O-C-H
                       |
                       H-C-OH
       ┌──────┐        |
  HOOC-C──────────── O-C-H
       |              |
       CH₂           HO-C-H
       |              |
       H-C-OH        H-C-O ─┘
       |              |
   AcNH-C-H          CH₂OH
       |
    ┕──O-C-H
       |
       CH₂OH
```

XXV

glycolipid as well as that of Tay–Sachs ganglioside is thus attributed to steric interference of the enzyme reaction by the neighboring N-acetyl-D-galactosamine. Such gangliosides were reported to react with neuraminidase from *Clostridium perfringens* in the presence of bile salts (Wenger and Wardell, 1973).

To our knowledge there has been no substantiated report of a naturally occurring β-ketoside linkage for sialic acid. However, the intriguing possibility that such a linkage may exist was raised by Liu *et al.* (1971) in their observation that group C meningococcal polysaccharide, a pure homopolymer of sialic acid, is not hydrolyzed by neuraminidase from *Vibrio cholera* or *Clostridium perfringens*. Other explanations were not excluded, however, and the substitution site within the polymer could conceivably be a critical factor.

The possibility of differentiating C_2-stereoisomers by circular dichroism was suggested in a study (Ledeen, 1970) which showed Cotton effects of opposite sign in methanol for the two methyl ketosides of NANA. Various gangliosides resembled the methyl α-ketoside in having a strongly negative Cotton effect in methanol at low wavelength. The circular dichroism of various gangliosides was found by Stone and Kolodny (1971) to have negative ellipticity in aqueous solution, in contrast to free sialic acids which were positive. Their study suggested an α-configuration for the C_2 bond in CMP-NANA which differed from the β-assignment of an earlier study (Comb *et al.,* 1966). The sign and location of ellipticity bands for sialic acid and its derivatives were found to depend on the state of the carboxyl group (free versus esterified) and the nature of the solvent (Ledeen, 1970; Dickinson and Bush, 1975). Additional influence is exerted by the grouping to which sialic acid is ketosidically linked, as shown in the distinguishable CD patterns obtained for the (2→3) and (2→6) isomers of neuraminosyl lactose (Dickinson and Bush, 1975).

C. Chemical Reactions and Derivatives

Sialic acids are highly reactive toward many reagents, and the diversity of products can be understood in terms of the nature and orientation of the various functional groups. Carboxyl, carbonyl-hemiketal, amide, and hydroxyl groups are common to all sialic acids, whereas ester and ether groups are more selectively distributed. The fact that the pKa values are in the range 2.6–2.75 (Blix *et al.*, 1956; Svennerholm, 1956) means the sialic acids are considerably more acidic than the common aliphatic acids. This is undoubtedly the result of the strongly electronegative grouping adjacent to the carboxyl. The latter is thus entirely ionized at physiological pH as well as the more acidic microenvironments believed to exist in some substructures. This strong acidity explains the capacity of sialic-acid containing compounds to undergo autohydrolysis on heating in aqueous solution.

1. Reactions with Acid and Alkali. Base-induced transformations that were instrumental in determining sialic acid structure have been discussed above. These included intramolecular condensation to pyrrole-2-carboxylic acid and reverse aldolization to form *N*-acylhexosamine and pyruvate. Ohkuma and Miyauchi (1966) heated NANA with 0.02 N barium hydroxide at 80° and observed several products including *N*-acetyl-D-mannosamine, *N*-acetyl-D-glucosamine, furan chromogens, pyrrole-2-carboxylic acid, and various other pyrrole chromogens. Base degradation occurs only with free sialic acid since ketoside derivatives are relatively stable to alkali. The base treatment needed to remove *O*-acetyl groups from sialic acids is sufficiently mild to avoid transformations of the above kind. Vigorous base treatment will, of course, result in de-*N*-acylation of ketosidically bound as well as free sialic acid.

Sialic acids undergo a series of complex reactions in hot acid medium, the degree of alteration being a function of acidity, temperature, and reaction period. Heating at pH 3 for 1 h at 100° appeared to have little effect, but when the acidity was increased to 0.1 N HCl there was considerable loss of color obtained with the thiobarbituric acid assay (Karkas and Chargaff, 1964). This sensitivity to mild acid treatment is reminiscent of the behavior of 2-deoxyhexoses, whose enhanced capacity to convert from the cyclic to open-chain form has been cited as a chief cause of their acid sensitivity (Tuppy and Gottschalk, 1972). This presumes that degradation is more rapid in the open-chain than in the cyclic conformation. Gottschalk (1966) has contended that sialic acids have similar tautomeric mobility by virtue of their 2-deoxycarbonyl grouping. Destruction by acid becomes an important consideration in quantitative assays which depend on prior acid hydrolysis (see below).

The mechanism of acid degradation and humin formation is undoubtedly complex. A Δ'-pyrroline derivative was proposed (Gottschalk, 1960) as an initial cyclization product, and this was later substantiated by identification of 4-hydroxy-5-(1,2,3,4,-tetrahydroxybutyl)-Δ'-pyrroline-2-carboxylic acid (XXVI) (Gielen, 1967a). This compound is

$$\text{HOOC}-\underset{N}{\overset{CH_2-CHOH}{C}}\diagdown\overset{|}{CH}-CHOH-CHOH-CHOH-CH_2OH$$

XXVI

dehydrated in part to one or more substituted pyrrole-2-carboxylic acids. Humin formation, which occurs for example when sialic acids are heated in 1 N HCl at 100°, is believed to be the result of condensation of unsaturated degradation products from such pyrrole derivatives. More vigorous treatment, such as refluxing with 12% hydrochloric acid, leads to decarboxylation (Blix *et al.*, 1956) while concentrated sulfuric acid causes release of carbon monoxide. As discussed above, reaction with methanolic hydrogen chloride gives methoxyneuraminic acid methyl ester, a product that is relatively stable to further decomposition in the methanolic medium.

2. Reactions of the Carboxyl Groups. The carboxyl group undergoes esterification quite readily, as illustrated by the fact that mere crystallization of sialic acid from anhydrous methanol leads to some ester formation (Blix *et al.*, 1956). Excellent yields of the methyl ester of NANA were prepared by reaction with anhydrous methanol at room temperature using Dowex 50 (H^+) as acid catalyst (Stephen and Jeanloz, 1966; Kuhn *et al.*, 1966). More vigorous conditions, such as refluxing the above mixture, led to the methylketoside methyl ester (see below), but a small amount of this product formed even at room temperature. This side reaction can be avoided by employing diazomethane, which has been used to esterify many synthetic ketosides of NANA (Meindl and Tuppy, 1966*b,c*; 1967) as well as sialic acid in naturally occurring saccharides (Kuhn and Brossmer, 1956*b*; Gottschalk, 1962). The diphenylmethyl group was introduced as an ester by reaction of NANA with diphenyldiazomethane and could be subsequently removed by hydrogenolysis (Khorlin and Privalova, 1966).

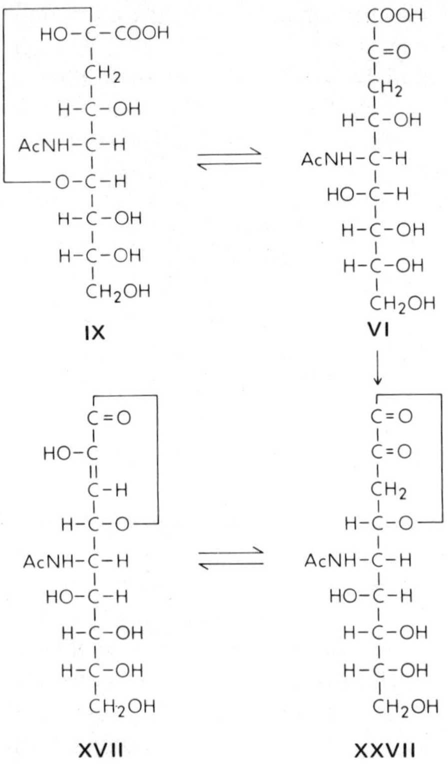

FIGURE 8. Lactonization of NANA: enol (XVII) and keto (XXVII) forms of γ-lactone.

The γ-lactone is an internal ester that also forms with relative ease, provided the sialic acid can assume the open-chain conformation* (Figure 8). Reaction of NANA with dicyclohexyl carbodiimide in cold pyridine, for example, produced the lactone in crystallizable form (Derevitskaya et al., 1966). Small quantities of the 2,6,7,8,9-penta-O-acyl derivatives of NANA-γ-lactone were produced by acylation of NANA in cold pyridine with acetic anhydride or benzoylchloride (Khorlin and Privalova, 1967). The C_2-carbonyl of the lactone can exist in both keto (XXVII) and enol (XVII) forms, the latter being susceptible to ozonolysis (Figure 6).

* Lactonization of the same groups is possible in the pyranose ring form but is thermodynamically less favorable due to the necessity of converting to the boat conformation (Yu and Ledeen, 1969).

Sialic acid esters are saponified with relative ease, brief treatment with dilute alkali at room temperature being sufficient. Acid-catalyzed hydrolysis requires only slightly stronger conditions (e.g., 0.1 N acid at 100° for 1 hr). The acid- and base-catalyzed reactions were shown (Karkas and Chargaff, 1964) to be retarded by the presence of glycosidic alkyl groups, and conversely the stability of the ketosidic linkage toward acid hydrolysis is increased by esterification of the neighboring carboxyl (Karkas and Chargaff, 1964; Ledeen and Salsman, 1965).

3. **Ketoside Derivatives.** A large number of synthetic ketosides have been prepared in recent years. Some of these provide useful models for studying the behavior of naturally occurring sialic acid ketosides. Two groups of anomers were synthesized by methods that are essentially the same as those used to prepare glycosides of simpler sugars. One group was prepared by heating sialic acid with an alcohol and an acid catalyst, while the anomeric forms were synthesized by the Koenigs–Knorr procedure (Figure 9). Although the various anomeric pairs could be differentiated by physical and enzymatic means, the precise structures remained uncertain until it was established that these reactions followed

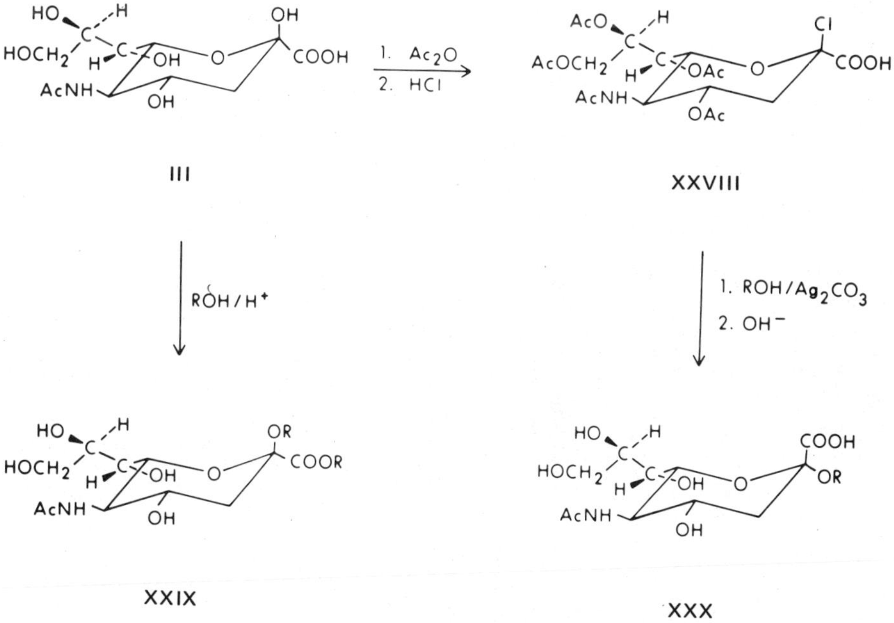

FIGURE 9. General procedures for synthesis of β-ketosides (XXIX) and α-ketosides (XXX) of sialic acids.

TABLE II. β-Ketosides of Neuraminic Acid and N-Acylneuraminic Acids and Their Methyl Esters

N-Acyl group	Ketoside group	4,7,8,9-O-Substitution	Melting or decomposition point, °C	$[\alpha]_D$ (solvent)	Reference
H	Methyl	—	199–200	−55(H_2O)	Böhm and Baumeister (1955); Gielen (1965; 1967a); Klenk and Lauenstein (1952)
Acetyl	Methyl	—	—	−44	Kuhn et al. (1966)
Acetyl[a]	Methyl	—	97–98	−39(CH_3OH)	McGuire and Binkley (1964)
			115–130	−46(CH_3OH)	Kuhn et al. (1966)
			111–115	−45(H_2O)	Yu and Ledeen (1969)
Acetyl[a]	Methyl	Acetyl	137–138	−11(CH_3OH)	Blix and Jeanloz (1969)
Acetyl	n-Amyl	—	124–125	−23(H_2O)	Meindl and Tuppy (1965a,b)
				−42($DMSO$)[b]	Meindl and Tuppy (1965a,b)
				−27(AcOH)	Meindl and Tuppy (1965a,b)
Acetyl	n-Amyl	Acetyl	199–200	−12(AcOH)	Meindl and Tuppy (1965a,b)
Acetyl	n-Hexyl	—	—	−20(H_2O)	Meindl and Tuppy (1965a,b)
Acetyl	n-Hexyl	Acetyl	175–177	—	Meindl and Tuppy (1965a,b)
Acetyl	Carboxymethyl	—	193–195	−24(H_2O)	Holmquist and Brossmer (1972a)
Acetyl	Benzyl	—	—	—	Faillard et al. (1966)

Chemistry and Analysis of Sialic Acid

Compound	Ester	M.p. (°C)	[α] (solvent)	Reference
Acetyl[a]	Benzyl	136–139	−13(CH$_3$OH)	Lutz et al. (1968)
Acetyl[a]	Benzyl	155–157	−12(CH$_3$OH)	Lutz et al. (1968)
Acetyl	Phenyl	—	−69(DMSO)[b]	Meindl and Tuppy (1967)
Acetyl	Phenyl	126–136	−59(AcOH)	Meindl and Tuppy (1967)
Acetyl[a]	Phenyl	189–194	−69(DMSO)[b]	Meindl and Tuppy (1967)
Acetyl	Pyridyl	—	—	Holmquist and Brossmer (1972b)
Acetyl[a]	Pyridyl	198–199	−43(CHCl$_3$)	Holmquist and Brossmer (1972b)
Trifluoracetyl	Methyl	154–158	—	Klenk et al. (1956)
Glycolyl	Methyl	—	—	Wirtz-Peitz et al. (1969)
Glycolyl[a]	Methyl	203–206	—	Blix et al. (1956)
		153–155 (oxalate)	−17	Yu and Ledeen (1970)
Glycyl[a]	Methyl	189–190	−25(H$_2$O)	Derevitskaya et al. (1965b)
Benzyloxycarbonylglycyl[a]	Methyl	—	—	Derevitskaya et al. (1965b)
Benzyloxycarbonyl	Methyl	—	—	Gielen (1965; 1967a)
Benzyloxycarbonyl[a]	Methyl	191–194	−39(H$_2$O)	Gielen (1967b)
		194–196	−43(CH$_3$OH)	Wesemann and Zilliken (1966)
Benzoyl	Methyl	145–146	−42(H$_2$O)	Meindl and Tuppy (1966c)
Benzoyl[a]	Methyl	213–215	−33(H$_2$O)	Meindl and Tuppy (1966c)

[a] Methyl ester replaces free carboxyl group.
[b] DMSO = dimethylsulfoxide.

the same stereochemical course previously demonstrated for the simple sugars. Reaction of the first type—heating with alcohol and an acid catalyst—gives the thermodynamically preferred product which, in the case of simple sugars, is generally the anomer whose glycosidic group is axial to the pyranose ring. This was found to be true of NANA as well (Yu and Ledeen, 1969). Thus, neither the presence of the carboxyl group nor the fact that the pyranose ring is in the 1C conformation served to alter the steric relationship. The anomer XXIX is designated the β-ketoside, according to nomenclature rules.

Table II, which, with Table III, represents an updating of the comprehensive list compiled by Tuppy and Gottschalk (1972), summarizes a number of sialic acid β-ketosides obtained synthetically. Since the alcohol incorporated into the ketoside grouping is generally the solvent for the reaction, the initial product is the ketoside-ester which may be converted to the corresponding carboxylate by mild base treatment. All such products with the β-configuration were found to be unreactive toward neuraminidase. Another route to β-ketosides is acylation of methoxyneuraminic acid, the product of vigorous acid methanolysis of sialic acids; a variety of N-acylated methyl β-ketosides have been synthesized by this route, including the N-trifluoroacetyl (Klenk et al., 1956), N-benzoyl (Meindl and Tuppy, 1966c), N-fluoroacetyl and N-chloroacetylneuraminic acids (Schauer et al., 1970).

The first stage of the Koenigs–Knorr procedure is reaction of the peracetylated derivative of sialic acid with hydrogen chloride to yield an unstable acetylglycosyl chloride XXVIII. This is heated with an alcohol or phenol in the presence of silver carbonate, silver oxide, or mercuric cyanide to form the α-ketoside by displacement of the chloride group, and O-acetyl groups are then removed with base. The configuration at C_2 is inverted during chloride displacement, which results in the thermodynamically less stable anomer (XXX) with the ketoside grouping equatorial to the pyranose ring (Figure 9). This sequence can be carried out with either sialic acid itself or its methyl ester. The relative instability of α-ketosides was indicated by rapid isomerization of the α-methyl anomer to the β-form on passing through a Dowex 50(H^+) column (Yu and Ledeen, 1969), and also from the observation that α-ketosides of NANA underwent acid hydrolysis more rapidly than the corresponding β-ketosides (Meindl and Tuppy, 1965b).

Synthetically prepared α-ketosides are summarized in Table III. The majority of these were shown to be susceptible to neuraminidase, provided the substrate was in the carboxylate rather than ester form. A wide variety of ketoside substituents were compatible with enzyme reactivity, but the presence of certain N-acyl groups inhibited this reaction. Thus, the N-butyryl, N-succinyl, N-benzoyl, and N-benzyloxycarbonylneuraminic acid α-ketosides were completely resistant to the enzyme, while the N-formyl and N-propionylneuraminic acid benzyl α-ketosides were cleaved, though more slowly than the corresponding benzyl α-ketoside of NANA (Meindl and Tuppy, 1966c; Brossmer and Nebelin, 1969; Faillard et al., 1969). Certain substituted phenolic α-ketosides were found useful as chromogenic neuraminidase substrates (Meindl and Tuppy, 1969; Tuppy and Palese, 1969; Privalova and Khorlin, 1969; Palese et al., 1973). Some additional ketosides were synthesized by Brossmer and coworkers to study the role of the anionic site in neuraminidase substrates. These included the carboxymethyl (Holmquist and Brossmer, 1972a) and 2-aminoethyl and 2-pyridyl (Holmquist and Brossmer, 1972b) ketosides of NANA; from their behavior the authors inferred the necessity for an ionic bond between the enzyme and the carboxylate group of the substrate.

A number of S- and N-ketosyl derivatives of NANA were prepared by Privalova and Khorlin (1969). The use of p-nitroaniline in a Koenigs–Knorr type reaction with the acetylglycosyl chloride methyl ester of NANA gave (after subsequent deacylation) the 2-deoxy-2-p-nitrophenylamino N-glycoside of NANA, assigned structure XXXI (Figure 10).

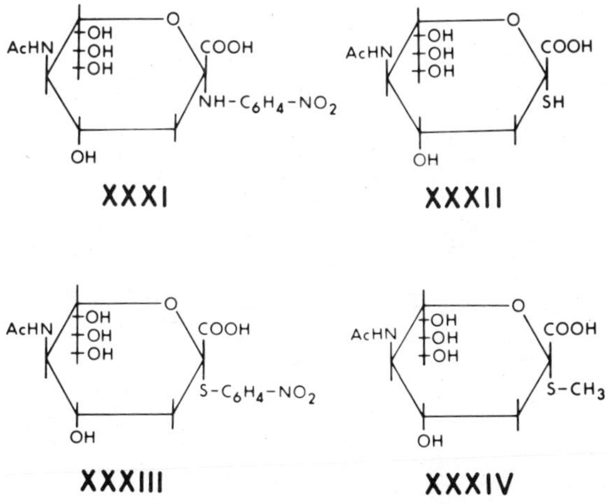

FIGURE 10. S-Ketosyl and N-ketosyl derivatives of NANA (Privalova and Khorlin, 1969).

TABLE III. α-Ketosides of *N*-Acylneuraminic Acids and Their Methyl Esters

N-Acyl group	Ketoside group	4,7,8,9-*O*-Substitution	Melting or decomposition point, °C	$[\alpha]_D$ (Solvent)	Reference
Formyl	Benzyl	—	—	$-15(H_2O)$	Brossmer and Nebelin (1969)
Formyl	Benzyl	Acetyl	202	—	Brossmer and Nebelin (1969)
Acetyl	Methyl	—	—	$-13(H_2O)$	Meindl and Tuppy (1965a,b)
				$-10(H_2O)$	Kuhn et al. (1966)
				$-1(CH_3OH)$	Kuhn et al. (1966)
Acetyl	Methyl	Acetyl	184–187	$-30(AcOH)$	Meindl and Tuppy (1965a,b)
Acetyl[a]	Methyl	—	167–169	$-12(H_2O)$	Meindl and Tuppy (1966b)
			160–164	$-23(H_2O)$	Yu and Ledeen (1969)
				$-6(CH_3OH)$	Kuhn et al. (1966)
Acetyl	*n*-Amyl	—	—	$-11(H_2O)$	Meindl and Tuppy (1965a,b)
Acetyl	*n*-Amyl	Acetyl	185–187	$-22(AcOH)$	Meindl and Tuppy (1965a,b)
Acetyl	*n*-Hexyl	—	—	$-11(H_2O)$	Meindl and Tuppy (1965a,b)
Acetyl	*n*-Hexyl	Acetyl	174–175	—	Meindl and Tuppy (1965a,b)
Acetyl	*n*-Decyl	—	—	$-10(H_2O)$	Meindl and Tuppy (1965a)
Acetyl	*n*-Decyl	Acetyl	158–162	$-17(AcOH)$	Meindl and Tuppy (1965a)
Acetyl	Cyclohexyl	—	—	$-8(H_2O)$	Meindl and Tuppy (1965a)
Acetyl	Cyclohexyl	—	186–188	—	Meindl and Tuppy (1965a)
Acetyl	3′-Hydroxypropyl	—	—	$-6(H_2O)$	Meindl and Tuppy (1965a)
Acetyl	3′-Acetoxypropyl	Acetyl	147–154	—	Meindl and Tuppy (1965a)
Acetyl	5′-Hydropentyl	—	—	$-6(H_2O)$	Meindl and Tuppy (1965a)
Acetyl	5′-Acetoxypentyl	Acetyl	160–162	—	Meindl and Tuppy (1965a)
Acetyl	2-Aminoethyl	—	—	$-17(H_2O)$	Holmquist and Brossmer (1972b)
Acetyl	Carboxymethyl	—	—	$-9(H_2O)$	Holmquist and Brossmer (1972a)
Acetyl	Butoxycarbonylmethyl	Acetyl	184–186	$-10(CH_3OH)$	Holmquist and Brossmer (1972a)
Acetyl	Benzyl	—	162–164	$-8(H_2O)$	Meindl and Tuppy (1965a; 1966a)
Acetyl	Benzyl	Acetyl	194–195	$-14(AcOH)$	Meindl and Tuppy (1965a)
Acetyl[a]	Benzyl	Acetyl	84–89	$-3(CH_3OH)$	Lutz et al. (1968)

Chemistry and Analysis of Sialic Acid

Acetyl[a]	Benzyl	Methyl	180–184	+18(CH_3OH)	Lutz et al. (1968)
Acetyl	m-Bromobenzyl				Meindl and Tuppy (1965a)
Acetyl	m-Bromobenzyl	Acetyl	186–187		Meindl and Tuppy (1965a)
Acetyl	m-Chlorobenzyl				Meindl and Tuppy (1965a)
Acetyl	m-Chlorobenzyl	Acetyl	195–200		Meindl and Tuppy (1965a)
Acetyl	m-Iodobenzyl				Meindl and Tuppy (1965a)
Acetyl	m-Iodobenzyl	Acetyl	180–182		Meindl and Tuppy (1965a)
Acetyl	m-Nitrobenzyl		177–180	−7(H_2O)	Meindl and Tuppy (1965a)
Acetyl	m-Nitrobenzyl	Acetyl	182–183		Meindl and Tuppy (1965a)
Acetyl	p-Methoxybenzyl			−16(H_2O)	Meindl and Tuppy (1965a)
Acetyl	p-Methoxybenzyl	Acetyl	191–192		Meindl and Tuppy (1965a)
Acetyl	Hexahydrobenzyl			−6(H_2O)	Meindl and Tuppy (1965a)
Acetyl	Hexahydrobenzyl	Acetyl	188–191		Meindl and Tuppy (1965a)
Acetyl	Phenyl		132–135	−14(DMSO)[b]	Meindl and Tuppy (1967)
Acetyl	Phenyl	Acetyl	182–185	O(AcOH)	Meindl and Tuppy (1967)
Acetyl[a]	Phenyl		95–100	−18(DMSO)[b]	Meindl and Tuppy (1967)
Acetyl	m-Methoxyphenyl		127–129	−31(DMSO)[b]	Tuppy and Palese (1969)
Acetyl	m-Methoxyphenyl	Acetyl	169–175	+69(AcOH)	Tuppy and Palese (1969)
Acetyl	p-Nitrophenyl			−1(CH_3OH)	Privalova and Khorlin (1969)
Acetyl	Pyridyl				Holmquist and Brossmer (1972b)
Acetyl[a]	Pyridyl	Acetyl		+43($CHCl_3$)	Holmquist and Brossmer (1972b)
n-Propionyl	Benzyl		162–163	−16(H_2O)	Meindl and Tuppy (1966c)
n-Propionyl	Benzyl	Acetyl	199–201	−17(AcOH)	Meindl and Tuppy (1966c)
n-Butyryl	Benzyl		163–164	−19(H_2O)	Meindl and Tuppy (1966c)
n-Butyryl	Benzyl	Acetyl	182–183	−15(AcOH)	Meindl and Tuppy (1966c)
Glycolyl	Methyl			−9(H_2O)	Meindl and Tuppy (1966a)
Glycolyl[a]	Methyl		165–168	−13(H_2O)	Meindl and Tuppy (1966a)
Acetoglycolyl	Methyl	Acetyl	183–187	−30(AcOH)	Meindl and Tuppy (1966a)
Glycolyl	n-Amyl			−9(H_2O)	Meindl and Tuppy (1966a)
Acetoglycolyl	n-Amyl	Acetyl	183–185	−22(AcOH)	Meindl and Tuppy (1966a)
Glycolyl	n-Decyl			−9(H_2O)	Meindl and Tuppy (1966a)

(continued)

TABLE III. (Continued)

N-Acyl group	Ketoside group	4,7,8,9-O-Substitution	Melting or decomposition point, °C	$[\alpha]_D$ (Solvent)	Reference
Acetoglycolyl	n-Decyl	Acetyl	169–171	−17(AcOH)	Meindl and Tuppy (1966a)
Glycolyl	Benzyl	—	—	−15(H_2O)	Meindl and Tuppy (1966a)
Acetoglycolyl	Benzyl	Acetyl	186–189	−14(AcOH)	Meindl and Tuppy (1966a)
Glycolyl	m-Nitrobenzyl	—	165	−9(H_2O)	Meindl and Tuppy (1966a)
Acetoglycolyl	m-Nitrobenzyl	Acetyl	202–205	−15(AcOH)	Meindl and Tuppy (1966a)
Succinyl	Benzyl	—	—	−10(H_2O)	Brossmer and Nebelin (1969)
Benzyloxycarbonyl	Benzyl	—	155–157	—	Faillard et al. (1969)
Benzoyl	Methyl	—	179–185	−3(H_2O)	Meindl and Tuppy (1966c)
Benzoyl	Methyl	Acetyl	110	−19(AcOH)	Meindl and Tuppy (1966c)
Benzoyl[a]	Methyl	—	174–175	−10(H_2O)	Meindl and Tuppy (1966c)
Benzoyl	Benzyl	—	190–199	−19(DMSO)[b]	Meindl and Tuppy (1966c)
Benzoyl	Benzyl	Acetyl	170–171	−9(AcOH)	Meindl and Tuppy (1966c)

[a] Methyl ester replaces free carboxyl group.
[b] DMSO = dimethylsulfoxide.

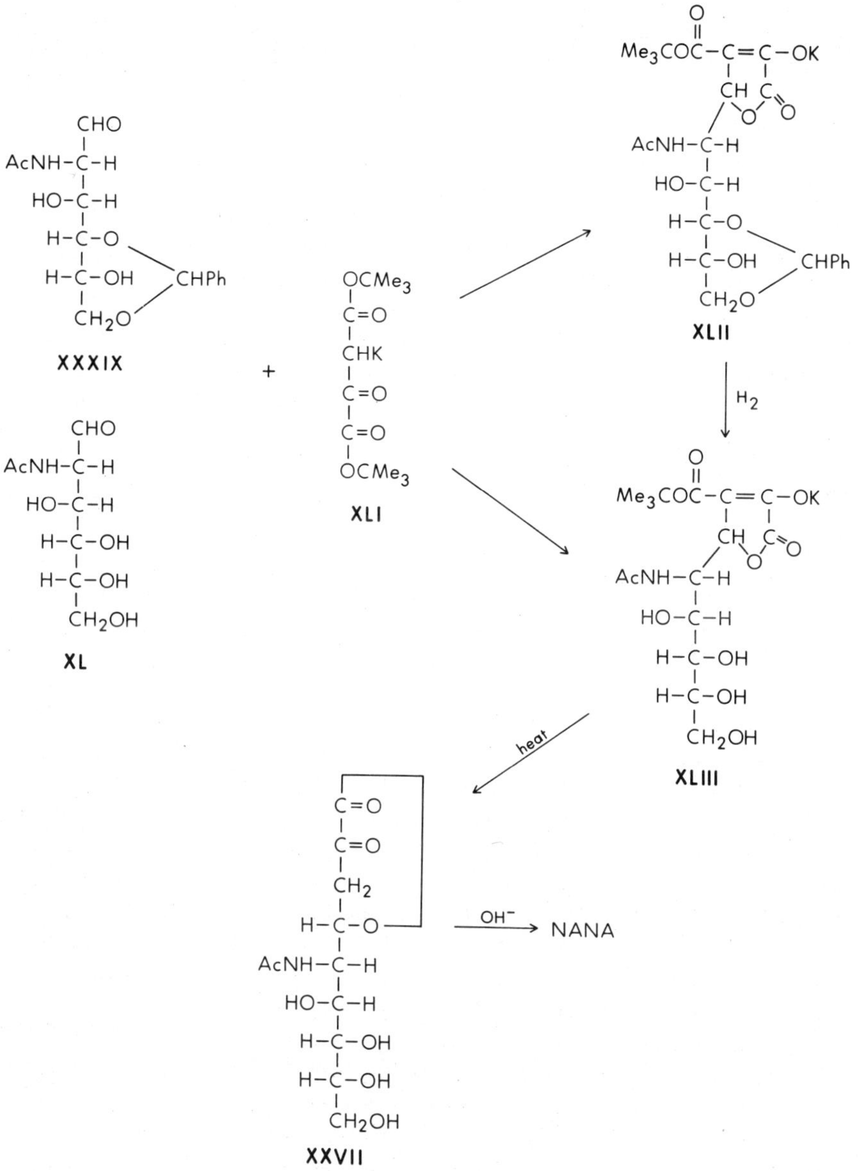

FIGURE 12. Sialic acid synthesis by the procedure of Kuhn and Baschang (1962b) (Ph = C_6H_5).

Interestingly, this substance was found to be resistant to N-acylneuraminate pyruvate-lyase.

The synthetic procedure developed by Kuhn and Baschang (1962b) employed the potassium salt of oxaloacetic acid di-*tert*-butyl ester (XLI, Figure 12) and resulted in greatly improved yields of sialic acid. Condensation of this activated form of oxaloacetate with N-acetyl-D-mannosamine gave lactone XLIII which, when heated to 90–100°C, formed the γ-lactone of NANA (XXVII) simultaneously with release of carbon dioxide and isobutylene. Following hydrolysis, NANA was obtained in 34% yield. Stephan and Jeanloz (1966) were able to improve the yield by using the 4,6-O-benzylidene derivative of N-acetyl-D-mannosamine (XXXIX), the benzylidene group being subsequently removed by hydrogenation. Somewhat lower yields of NANA had been obtained earlier with the 4,6-O-benzylidene derivative of N-acetyl-D-glucosamine, and in this case reaction temperature was found to be critical: condensation at 20–25° gave mostly NANA, whereas at 70° the C_5 epimer with the D-*glycero* configuration predominated (Kuhn and Baschang, 1962b).

The procedure of Kuhn and Baschang provided a convenient route to the preparation of several N-acylated neuraminic acids including the N-benzoyl, N-propionyl and N-butyryl (Meindl and Tuppy, 1966c), N-formyl and N-succinyl (Brossmer and Nebelin, 1969), and N-ethoxycarbonyl and N-benzyloxycarbonyl (Wesemann and Zilliken, 1966; Gielen, 1965; 1967a) derivatives. NGNA could be synthesized from 4,6-O-benzylidene-N-glycolylglucosamine, the latter being derived from D-glucosamine and 1,3-dioxolane-2,4-dione (Faillard, 1965; Faillard and Blohm, 1965). The 8-O-methyl ether of NANA was also synthesized by this route (Khorlin and Privalova, 1970) prior to its discovery as a naturally occurring sialic acid in starfish (Kochetkov *et al.*, 1973). Use of 2-acetamido-2-deoxy-D-lyxose in the condensation gave rise to 5-acetamido-3,5-dideoxy-D-*galacto*-octulosonic acid, the 8-carbon analog of NANA (McLean and Beidler, 1969).

Despite discouraging results in the attempt to condense pyruvate with hexosamine, some success was achieved with pyruvate derivatives containing electronegative groups on the methyl carbon. Reaction of 3-fluoropyruvate with N-acetyl-D-glucosamine or N-acetyl-D-mannosamine gave low yields of a product believed to be N-acetyl-3-fluoroneuraminic acid, partially on the basis of its inhibitory effect on N-acylneuraminate pyruvate-lyase (Gantt *et al.*, 1964). The use of bromopyruvate or hydroxypyruvate gave a product described as 3-hydroxy-N-acetylneuraminic acid (DeVries and Binkley, 1972b). The stereochemistry of these

B. Enzymatic Assay

Brunetti *et al.* (1963) have described an enzymatic method for quantification of sialic acid based on cleavage of *N*-acylneuraminic acid by *N*-acylneuraminate pyruvate-lyase to pyruvate and the corresponding *N*-acylmannosamine. This reaction was coupled to reduction of pyruvate by lactate dehydrogenase and NADH, this second reaction driving the aldolase cleavage to completion. Sialic acid concentration was estimated in one of the following ways: (a) determination of the quantity of NADH oxidized at the end of the reaction by spectrophotometry or fluorometry; (b) determination of the initial rate of NADH oxidation (which is a function of sialic acid concentration); (c) determination of *N*-acylhexosamine by a modified Morgan–Elson reaction (Brunetti *et al.*, 1962). The method requires free NANA or NGNA, and since *O*-acetylated sialic acids are sometimes resistant to the aldolase, they must be deacylated by mild acid or alkali treatment prior to the assay. Although methods for preparation and purification of the *Clostridium perfringens* aldolase have been described in detail (Comb and Roseman, 1962; Brunetti *et al.*, 1963; DeVries and Binkley, 1972*a*), the requirement of a purified enzyme free of NADH oxidase activity has imposed a limitation. Nevertheless, it remains one of the most specific methods for sialic acid assay, and the fact that 15 ng or less can be determined also makes it the most sensitive.

C. Gas–Liquid Chromatography

Use of GLC for analysis of sialic acids requires conversion of these highly polar, ionic substances to nonionic derivatives of sufficient volatility. Sweeley and coworkers (Sweeley and Walker, 1964; Vance and Sweeley, 1967) demonstrated the utility of trimethylsilyl ether derivatives for analyzing sialic acid as well as other sugar constituents of glycolipids. The ganglioside is subjected to prolonged heating in 0.5 N methanolic-HCl to form the methylketoside methyl ester of neuraminic acid and the methyl glycosides of the other carbohydrates. Reaction with a mixture of pyridine-trimethylchlorosilane-hexamethyldisilazane then converts all free hydroxyls to trimethylsilyl ethers. A somewhat similar procedure has been described for sialic acid for glycopeptides and glycoproteins (Clamp *et al.*, 1967; 1972). Both methods employ rather strong acidic conditions for methanolysis which cause complete cleavage of the *N*-acyl groups from sialic acids. *N*-acetylation can be carried out selectively (to the exclusion of *O*-acetylation) in methanolic solvent prior to trimethylsilylation.

A procedure employing mild methanolysis was subsequently developed in order to preserve the identity of the N-acyl group and thus provide some differentiation of sialic acid species during the course of analysis (Yu and Ledeen, 1970). It has proved useful for bound sialic acid in general and was first applied to gangliosides. Heating in 0.05 N methanolic-HCl for 1 hr at 80° releases most of the bound sialic acid in the form of methylketoside methyl ester, and this is subsequently derivatized to the trimethylsilyl ether. Only small amounts of de-N-acylation product are detected. Both the α- and β-ketosides of sialic acid form in this reaction, but since the latter is considerably more abundant, quantification is based empirically on the β-peak alone. Phenyl N-acetyl-α-D-glucosaminide was found to be a convenient internal standard as its trimethylsilyl ether derivative emerges midway between NANA and NGNA on a OV-1 column (see below). For maximum accuracy it is desirable to use a calibration standard similar in composition to the ganglioside being assayed. The method was recently adapted with minor modifications for determining protein-bound sialic acid (Ledeen and Yu, 1973b).

In addition to enabling separate analysis of at least two types of sialic acid there are the additional advantages inherent in GLC assay: sensitivity and specificity. As little as 1 μg of lipid-bound sialic acid could be conveniently analyzed and this could be reduced to 0.3μg and perhaps less with suitable controls for background peaks. For routine assays OV-1 has been the column of choice, but the somewhat more polar OV-225 has proved quite useful in confirming peak identification where required. A typical analysis with both columns is shown in Figure 14: the sample was purified ganglioside mixture from ovine adrenal medulla (Price and Yu, 1976), which was found to contain NANA and NGNA in a ratio of approximately 3:2, respectively.

Although artifactual peaks are sometimes seen, these are minimized by employing ganglioside samples of at least moderate purity. The use of two columns enables sialic acid identification to be made with a high degree of certainty, although the OV-1 column alone is usually sufficient for routine assays. O-acetylated sialic acids lose some or all of their ester groups despite the mild methanolysis conditions, so that quantification of such species is not yet feasible; these may, however, be detected qualitatively when deacetylation is not complete (Yu and Ledeen, 1970).

Glycolipid-sialic acid has also been analyzed by GLC of the N-,O-trifluoroacetyl derivative after first cleaving sialic acid by heating 6 hr at 75–80° in 3% methanolic-HCl (Ando and Yamakawa, 1971). It should be emphasized that this GLC procedure as well as those described above have been applied only to ketosidically linked NANA. However, a

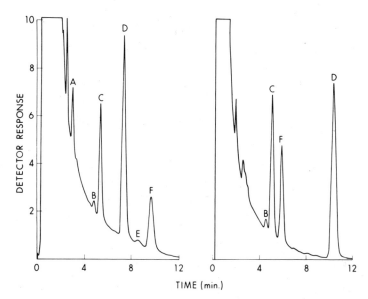

FIGURE 14. Gas–liquid chromatograms of sialic acid liberated by mild methanolysis from purified ovine adrenal medulla and converted to trimethylsilyl ether derivatives. Left, 3% OV-1, 225°C; right, 3% OV-225, 225°C. Peak identification: A, methoxyneuraminic acid methyl ester (de-N-acylated product); B, methyl α-ketoside methyl ester of NANA; C, methyl β-ketoside methyl ester of NANA; D, phenyl N-acetyl-α-D-glucosaminide (internal standard); E, methyl α-ketoside methyl ester of NGNA; F, methyl β-ketoside methyl ester of NGNA. (Price and Yu, 1976).

method for GLC analysis of free sialic acid was described by Craven and Gehrke (1968) which would presumably allow differentiation of NANA and NGNA (although this was not attempted by the authors). Glycoprotein was subjected to mild aqueous hydrolysis and the liberated sialic acid derivatized to the trimethylsilyl ether–ester by heating with bis-(trimethylsilyl) trifluoroacetamide (BSTFA) in acetonitrile. Linear detection response was observed between 200 and 800 μg NANA, but greater sensitivity was implied. It is to be hoped that GLC methods will soon appear that permit assay of the several O-acylated sialic acids in addition to those previously described.

ACKNOWLEDGMENTS

This work was supported by grants NS-04834, NS-10931, NS-03356, and NS-11853 from the National Institutes of Health.

VII. REFERENCES

Abrahamson, S., Fischmeister, I., and Svennerholm, L., 1962, X-ray powder diagrams of sialic acids, *Ark. Kemi* **18**:435.
Aminoff, D., 1959, The determination of free sialic acid in the presence of the bound compound, *Virology* **7**:355.
Aminoff, D., 1961, Method for the quantitative estimation of N-acetylneuraminic acid and their application to hydrolysates of sialomucoids, *Biochem. J.* **81**:384.
Ando, S., and Yamakawa, T., 1971, Application of trifluoroacetyl derivatives to sugars and lipid chemistry. I. Gas chromatographic analysis of common constituents of glycolipids, *J. Biochem.* (Tokyo) **70**:335.
Ayala, W., Moore, L. V., and Hess, E. L., 1951, The purple color reaction given by diphenylamine reagent. I. With normal and rheumatic fever sera, *J. Clin. Invest.* **30**:781.
Barry, G. T., 1958, Colominic acid, a polymer of N-acetylneuraminic acid, *J. Exp. Med.* **107**:507.
Barry, G. T., 1959, Detection of sialic acid in various *Escherichia coli* strains and in other species of bacteria, *Nature* **183**:117.
Barry, G. T., and Goebel, W. F., 1957, Colominic acid, a substance of bacterial origin related to sialic acid, *Nature* **179**:206.
Barry, G. T., Abbot, V., and Tsai, T., 1962, Relationship of colominic acid (poly N-acetylneuraminic acid) to bacteria which contain neuraminic acid, *J. Gen. Microbiol.* **29**:335.
Barry, G. T., Hamm, J. D., and Graham, M. G., 1963, Evaluation of colorimetric methods in the estimation of sialic acid in bacteria, *Nature* **200**:806.
Bentley, R., 1972, Configurational and conformational aspects of carbohydrate biochemistry, *Ann. Rev. Biochem.* **41**:953.
Bertolini, M., and Pigman, W., 1967, Action of alkali on bovine and ovine submaxillary mucins, *J. Biol. Chem.* **242**:3776.
Blix, G., 1936, Üder die Kohlenhydratgruppen des Submaxillarismucins, *Hoppe-Seyl. Z.* **240**:43.
Blix, G., 1938, Einige Beobachtungen über eine hexosaminhaltige Substanz in der Protagonfraktion des Gehirns, *Skand. Arch. Physiol.* **80**:46.
Blix, G., 1959, Sialic acids, in: *Proceedings of the Fourth International Congress on Biochemistry*, Vienna, **1**:94.
Blix, G., 1962, Sialic Acids, *Methods Carbohyd. Chem.* **1**:246.
Blix, G., and Jeanloz, R. W., 1969, Sialic acids and muramic acid, in: *The Amino Sugars* (R. W. Jeanloz and E. A. Balazs, eds.), pp. 213–265, Academic Press, New York.
Blix, G., and Lindberg, E., 1960, The sialic acids of bovine and equine submaxillary mucins, *Acta Chem. Scand.* **14**:1809.
Blix, G., and Odin, L., 1955, Isolation of sialic acid from gangliosides, *Acta Chem. Scand.* **9**:1541.
Blix, G., Svennerholm, L., and Werner, I., 1952, The Isolation of chondrosamine from gangliosides and from submaxillary mucin, *Acta Chem. Scand.* **6**:358.
Blix, G., Lindberg, E., Odin, L., and Werner, I., 1955, Sialic acids, *Nature* **175**:340.
Blix, G., Lindberg, E., Odin, L., and Werner, I., 1956, Studies on sialic acids, *Acta Soc. Med. Upsalien* **61**:1.
Blix, G., Gottschalk, A., and Klenk, E., 1957, Proposed nomenclature in the field of neuraminic and sialic acids, *Nature* **175**:340.

Böhm, P., and Baumeister, L., 1955, über die Isolierung der Methoxyneuraminisäure als Spaltprodukt des Serumeiweisses, *Hoppe-Seyl. Z.* **300:**153.
Bose, A. K., and Chatterjee, B. G., 1958, Molecular rotation and absolute configuration. II. Sugars, *J. Org. Chem.* **23:**1425.
Brossmer, R., and Nebelin, E., 1969, Synthesis of N-formyl- and N-succinyl-D-neuraminic acid and the specificity of neuraminidase, *FEBS Lett.* **4:**335.
Brug, J., and Paerels, G. B., 1958, Configuration of N-acetylneuraminic acid, *Nature* **182:**1159.
Brunetti, P., Jourdian, G. W., and Roseman, S., 1962, The sialic acids. III. Distribution and properties of animal N-acetylneuraminic acid aldolase, *J. Biol. Chem.* **237:**2447.
Brunetti, P., Swanson, A., and Roseman, S., 1963, Enzymatic Determination of sialic acids, *Methods Enzymol.* **6:**465.
Brunngraber, E. G., and Brown, B. D., 1967, Preparation and properties of sialomucopolysaccharides obtained from rat brain, *Biochem. J.* **103:**65.
Buscher, H., Casals-Stenzel, J., and Schauer, R., 1974, New sialic acids, *Eur. J. Biochem.* **50:**71.
Cabezas, J. A., and Feo, F., 1969, Sialic Acids. XI. On the thiobarbituric acid positive reaction in several materials from the vegetal kingdom, *Rev. Esp. Fisiol.* **25:**153.
Cabezas, J. A., and Ramos, M., 1972, The type and content of sialic acid of bile from several animal sources, *Carbohyd. Res.* **24:**486.
Carrion, A., Bourrillon, R., and Cabezas, J. A., 1969, N-Acetyl- and N, O-diacetylneuraminic acids in a sialoglycopeptide from normal human urine, *Clin. Chim. Acta* **26:**481.
Carroll, P. M., and Cornforth, J. W., 1960, Preparation of N-acetylneuraminic acid from N-acetyl-D-mannosamine, *Biochim. Biophys. Acta* **39:**161.
Carubelli, R., and Griffin, M. J., 1968, On the presence of N-glycolylneuraminic acid in HeLa cells, *Biochim. Biophys. Acta* **170:**446.
Clamp, J. R., Dawson, G., and Hough, L., 1967, The simultaneous estimation of 6-deoxy-L-galactose (L-fucose), D-mannose, D-galactose, 2-acetamido-2-deoxy-D-glucose (N-acetyl-D-glucosamine) and N-acetyl-neuraminic acid (sialic acid) in glycopeptides and glycoproteins, *Biochim. Biophys. Acta* **148:**342.
Clamp, J. R., Bhatti. T., and Chambers, R. E., 1972, The examination of carbohydrate in glycoproteins by gas–liquid chromatography, in: *Glycoproteins* (A. Gottschalk, ed.), 2nd ed., pp. 300–321 Elsevier Publishing Co., Amsterdam.
Clark, W. R., Jackson, R. H., and Pallansch, M. J., 1962, Isolation of sialic acid in high yield from colostrum, *Biochim. Biophys. Acta* **58:**129.
Comb, D. G., and Roseman, S., 1958, Composition and enzymatic synthesis of N-acetylneuraminic acid (sialic acid), *J. Am. Chem. Soc.* **80:**497.
Comb, D. G., and Roseman, S., 1960, Sialic acids. I. Structure and enzymic synthesis of N-acetylneuraminic acid, *J. Biol. Chem.* **235:**2529.
Comb, D. G., and Roseman, S., 1962, N-acetylneuraminic acid aldolase, *Methods Enzymol.* **5:**391.
Comb, D. G., Watson, D. R., and Roseman, S., 1966, The sialic acids. IX. Isolation of cytidine 5'-monophospho-N-acetylneuraminic acid from *Escherichia coli* K-235, *J. Biol. Chem.* **241:**5637.
Cornforth, J. W., Firth, M. E., and Gottschalk, A., 1958, The synthesis of N-acetylneuraminic acid, *Biochem. J.* **68:**57.
Craven, D. A., and Gehrke, C. W., 1968, Quantitative determination of N-acetylneuraminic acid by gas–liquid chromatography, *J. Chromatography* **37:**414.
Dawson, G., and Sweeley, C. C., 1971, Mass spectrometry of neutral, mono- and disialoglycosphingolipids, *J. Lipid Res.* **12:**56.

Delmotte, P., 1968, The automatic determination of blood serum sialic acid levels, *Z. Klin. Chem. Klin. Biochim.* **6**:46.
Derevitskaya, V. A., Kalinevich, V. M., and Kochetkov, N. K., 1965a, Synthesis of methyl ester of 9-*O*-glycyl-*N*-acetylneuraminic acid, *Dokl. Akad. Nauk SSSR* **160**:596.
Derevitskaya, V. A., Kalinevich, V. M., and Kochetkov, N. K., 1965b, Glycopeptides. XVI. Synthesis of methyl ester of *N*-glycylmethoxyneuraminic acid, *Khim. Prirodn. Soedin* **1965**:241.
Derevitskaya, V. A., Kalinevich, V. M., and Kochetkov, N. K., 1966, Lactone of *N*-acetylneuraminic acid, *Dokl. Akad. Nauk SSSR* **169**:1087.
DeVries, G. H., and Binkley, S. B., 1972a, *N*-acetylneuraminic acid aldolase of *Clostridium perfringens:* Purification, properties and mechanism of action, *Arch. Biochem. Biophys.* **151**:234.
DeVries, G. H., and Binkley, S. B., 1972b, 3-Hydroxy-*N*-acetylneuraminic acid. Synthesis and inhibitory properties, *Arch. Biochem. Biophys.* **151**:243.
Dickinson, H. R., and Bush, C. A., 1975, Circular dichroism of oligosaccharides containing neuraminic acid, *Biochem.* **14**:2299.
Dimitrov, G. D., 1973, Spectrophotometric method for qualitative and quantitative determination of sialic acid in glycoproteins and glycopeptides, *Hoppe-Seyl. Z.* **354**:121.
Diringer, H., 1972, The thiobarbituric acid assay of sialic acids in the presence of large amounts of lipids, *Hoppe-Seyl. Z.* **353**:39.
Faillard, H., 1965, The synthesis of *N*-glycolylneuraminic acid and the biochemistry of glycoproteins containing neuraminic acid, *Angew. Chem. (Intern. Ed.)* **4**:445.
Faillard, H., and Blohm, M., 1965, Synthese der *N*-Glycolyl-neuraminsäure, *Hoppe-Seyl. Z.* **341**:167.
Faillard, H., Kirchner, G., and Blohm, G., 1966, Anomere Benzylglykoside der *N*-Acetylneuraminsäure, *Hoppe-Seyl. Z.* **347**:87.
Faillard, H., Ferreira Do Amaral, C., and Blohm, M., 1969, Untersuchungen zur enzymatischen Spezifität der Neuraminidaze und *N*-Acyl-neuraminat-Lyase in bezug auf die *N*-Substitution, *Hoppe-Seyl. Z.* **350**:798.
Fidgen, K. J., 1973, An improved automated method for the estimation of sialic acid released in the neuraminidase assay, *Anal. Biochem.* **54**:349.
Fischmeister, I., 1958, Infrarotspektren von Sialinsäuren und Sialinsäuremethylestern, *Arkiv Kemi.* **13**:247.
Flippen, J. L., 1973, The crystal structure of β-D-*N*-acetylneuraminic acid dihydrate (sialic acid), $C_{11}H_{19}NO_9 \cdot 2H_2O$, *Acta Cryst.* **B29**:1881.
Folch, J., Arsove, S., and Meath, J. A., 1951, Isolation of brain strandin, a new type of large molecule tissue component, *J. Biol. Chem.* **191**:819.
Gantt, R., Millner, S., and Binkley, S. B., 1964, Inhibition of *N*-acetylneuraminic acid aldolase by 3-fluorosialic acid, *Biochemistry* **3**:1952.
Gerbant, L., Rey, E., and Lombart, C., 1973, Improved automated determination of bound *N*-acetylneuraminic acid in serum, *Clin. Chem.* **19**:1285.
Gibbons, R. A., 1963, The sensitivity of the neuraminosidic linkage in mucosubstances towards acid and towards neuraminidase, *Biochem. J.* **89**:380.
Gielen, W., 1965, Beitrag zur Chemie der Neuraminsäure, *Hoppe-Seyl. Z.* **342**:170.
Gielen, W., 1967a, Beitrag zur Chemie der Neuraminsäure, *Hoppe-Seyl. Z.* **348**:329.
Gielen, W., 1967b, Die Synthese der Methoxyneuraminsäure, *Hoppe-Seyl. Z.* **348**:378.
Gielen, W., 1968, Neuraminsäure in Pflanzen? Die 2-Keto-3-desoxyaldonsäuren in Pflanzen und die Synthese der 3-Deoxy-D-*glycero*-β-D-*galakto*-nonulosonäure, *Z. Naturforsch.* **23b**:1598.

Gottschalk, A., 1955, 2-carboxypyrrole: its preparation from and its precursor in mucoproteins, *Biochem. J.* **61**:298.
Gottschalk, A., 1958, Neuraminidase: its substrate and mode of action, *Adv. Enzymol.* **20**:135.
Gottschalk, A., 1960, *The Chemistry and Biology of Sialic Acids and Related Substances*, Cambridge University Press, London.
Gottschalk, A., 1962, The relation between structure and function in some glycoproteins, *Perspectives Biol. Med.* **5**:327.
Gottschalk, A., 1966, in: *Glycoproteins. Their Composition, Structure and Function* (A. Gottschalk, ed.) 1st ed., p. 173, Elsevier Publishing Co., Amsterdam.
Gottschalk, A., and Lind, P. E., 1949, Product of interaction between influenza virus enzyme and ovomucin, *Nature* **164**:232.
Gutteridge, J. M. C., Stocks, J., and Dormandy, T. L., 1974, Thiobarbituric acid reacting substances derived from autoxidizing linoleic and linolenic acids, *Anal. Chim. Acta* **70**:107.
Granzer, E., 1962, Dünnschichtchromatographie von Neuraminsäure-Derivaten, *Hoppe-Seyl. Z.* **328**:277.
Hahn, H., Hellman, B., Lernmark, A., Sehlin, J., and Tüljedal, I., 1974, The pancreatic β-cell recognition of insulin secretagogues, *J. Biol. Chem.* **249**:5275.
Hakomori, S., and Saito, T., 1969, Isolation and characterization of glycosphingolipid having a new sialic acid, *Biochemistry* **8**:5082.
Harris, J. U., and Klingman, J. D., 1972, Detection, determination, and metabolism *in vitro* of gangliosides in mammalian sympathetic ganglia, *J. Neurochem.* **19**:1267.
Haverkamp, J., Schauer, R., Wember, M., Kamerling, J. P., and Vliegenthart, J. F. G., 1975, Synthesis of 9-*O*-acetyl- and 4,9-di-*O*-acetyl derivatives of the methyl ester of *N*-acetyl-β-D-neuraminic acid methylglycoside. Their use as models in periodate oxidation studies, *Hoppe-Seyl. Z.* **356**:1575.
Heimer, R., and Meyer, K., 1956, Studies on sialic acid of submaxillary mucoid, *Proc. Nat. Acad. Sci.* **42**:728.
Hess, E. L., Coburn, A. F., Bates, R. C., and Murphy, P., 1957, A new method for measuring sialic acid levels in serum and its application to rheumatic fever, *J. Clin. Invest.* **36**:449.
Hess, H. H., and Rolde, E., 1964, Fluorometric assay of sialic acid in brain gangliosides, *J. Biol. Chem.* **239**:3215.
Hof, L., and Faillard, H., 1973, The serum dependence on the occurrence of *N*-glycolylneuraminic acid in HeLa cells, *Biochim. Biophys. Acta* **297**:561.
Holmquist, L., and Brossmer, R., 1972*a*, On the specificity of neuraminidase. The carboxymethyl α-ketoside of *N*-acetyl-D-neuraminic acid, a *Vibrio cholerae* neuraminidase substrate having two anionic sites, *FEBS Lett.* **22**:46.
Holmquist, L., and Brossmer, R., 1972*b*, On the specificity of neuraminidase. Synthesis and properties of the 2-aminoethyl α- and the 2-pyridyl α- and β-ketosides of N-acetyl-D-neuraminic acid, *Hoppe-Seyl. Z.* **353**:1346.
Horvat, A., and Touster, O., 1968, On the lysosomal occurrence and the properties of the neuraminidase of rat liver and of Ehrlich ascites tumor cells, *J. Biol. Chem.* **243**:4380.
Hotta, K., Kurokawa, M., and Isaka, S., 1970, Isolation and identification of two sialic acids from the jelly coat of sea urchin eggs, *J. Biol. Chem.* **245**:6307.
How, M. J., Halford, M. D. A., Stacey, M., and Vickers, E., 1969, An improved synthesis of *N*-acetylneuraminic acid, *Carbohyd. Res.* **11**:313.
Huang, R. T. C., and Klenk, E., 1972, α-Ketosidic linkage of the neuraminidase-resistant neuraminic acid in brain gangliosides, *Hoppe-Seyl. Z.* **353**:679.

Isemura, M., Zahn, R. K., and Schmid, K., 1973, A new neuraminic acid derivative and three types of glycopeptides isolated from the cuvierian tubules of the sea cucumber *Holothuria forskali, Biochem. J.* **131**:509.
Ishizuka, I., Kloppenburg, M., and Wiegandt, H., 1970, Characterization of gangliosides from fish brain, *Biochim. Biophys. Acta* **210**:299.
Jann, K., 1969, Neuraminsäurehaltige Polysaccharide in *Escherichia coli, Hoppe-Seyl. Z.* **350**:666.
Jeanloz, R. W., 1963, Recent developments in the biochemistry of amino sugars, *Adv. Enzymol.* **25**:433.
Jochims, J. C., Taigel, G., Seeliger, A., Lutz, P., and Drieson, H. E., 1967, Stereospecific long-rage couplings of hydroxyl protons of pyranoses, *Tetrahed. Lett.* **44**:4363.
Jourdian, G. W., Dean, L., and Roseman, S., 1971, The sialic acids. XI. A periodate-resorcinol method for the quantitative estimation of free sialic acids and their glycosides, *J. Biol. Chem.* **246**:430.
Kamerling, J. P., Vliegenthart, J. F. G., and Vink, J., 1974, Mass spectrometry of pertrimethylsilyl neuraminic acid derivatives, *Carbohyd. Res.* **33**:297.
Kamerling, J. P., Vliegenthart, J. F. G., Versluis, C., and Schauer, R., 1975a, Identification of O-Acetylated N-acylneuraminic acids by mass spectrometry, *Carbohyd. Res.* **41**:7.
Kamerling, J. P., Vliegenthart, J. F. G., Schauer, R., Strecker, G., and Montreuil, J., 1975b, Isolation and identification of 2-deoxy-2,3-dehydro-N-acetylneuraminic acid from the urine of a patient with sialuria, *Eur. J. Biochem.* **56**:253.
Karkas, J. D., and Chargaff, E., 1964, Studies on the stability of simple derivatives of sialic acid, *J. Biol. Chem.* **239**:949.
Karlsson, K. A., 1973, Carbohydrate composition and sequence analysis of cell surface components by mass spectrometry: characterization of the major monosialoganglioside of brain, *FEBS Lett.* **32**:317.
Karlsson, K. A., 1974, Carbohydrate composition and sequence analysis of a derivative of brain disialoganglioside bt mass spectrometry, with molecular weight ions at m/e 2245. Potential use in the specific microanalysis of cell surface components, *Biochemistry* **13**:3643.
Karlsson, K. A., Pascher, I., and Samuelsson, B. E., 1974, Analysis of intact gangliosides by mass spectrometry. Comparison of different derivatives of a hematoside of a tumour and the major monosialoganglioside of brain, *Chem. Phys. Lipids* **12**:271.
Kendal, A. P., 1968, An automated determination of sialic acids, *Anal. Biochem.* **23**:150.
Khorlin, A. Ya., and Privalova, I. M., 1966, The benzhydryl ester of N-acetylneuraminic acid, *Izv. Akad. Nauk SSSR Ser. Khim.* **1966**:1261.
Khorlin, A. Ya., and Privalova, I., 1967, Acylation of N-acetylneuraminic acid, *Khim, Prirodn. Soedin.* **1967**:191.
Khorlin, A. Ya., and Privalova, I. M., 1968, Glycosylation of N-acetylneuraminic acid, *Izv. Akad. Nauk SSSR, Ser. Khim.* **1968**:215.
Khorlin, A. Ya., and Privalova, I. M., 1970, Synthesis of N-acetylneuraminic acid 8-methyl ether, *Carbohyd. Res.* **13**:373.
Klenk, E., 1935, Über die Natur der Phosphatide und anderer Lipoide des Gehirns und der Leber in Niemann–Pickscher Krankheit, *Hoppe-Seyl. Z.* **235**:24.
Klenk, E., 1941, Neuraminsäure, das Spaltproduct eines neuen Gehirnlipoids, *Hoppe-Seyl. Z.* **268**:50.
Klenk, E., 1942, Über die Ganglioside, eine neue Gruppe von zuckerhaltigen Gehirnlipoiden, *Hoppe-Seyl. Z.* **273**:76.

Klenk, E., and Faillard, H., 1954, Zur Kenntnis der Kohlenhydratgruppen der Mucoproteide, *Hoppe-Seyl. Z.* **298**:230.

Klenk, E., and Langerbeins, H., 1941, Verteilung von Neuraminsäure in Gehirn. Mit einer Mikromethode für die Bestimmung dieser Substanz im Nervengewebe, *Hoppe-Seyl. Z.* **270**:185.

Klenk, E., and Lauenstein, K., 1952, Zur Kenntnis der Kohlenhydratgruppen des Submaxilleris mucins und Harnmucoproteids. Die Isolierung von Neuraminsaüren als Spaltprodukt, *Hoppe-Seyl. Z.* **291**:147.

Klenk, E., Faillard, H., Weygand, F., and Schöne, H. H., 1956, Untersuchungen über die Konstitution der Neuraminsäure, *Hoppe-Seyl. Z.* **304**:35.

Kochetkov, N. K., Chizhov, O. S., Kadentsev, V. I., Smirnova, G. P., and Zhukova, I. G., 1973, Mass spectra of acetylated derivatives of sialic acid, *Carbohyd. Res.* **27**:5.

Kuhn, R., and Baschang, G., 1962a, Die Konfiguration der Sialinsäuren am C-Atom 4, *Chem. Ber.* **95**:2384.

Kuhn, R., and Baschang, G., 1962b, Synthese der lactaminsäure, *Ann. Chem.* **659**:156.

Kuhn, R., and Brossmer, R., 1956a, Abbau der Lactaminsäure zu *N*-Acetyl-D-glucosamin, *Chem. Ber.* **89**:2471.

Kuhn, R., and Brossmer, R., 1956b, Über *O*-Acetyl-lactaminsäurelactose aus Kuh-Colostrum und ihre Spaltbarkeit durch Influenza-Virus, *Chem. Ber.* **89**:2013.

Kuhn, R., and Brossmer, R., 1957, Die Konfiguration der Lactaminsäure, *Angew. Chem.* **69**:534.

Kuhn, R., and Brossmer, R., 1958a, Zur Konfiguration der Lactaminsäure, *Ann. Chem.* **616**:221.

Kuhn, R., and Brossmer, R., 1958b, Die Konstitution der Lactaminsäurelactose; α-Ketosidase-Wirkung von Viren der Influenza-Gruppe, *Angew, Chem.* **70**:25.

Kuhn, R., and Brossmer, R., 1959, Über das durch Viren der Influenza-Gruppe spaltbare Trisaccharid der Milch, *Chem. Ber.* **92**:1667.

Kuhn, R., and Brossmer, R., 1962, The configuration of sialic acids, *Angew. Chem. Intern. Ed.* **1**:218.

Kuhn, R., and Lutz, P., 1963, Über Formyl-brenztraubensäure und den Farbstoff der Warren-Reaktion, *Biochem. Z.* **338**:554.

Kuhn, R., Brossmer, R., and Schulz, W., 1954, Über die prosthetische Gruppe der Mucoproteine des Kuh-Colstrums, *Chem. Ber.* **87**:123.

Kuhn, R., Lutz, P., and MacDonald, D. L., 1966, Synthese anomerer Sialinsäuremethylketoside, *Chem. Ber.* **99**:611.

Labat, J., and Schmid, K., 1969, Neuraminidase-resistant sialyl residues of α_1-acid glycoprotein, *Experientia* **25**:701.

Ledeen, R., 1970, New developments in the study of ganglioside structures, *Chem. Phys. Lipids* **5**:205.

Ledeen, R., and Salsman, K., 1965, Structure of the Tay–Sachs ganglioside, *Biochemistry* **4**:2225.

Ledeen, R. W., and Yu, R. K., 1973a, Structure and enzymic degradation of sphingolipids, in: *Lysosomes and Storage Diseases* (H. G. Hers and F. Van Hoof, eds.), pp. 105–145, Academic Press, New York and London.

Ledeen, R. W., and Yu, R. K., 1973b, GLC assay of sialic acid, in: *Biological Diagnosis of Brain Disorders* (S. Bogoch, ed.), pp. 377–9, Spectrum Publications, New York.

Ledeen, R., Salsman, K., and Cabrera, M., 1968, Gangliosides of bovine adrenal medulla, *Biochemistry* **7**:2287.

Ledeen, R. W., Kundu, S. K., Price, H. C., and Fong, J. W., 1974, Mass spectra of permethyl derivatives of glycosphingolipids, *Chem. Phys. Lipids* **13**:429.
Liao, T.-H., Gallop, P., and Blumenfeld, O. O., 1973, Modification of sialyl residues of sialoglycoprotein(s) of the human erythrocyte surface, *J. Biol. Chem.* **248**:8247.
Liu, T. Y., Gotschlich, E. C., Dunne, F. T., and Jonssen, E. K., 1971, Studies on the menigococcal polysaccharides. II. Composition and chemical properties of the group B and group C polysaccharide, *J. Biol. Chem.* **246**:4703.
Lutz, P., Lochinger, W., and Taigel, G., 1968, Zur Konformation der N-Acetylneuraminsäure, *Chem. Ber.* **101**:1089.
Martensson, E., Raal, A., and Svennerholm, L., 1958, Sialic acids in blood serum, *Biochim. Biophys. Acta* **30**:124.
McGuire, E. J., and Binkley, S. B., 1964, The structure and chemistry of colominic acid, *Biochemistry* **3**:247.
McLean, R., and Beidler, J., 1969, 5-Acetamido-3,5-dideoxy-D-*galacto*-octulosonic acid, an eight-carbon analog of N-acetylneuraminic acid, *J. Am. Chem. Soc.* **91**:5388.
Meindl, P., and Tuppy, H., 1965a, Über synthetische Ketoside der N-Acetyl-D-neuraminsäure. 1. Mitt.: Darstellung einer Reihe durch Neuraminidase spaltbarer Ketoside, *Monatsh. Chem.* **96**:802.
Meindl, P., and Tuppy, H., 1965b, Über synthetische Ketoside der N-Acetyl-D-neuraminsäure, 2. Mitt.: Anomere n-Amyl-und n-Hexylketoside der N-Acetyl-D-neuraminsäure, *Monatsh, Chem.* **96**:816.
Meindl, P., and Tuppy, H., 1966a, Über synthetische Ketoside der N-Glykolyl-D-neuraminsäure, *Monatsh. Chem.* **97**:654.
Meindl, P., and Tuppy, H., 1966b, Über die Spaltung synthetischer Sialinsäure-α-Ketoside durch Neuraminidase, *Monatsh. Chem.* **97**:990.
Meindl, P., and Tuppy, H., 1966c, Darstellung und enzymatische Spaltbarkeit von α-Ketosiden der N-Propionyl-, N-Butyryl-, und N-Benzoyl-D-neuraminsäure, *Monatsh. Chem.* **97**:1628.
Meindl, P., and Tuppy, H., 1967, Über synthetische Ketoside der N-Acetyl-D-neuraminsäure, 3. Mitt.: Darstellung des Phenyl-α-Ketosides der N-Acetyl-D-neuraminsäure und seines β-Anomeren, *Monatsh. Chem.* **98**:53.
Meindl, P., and Tuppy, H., 1969, Uber 2-Desoxy-2.3-dehydro-sialinsäueren, II. Kompetitive Hemmung der Vibriocholerae-Neuraminidase durch 2-Desoxy-2.3-dehydro-N-acyl-neuraminsäueren, *Hoppe-Seyl. Z.* **350**:1088.
Meindl, P., and Tuppy, H., 1973, 2-Deoxy-2,3-dehydrosialic acid. III. Synthesis and properties of 2-deoxy-2,3-dehydroneuraminic acid and of new N-acyl derivatives, *Monatsh. Chem.* **104**:402.
Meindl, P., Bodo, G., Palese, P., Schulman, J., and Tuppy, H., 1974, Inhibition of neuraminidase activity by derivatives of 2-deoxy-2,3-dehydro-N-acetylneuraminic acid, *Virology* **58**:457.
Mesnard, P., and Devaux, G., 1964, Nouvelle méthode de dosage colorimétrique de l'acide quinique et de ses esters naturels, *Bull. Soc. Chim.* (France) **1964**:43.
Miettinen, T., and Takki-Luukkainen, I.-T., 1959, Use of butyl acetate in determination of sialic acid, *Acta Chem. Scand.* **13**:856.
Mirzayanova, M. N., Davydova, L. P., and Samokhvalov, G. I., 1970, New synthesis of N-acetylneuraminic acid, *Z. Obshchei Khim.* **40**:693.
Neuberger, A., and Ratcliffe, W. A., 1972, The acid and enzymic hydrolysis of O-acetylacted sialic acid residues from rabbit Tamm–Horsfall glycoprotein, *Biochem. J.* **129**:683.

Neuberger, A., and Ratcliffe, W. A., 1973, Kinetic studies on the acid hydrolysis of the methyl ketoside of unsubstituted and O-acetylated N-acetylneuraminic acid, *Biochem. J.* **133**:623.

O'Connell, A. M., 1973, The crystal structure of N-acetylneuraminic acid methyl ester monohydrate, *Acta Cryst.* **B29**:2320.

Ohkuma, S., and Miyauchi, C., 1966, Chromogen formation and degradation of N-acetylhexosamines and N-acetylneuraminic acid by barium hydroxide treatment, *Nature* **211**:190.

Onodera, K., Hirano, S., and Hayashi, H., 1965, Sialic acid and related substances. II. A comparative assay of N-acetylneuraminic acid, *Carbohyd. Res.* **1**:44.

Paerels, G. B., and Schut, J., 1965, The mechanism of the periodate-thiobarbituric acid reaction of sialic acids, *Biochem. J.* **96**:787.

Palese, P., Bucher, D., and Kilbourne, E. D., 1973, Applications of a synthetic neuraminidase substrate, *Appl. Microbiol.* **25**:195.

Pepper, D. S., 1968, The sialic acids of horse serum with special reference to their virus inhibitory properties, *Biochim. Biophys. Acta* **156**:317.

Price, H. C., and Yu, R. K., 1976, Adrenal medulla gangliosides. A comparative study of some mammals, *Comp. Biochem. Physiol.* **54B**:451.

Price, H. C., Kundu, S., and Ledeen, R., 1975, Structures of gangliosides from bovine adrenal mudulla, *Biochemistry* **14**:1512.

Privalova, I. M., and Khorlin, A. Ya., 1969, Substrates and inhibitors of neuraminidases. Communication synthesis of O-, S-, and N-ketosides of N-acetyl-D-neuraminic acid, *Izv. Akad. Nauk SSSR, Ser. Khim.* **1969**:2785 (Eng. trans. p. 2614).

Roseman, S., Jourdian, G. W., Watson, D., and Rood, R., 1961, Enzymic synthesis of sialic acid 9-phosphates, *Proc. Nat. Acad. Sci. U.S.* **47**:955.

Rosenberg, S. A., and Einstein, Jr., A. B., 1972, Sialic acids on the plasma membrane of cultured human lymphoid cells, *J. Cell. Biol.* **53**:466.

Saifer, A., and Feldman, N. I., 1971, The photometric determination of gangliosides with the sulfo-phospho-vanillin reaction, *J. Lipid Res.* **12**:112.

Saito, Y., 1956, An improved method for the preparation of sialic acid or acetylneuraminic acid, *Nature* **178**:995.

Schauer, R., 1970, Biosynthese von N-acetyl-O-acetylneuraminsäuren, I., *Hoppe-Seyl. Z.* **351**:595.

Schauer, R., 1973, Chemistry and biology of acylneuraminic acids, *Angew. Chem. Intern. Ed.* **12**:127.

Schauer, R., and Buscher, H. P., 1974, An improved method for the synthesis of ^{14}C-labeled or ^{3}H-labeled N-acetylneuraminic acid, *Biochim. Biophys. Acta* **338**:369.

Schauer, R., and Faillard, H., 1968, Das Verhalten isomerer N,O-Diacetylneuraminsäure glykoside in Submaxillarismucin von Pferd und Rind bei Einwirkung bakterieller Neuraminidase, *Hoppe-Seyl. Z.* **349**:961.

Schauer, R., Wirtz-Peitz, F., and Faillard, H., 1970, Synthese von N-acylneuraminsäueren. II. N-[1-^{14}C]Glycolyl-N-Chloracetyl-und N-Fluoracetylneuraminsäure, *Hoppe-Seyl. Z.* **351**:359.

Schauer, R., Haverkamp, J., Wember, M., Vliegenthart, J. F., and Kamerling, J. P., 1976, N-Acetyl-9-O-L-lactylneuraminic acid. A new acylneuraminic acid from bovine submandibular gland, *Eur. J. Biochem.* **62**:237.

Schneir, M., Benya, P., and Buch, L., 1970, Determination of malonaldehyde in the presence of sialic acid, *Anal. Biochem.* **35**:46.

Seibert, F. B., Pfabb, M. L., and Seibert, M. V., 1948, *Arch. Biochem. Biophys.* **18**:279.

Spence, M. W., and Wolfe, L. S., 1967, Gangliosides in developing rat brain. Isolation and composition of subcellular membranes enriched in gangliosides, *Can. J. Biochem.* **45**:671.
Spiro, R. G., 1964, Periodate oxidation of the glycoprotein fetuin, *J. Biol. Chem.* **239**:567.
Spiro, R. G., 1966, Analysis of sugars found in glycoproteins, *Methods Enzymol.* **8**:3.
Stephen, A. M., and Jeanloz, R. W., 1966, Synthesis of N-acetylneuraminic acid and derivatives, *Fed. Proc.* **25**:409.
Stone, A. L., and Kolodny, E. H., 1971, Circular dichroism of gangliosides from normal and Tay–Sachs tissues, *Chem. Phys. Lipids* **6**:274.
Suttajit, M., and Winzler, R. J., 1971, Effect of modification of N-acetylneuraminic acid on the binding of glycoproteins to influenza virus and on susceptibility to cleavage by neuraminidase, *J. Biol. Chem.* **246**:3398.
Svennerholm, L., 1956, On the isolation and characterization of N-acetylsialic acid, *Acta Soc. Med. Upsalien* **61**:75.
Svennerholm, L., 1957a, Quantitative estimation of sialic acids. I. A colorimetric method with orcinol–hydrochloric acid (Bial's) reagent, *Arkiv Kemi.* **10**:577.
Svennerholm, L., 1957b, Quantitative estimation of sialic acids. II. A colorimetric resorcinol–hydrochloric acid method, *Biochim. Biophys. Acta* **24**:604.
Svennerholm, L., 1958, Quantitative estimation of sialic acids. III. An anionic exchange resin method, *Acta Chem. Scand.* **12**:547.
Svennerholm, L., 1963a, Sialic acids and derivatives: Preparation, *Methods Enzymol.* **6**:453.
Svennerholm, L., 1963b, Sialic acids and derivatives: Estimation by the ion-exchange method, *Methods Enzymol.* **6**:459.
Svennerholm, L., 1963c, Chromatographic separation of human brain gangliosides, *J. Neurochem.* **10**:613.
Sweeley, C. C., and Walker, B., 1964, Determination of carbohydrates in glycolipids and gangliosides by gas chromatography, *Anal. Chem.* **36**:1461.
Tettamanti, G., Bertona, L., Berra, B., and Zambotti, V., 1964, On the evidence of glycolylneuraminic acid in beef brain gangliosides, *Ital. J. Biochem.* **13**:315.
Tuppy, H., and Gottschalk, A., 1972, The structure of sialic acids and their quantitation, in: *Glycoproteins. Their Composition, Structure and Function* (A. Gottschalk, ed.), 2nd ed., pp. 403–449, Elsevier Publishing Co., Amsterdam.
Tuppy, H., and Palese, P., 1969, Chromogenic substrate for the investigation of neuraminidases, *FEBS Lett.* **3**:72.
Vaitukaitis, J. L., Sherins, R., Ross, G. T., Hickman, J., and Ashwell, G., 1971, A method for the preparation of radioactive FSH with preservation of biologic activity, *Endocrinology* **89**:1356.
Vance, D. E., and Sweeley, C. C., 1967, Quantitative determination of the neutral glycosyl ceramides in human blood, *J. Lipid Res.* **8**:621.
Van Lenten, L., and Ashwell, G., 1971, Studies on the chemical and enzymatic modification of glycoproteins. A general method for the tritiation of sialic-acid containing glycoproteins, *J. Biol. Chem.* **246**:1426.
Walkowiak, H., Kedzierska, B., and Starzynski, W., 1968, Horizontal circular chromatography of sialic acids from serum of certain mammal species, *Bull. Acad. Pol. Sci., Ser. Sci. Biol.* **16**:97.
Waravdekar, V. S., and Saslaw, L. D., 1957, A method of estimation of 2-deoxyribose, *Biochim. Biophys. Acta* **24**:439.
Warren, L., 1959, The thiobarbituric acid assay of sialic acids, *J. Biol. Chem.* **234**:1971.

Warren, L., 1963a, The distribution of sialic acids in nature, *Comp. Biochem. Physiol.* **10**:153.
Warren, L., 1963b, Thiobarbituric acid assay of sialic acid, *Methods Enzymol.* **6**:463.
Warren, L., 1964, *N*-Glycolyl-8-*O*-methylneuraminic acid. A new form of sialic acid in the starfish *Asterias forbesi*, *Biochim. Biophys. Acta* **83**:129.
Warren, L., and Felsenfeld, H., 1961, *N*-Acetylmannosamine-6-phosphate and *N*-acetylneuraminic acid -9-phosphate as intermediates in sialic acid biosynthesis, *Biochem. Biophys. Res. Comm.* **5**:185.
Warren, L., and Felsenfeld, H., 1962, The biosynthesis of sialic acids, *J. Biol. Chem.* **237**:1421.
Watson, D. F., Jourdian, G. W., and Roseman, S., 1966a, The sialic acids. VIII. Sialic acid 9-phosphate synthetase, *J. Biol. Chem.* **241**:5627.
Watson, D. R., Jourdian, G. W., and Roseman, S., 1966b, *N*-Acylneuraminic (Sialic) acid 9-phosphate synthetase, *Methods Enzymol.* **8**:201.
Watson, R. G., Marinetti, G. V., and Scherp, H. W., 1958, The specific hapten of group C (Group II α) Menigococcus. II. Chemical Nature, *J. Immunol.* **81**:337.
Weissbach, A., and Hurwitz, J., 1959, The formation of 2-keto-3-deoxyheptonic acid in extracts of *Escherichia coli* B, *J. Biol. Chem.* **234**:705.
Wenger, D. A., and Wardell, S., 1973, Action of neuraminidase (E.C.3.2.1.18) from *Clostridium perfringens* on brain gangliosides in the presence of bile salts, *J. Neurochem.* **20**:607.
Werner, I., and Odin, L., 1952, On the presence of sialic acid in certain glycoproteins and in gangliosides, *Acta Soc. Med. Upsalien* **57**:230.
Wesemann, W., and Zilliken, F., 1966, Synthesen von *N*-Acyl-neuraminsäuren, I. Rezeptoren der Neurotransmitter, *Ann. Chem.* **695**:209.
Wirtz-Peitz, F., Schauer, R., and Faillard, H., 1969, Synthese von *N*-Acyl-neuraminsäuren aus Neruaminsäure-β -methylglykosid, I., *Hoppe-Seyl. Z.* **350**:111.
Yamakawa, T., and Suzuki, S., 1952, The chemistry of the lipids of posthemolytic residue or stroma of erythrocytes. III. On the structure of hemataminic acid, *J. Biochem. (Tokyo)* **39**:175.
Yu, R. K., and Ledeen, R. W., 1969, Configuration of the ketosidic bond of sialic acid, *J. Biol. Chem.* **244**:1306.
Yu, R. K., and Ledeen, R. W., 1970, Gas–liquid chromatographic assay of lipid-bound sialic acid: measurement of gangliosides in brain of several species, *J. Lipid Res.* **11**:506.
Yu, R. K., and Ledeen, R. W., 1972, Gangliosides of human, bovine, and rabbit plasma, *J. Lipid Res.* **13**:680.
Zilliken, F., and Glick, M. C., 1956, Alkalischer Abbau von Gynaminsäure zu Brenztraubensäure und *N*-Acetyl-D-glucosamin, *Naturwissenschaften* **43**:536.
Zilliken, F., and O'Brien, P. J., 1960, *N*-Acetylneuraminic acid, *Biochem. Prep.* **7**:1.
Zilliken, F., and Whitehouse, M. W., 1958, The nonulosaminic acids. Neuraminic acids and related compounds (sialic acids), *Adv. Carbohyd. Chem.* **13**:237.
Zilliken, F., Braun, G. A., and György, P., 1955, Gynaminic acid. A naturally occurring form of neuraminic acid in human milk, *Arch. Biochem. Biophys.* **54**:564.

Chapter 2

The Natural Occurrence of Sialic Acids

Sai-Sun Ng and Joel A. Dain

I. INTRODUCTION

The sialic acids are widely distributed in nature, either free or as components of homo- and heterosaccharides, glycoproteins, and glycolipids (Tuppy and Gottschalk, 1972). Although a considerable literature has accumulated on sialic acids and the sialo compounds, a unified concept about the biological roles of sialic acids has not yet developed due to both their heterogeneous occurrence and their possible involvement in diverse cellular functions (Faillard and Schauer, 1972; Gottschalk, 1972; Marshall, 1972; Mehrishi, 1972; Curtis, 1973; Hughes, 1973; Schauer, 1973; Weiss, 1973). This chapter will approach a better understanding of the biological roles of sialic acids by examining their occurrence in nature. A possible correlation between the evolution and biological roles of sialic acids is considered.

II. THE NATURAL OCCURRENCE OF SIALIC ACIDS

A. Viruses

Sialic acid has been reported in an arborvirus, the Sindbis virus. The envelope membrane of this virus consists of a lipid and a protein containing 14% carbohydrate. The sialic acid present probably occupies the terminal positions of at least the largest carbohydrate chains (Strass et al., 1970; Burge and Strass, 1970). While the membrane protein core

---

SAI-SUN NG and JOEL A. DAIN • Department of Biochemistry, University of Rhode Island, Kingston, Rhode Island 02881.

of Sindbis virus is coded by the viral genome, the addition of the carbohydrate moieties may be partially or entirely mediated by glycosyltransferases of the cells infected by the virus (Burge and Strass, 1970).

Recent reports on Sindbis virus have demonstrated that there are two sialoglycoproteins located on its membrane and at least two more virus-specific glycoproteins in the infected cells (Compans, 1971; Sefton and Burge, 1973). Gas chromatographic analysis of the two virion glycoproteins revealed a carbohydrate content of 8% by weight, with mannose as the major constituent and with lesser amounts of glucosamine, galactose, sialic acid, and fucose. Their oligosaccharide structures appeared to be unrelated (Sefton and Keegstra, 1974). Sindbis virus grown in chick embryo fibroblasts also contained four sialoglycolipids which were apparently the same as those of the host cell. The principal glycolipid was hematoside (G_{M3}). This finding further supports the hypothesis that the virus probably incorporates a nonspecific population of host cell plasma membrane components during viral morphogenesis (Hirschberg and Robbins, 1974).

Two other viruses, Rous sarcoma and vesicular stomatitis, have also been shown to possess sialic acid. The major glycoprotein component of Rous sarcoma virus, g2, contained 40% by weight carbohydrate, with all the sialic acid (0.4 mole/mg isolated peptide) susceptible to *Clostridium perfringens* neuraminidase (Krantz *et al.*, 1974). Both sialoglycoproteins and sialoglycolipids were found in vesicular stomatitis virus. Interestingly, removal of sialic acid by neuraminidase treatment reduced the infectivity of the virus, and resialylation of desialylated virions substantially restored infectivity (Schloemer and Wagner, 1974). Furthermore, differences were evident in the composition of component sugars of glycoproteins when the virus was grown in different host cells (Etchison and Holland, 1974). Detailed analyses on viral membrane components will prove extremely useful in elucidating the mechanisms of interactions of viruses with host cells.

B. Bacteria

The occurrence of sialic acids in bacteria and their relationship to pathogenicity have been studied in some detail. Aaronson and Lessie (1960) examined the presence of sialic acids in several classes of bacteria and concluded that sialic acids are limited to the gram-negative bacteria. A later study, however, has indicated that some gram-positive bacteria also contain sialic acids, and the species examined all are pathogenic (Irani and Ganapathi, 1962; Luppi and Cavazzini, 1966). In *Escherichia coli*, both *N*-acetylneuraminic and *N*-acetyl-7-*O*-acetylneuraminic acids are found (DeWitt and Rowe, 1958; 1959; 1961). Some *E. coli* strains, however, do not contain any sialic acids (Barry *et al.*, 1963).

Most of the sialic acids in bacteria are in a bound form as components of membrane structures (Irani and Ganapathi, 1962). The *E. coli* endotoxin contains sialic acids, and in *Neisseria meningitidis*, strain 1908, the sialic acids are the main structural units of capsular material (Watson *et al.*, 1958).

Bacteria of the O serotype, such as *Salmonella* O48 and *Arizona* O29, contain sialic acids as part of their O antigens (Barry *et al.*, 1962; Barry, 1965; Kauffmann *et al.*, 1962). The nature of the O serotype and the production of hemolysins or bacteriocines are related to the presence of sialic acids. The O antigens presumably are part of the membrane glycoproteins (Barry *et al.*, 1960; Barry, 1965) and may contribute to the pathogenic properties of the organisms (Frits *et al.*, 1971).

E. coli K-235 of K-1 serotype produces colominic acid, a homopolymer of N-acetylneuraminic acid with (2→8) glycosidic linkages (Barry and Goebel, 1957; Barry, 1958). In addition, an ester linkage between the carboxyl group of one N-acetylneuraminic acid residue and the C_7- or C_9-hydroxyl group of an adjacent residue (or a lactone linkage) presumably exists, since the complete hydrolysis of the homopolymer by *Clostridium perfringens* neuraminidase has a very long lag period (McGuire and Binkley, 1964). Two aspects of the biosynthesis of the colominic acid polymer are of interest: (i) the N-acetylneuraminic acid moiety of CMP-N-acetylneuraminic acid is transferred to a lipid molecule to form a lipid–N-acetylneuraminic acid complex which then transfers the sialic acid moiety to the elongating polymer chain (Troy *et al.*, 1973), and (ii) chain elongation proceeds comparably to the formation of glycogen at the nonreducing terminal of the homopolymer rather than at the reducing end as in bacterial lipopolysaccharide biosynthesis (Kundig *et al.*, 1971).

Strains of other K serotypes do not contain significant amounts of sialic acids (Barry *et al.*, 1962; Barry, 1959). Interestingly, there is no serological relationship between *E. coli* of the K serotypes and other bacteria of O serotype (Barry *et al.*, 1962; Kedzierska *et al.*, 1968). Furthermore, the existence of sialic acids in *E. coli*, in contrast to bacteria of O serotype, is not related to the presence of bacteriocine or hemolysin (Barry, 1959).

C. Plants

The presence of sialic acids could not be demonstrated in fungi and yeast (Aaronson and Lessie, 1960). Nor, in an extensive survey of sialic acid distribution in nature, could they be detected in slime mold, lichens, mosses, ferns, and the flowering plants (Warren, 1963). In fact, it was not until 1964 that hexosamine was first reported in higher plant glycoproteins (Pustzai, 1964).

Several studies, nevertheless, more recently have suggested the presence of sialic acids in plants. Correll (1964) observed a sialic-acid-containing glycopeptide in the plantlike protist, *Chlorella*. Mayer et al. (1964) reported N-glycolylneuraminic acid in the mild acid hydrolyzate of the delipidated seeds of soya bean (*Glycine max*) and alfalfa (*Medicago sativa*). Onodera et al. (1966) claimed that most of the 64 plant materials that they studied contained more than 1 mg% sialic acids, and there were some that contained greater than 10 mg%. More recently Faillard (1969) has detected sialic acids in the leguminoses.

Despite these reports, however, the presence of sialic acids in plants remains doubtful. Gielen (1968a) has shown that the substance isolated from soya beans and bananas belongs to the group of 2-keto-3-deoxyaldonic acids which do not contain nitrogen. These compounds give a positive reaction with periodate-thiobarbiturate but not with Bial's reagent. This observation correlates well with studies by Cabezas (1968). Detailed analyses (Cabezas and Feo, 1969) revealed that the periodate-thiobarbiturate positive compound from several plant materials was negative towards resorcinol-butyl-acetate, direct Ehrlich, Bial's orcinol, and Hestrin's O-acetyl group reagents. The compound migrated close to glucosamine on paper chromatography and was split by N-acetylneuraminic acid aldolase to give neither N-acetyl-D-mannosamine nor N-acetyl-D-galactosamine. It also had a gas chromatographic property distinct from N-acetylneuraminic acid, hexoses, osamines, and several acids that were tested. From these studies it has become clear that the characterization of sialic acid by the thiobarbituric acid method (Warren, 1959) is not sufficient, a point that was stressed earlier by Barry et al. (1963) and Warren (1963).

D. Invertebrates

In early articles, sialic acids were reported to be absent in protozoa (Aaronson and Lessie, 1960; Warren, 1963). Some representative sponges, ctenophores, coelenterates, and the bryozoans also were devoid of sialic acids (Warren, 1963).

More recent reports, however, suggest that some protozoans do contain sialic acid components. The marine diatom, *Nitzschia alba,* was found to possess components containing sialic acid, with the highest concentration in a light smooth membrane preparation (Sullivan and

Volcani, 1974). The extracellular malaria parasites, *Plasmodium berghei,* were shown to contain about half the amount of sialic acid per unit weight as the control red cell extract, although the sialic acid residues were not exposed to the outer surface of the cell membrane (Seed *et al.,* 1974). Still to be investigated, as has been done with the Sindbis virus, is the extent of host dependency on the presence of sialic acids in this and other parasites which have extensive intracellular phases in their life cycles.

A primitive platyhelminth, *Polychoerus carmelensis,* an acoel turbellarian, contains bound N-acetylneuraminic acid. Other turbellarians tested were found to be devoid of sialic acids. A trematode, *Fascioloides magna,* contains sialic acids which may be of dietary blood origin (Warren, 1963).

Sialic acids have not been detected in annelids, nemertines, sipunculids, or chaetognaths (Warren, 1963).

The presence of sialic acids in molluscs was demonstrated in the digestive glands of the squid, *Loligo pealii,* at least 80% of which is in a free state (Warren, 1963). Sialic acids are also present in a snail (Eldredge *et al.,* 1963). Sialoglycolipids (gangliosides) have been reported present in the nervous tissues of squid and octopus (Ishizuka *et al.,* 1970; Noren and Svennerholm, 1970). Purified hemagglutinin subunits from the oyster, *Crassostrea virginica,* contain one sialic acid residue per molecule (Acton *et al.,* 1973).

A hemagglutinin containing sialic acid has also been purified from the lobster, *Panulirus argus* (Acton *et al.,* 1973). Free sialic acid has been found (Warren, 1963) in the digestive glands of the lobster, *Homarus americanus.* The eye stalks of crab contain gangliosides (Wiegandt, 1968; Ishizuka *et al.,* 1970; Noren and Svennerholm, 1970). However, the occurrence of gangliosides in the eye stalk of the crab, *Carcinus maenas,* could not be confirmed (Wiegandt, 1970). No gangliosides could be found in the blow fly, *Calliphora erythrocephala,* or the house fly, *Musca domestica* (Wiegandt, 1970). In fact no sialic acids were detected previously in three species of insects as well as two species of arachnoids (Warren, 1963).

In echinoderms, sialic acids are of common occurrence and are detected in most tissues of all species studied (Warren, 1963; Eldredge *et al.,* 1963). Most of the sialic acids are of the N-glycolyl type, but the starfish, *Asterias forbesi,* possesses a unique species, possibly 8-methoxyl-N-glycolylneuraminic acid (Warren, 1963). Another novel sialic acid has been identified in the sea urchin, *Pseudocentrotus depressus,* with the structure N-acetoglycolyl-4-methyl-4,9-dideoxyneuraminic acid (Hotta *et al.,* 1970*b*), cf. Chapter 1.

A sialoglycolipid was isolated from the sea urchin, *Strongylocentrotus intermedius*, which contained 2 moles each of glucose and sialic acid per mole of glycolipid. Both N-acetyl-(65%)- and N-glycolyl-(35%)-neuraminic acids were identified (Kochetkov et al., 1973).

While no N-acetylneuraminic acid was found in the egg jelly coat glycoprotein from *Pseudocentrotus depressus* (Hotta et al., 1970a), it is present in that of *Arbacia lixula, Paracentrotus lividus,* and *Sphaerechinus granularis* (Immers, 1968). In *Pseudocentrotus depressus* jelly coat glycoprotein, some if not all of the N-glycolylneuraminic acid is in a novel association with fucose as fucopyranosyl-(1→4)-N-glycolylneuraminic acid (Hotta et al., 1973).

The jelly coat of *Arbacia punctalata*, however, has no sialic acids. All sialic acids are in bound forms in the eggs, and most of the content is N-glycolylneuraminic acid associated with a lipid. The glycolipid does not resemble gangliosides of animal brain tissues (Warren, 1960). A different glycolipid was found in eggs and sperms of *Pseudocentrotus depressus*. It migrates as the monosialoganglioside on thin layer chromatography, but chemical analysis reveals only the presence of glucose and no galactose or N-acetylgalactosamine as in gangliosides (Isono and Nagai, 1966). Gonads of sea urchin, *Strongylocentrotus intermedius*, also contain sialoglycolipid; at least two species of sialic acids were detected (Kochetkov et al., 1968). Unidentified glycolipids, resembling gangliosides, were also reported in several species of echinoderms (Vaskovasky et al., 1970). So far, however, a definitive characterization of the gangliosides in invertebrates, including those of squid, octopus, and crab (Svennerholm, 1970), is still not available.

A previously unknown sialic acid was found in two glycopeptide fractions isolated from the Cuvierian tubules of the sea cucumber, *Holothuria forskali*. It is resistant to enzymic cleavage by neuraminidase, even after mild alkaline hydrolysis for the removal of O-acyl residues; however it is readily released by mild acid hydrolysis. Its chromatographic properties differ from other known sialic acids, but presumably the new species possesses neuraminic acid as its basic structure (Isemura et al., 1973). See Chapter 1.

E. Primitive Chordates

The hemichordate, *Dolichoglossus kowalevskii*, and the cephalochordate, *Branchiostoma (Amphioxus)*, contain sialic acids, mostly as N-glycolylneuraminic acid. In urochordates, including one urochordate larva, sialic acids could not be detected (Warren, 1963). The urochordates also do not contain any ganglioside-like glycolipids (Vaskovasky et al., 1970).

F. Vertebrates

1. Fishes. Sialic acids are present in cyclostomes (Turumi and Saito, 1953; Wessler and Werner, 1957), the lamprey (Eylar et al., 1962a; Cabezas and Frois, 1966), elasmobranchs, and teleosts (Warren, 1963). They are found in all tissues studied (Warren, 1963). The eggs of brook trout and rainbow trout have 50% of the sialic acid content in a free state; and both N-glycolyl- and N-acetylneuraminic acids are present (Warren, 1960). No N-glycolylneuraminic acid is detected in lamprey liver and eggs and in eggs from two teleosts. The sialic acid is chiefly N-acetylneuraminic acid with a small amount of N,O-diacetylneuraminic acid (Cabezas and Frois, 1966).

The fishes are the most primitive animals definitely known to possess the gangliosides containing sialic acid and fish brain tissues have been extensively studied for this class of glycolipid (Eldredge et al., 1963; Tettamanti et al., 1965; Tettamanti and Zambotti, 1967; Gielen, 1968b; Schengrund and Garrigan, 1969; Ishizuka et al., 1970; Seiter, 1970; Avrova, 1971; McCluer and Agranoff, 1972). Although fishes contain relatively low amounts of gangliosides when compared with the higher vertebrates, the ganglioside sialic acid content is high. This is attributed to the occurrence of tetrasialo- and pentasialogangliosides as well as to considerable amounts of monosialo- and disialogangliosides with shorter carbohydrate chain lengths (Tettamanti and Zambotti, 1967; Gielen, 1968b, Avrova, 1971). The polysialogangliosides may contribute about 50% of the total ganglioside content in most fishes (Ishizuka et al., 1970), while gangliosides of shorter carbohydrate chain lengths are more important in others (Tettamanti and Zambotti, 1967; Gielen, 1968b).

The structures of the polysialogangliosides have been elucidated by Ishizuka et al. (1972). Most of the sialic acid is N-acetylneuraminic acid. Also identified is N-acetyl-8-O-acetylneuraminic acid (Ishizuka et al., 1970; Seiter, 1970; McCluer and Agranoff, 1972).

2. Amphibians. Sialic acids are present in frogs (Bohm and Baumeister, 1956). In *Rana catesbeiana* they are found in muscle, skin, blood, and eggs (Warren, 1963). As in fishes, relatively high content of polysialogangliosides is characteristic of brains of the frogs (*R. clamitans, R. esculenta, R. pipiens, R. sylvaticus*, and *R. temporaria*) and the triton (*Triturus* sp.) (Tettamanti et al., 1965; Dain and Yip, 1969;

Yiamouyiannis and Dain, 1968; Avrova, 1968; 1971). Most, if not all, of the sialic acid is N-acetylneuraminic acid (Eldredge *et al.*, 1963).

In the oviduct mucoprotein of the toad, *Bufo arenarum,* four species of sialic acids were detected: N-acetyl-, $N,4$-O-diacetyl-, and two species of N-glycolylneuraminic acids (DeMartinez and Olavarria, 1973).

3. Reptiles. Turtle skin contains sialic acids (Warren, 1963). Turtle brain tissue contains N-acetylneuraminic acid, most of which is in gangliosides (Eldredge, 1963). The ganglioside pattern differs from that in fish and frog and resembles those of birds and mammals, with monosialo-, disialo-, and trisialogangliosides containing 80–90% of the total gangliosidic N-acetylneuraminic acid (Avrova, 1971). The content of brain gangliosides in reptiles in general is intermediate between that of fishes and amphibians on the one hand, and birds and mammals on the other (Avrova, 1968).

Histochemical studies have revealed the presence of sialic acid in cephalic glands of several reptilian species, *Ameiva ameiva* (Lacertilia, Teiidae), *Micrurus corallinus corallinus* (Ophiadea, Elapidae), and *Bothrops jararaca* (Ophiadia, Viperidae) (Lopes and Valeri, 1972; Lopes *et al.*, 1973; 1974).

A phylogenetic survey of the neuraminidase sensitivity of reptilian gonadotropins suggests that sialic acid is necessary for full biological activity of reptiles and is probably a primitive trait of reptilian gonadotropins. This trait has disappeared at least twice in reptilian evolution in one lineage of turtles and in some crocodilia (Licht and Papkoff, 1974).

4. Birds. Early studies showed the presence of sialic acids in the serum of hen (Bohm and Baumeister, 1956) and mucous from pheasant and goose (Wessler and Werner, 1957). Feeney *et al.* (1960) studied the chemical and the physical characteristics of avian egg white proteins and found sialic acids in the proteins of 25 species belonging to six orders of *Aves*. There are marked variations in sialic acid content between bird species, but they all appear to contain the same sialic acid, N-acetylneuraminic acid. Only a trace amount of free sialic acid was detected.

At least two species of sialic acid, N-acetyl- and N-glycolylneuraminic acids, have been identified in poultry serum, together with a very small amount of a third species, possibly N,O-diacetylneuraminic acid (Faillard and Cabezas, 1963; Dzulynska *et al.*, 1969). The nest-cementing substance of the Chinese swiftlet, collocalia mucoid, is a glycoprotein containing N-acetyl-4-O-acetylneuraminic acid (Kathan and Weeks, 1969).

As with all other vertebrates, the brain tissues of birds contain gangliosides. Qualitative and quantitative studies on chicken and pigeon have been given in several reports (Eldredge *et al.*, 1963; Garrigan and Chargaff, 1963; Tettamanti *et al.*, 1965; Avrova, 1968; 1971; Schengrund and Garrigan, 1969). The qualitative pattern in general resembles that of reptiles and mammals while the content is higher than that of the lower vertebrates (Avrova, 1968).

Gangliosides have also been identified as constituents of egg yolk. These gangliosides have shorter carbohydrate chains than the major brain gangliosides, and are predominantly of the monosialo type (Keenan and Berridge, 1973).

In mallard ducks, the salt gland secretory cells contain a relatively high content of sialic acid, the amount of which increases with saltwater adaptation. In this case, sialic acid is believed to confer to the salt gland cells certain properties conducive to osmoregulation in saltwater habitat (Martin and Philpott, 1974), a situation similar to that of the silver eel (Lemoine, 1974; Lemoine and Olivereau, 1973; 1974; Olivereau and Lemoine, 1972).

Krysteva *et al.* (1973) have studied the environment of the most exposed tyrosine of chicken ovomucoid. The glycoprotein was treated with tetranitromethane, and the nitrotyrosine peptide was isolated from the degraded fragments. The peptide which contained neither neutral sugar nor glucosamine was positive for sialic acid. The authors proposed that the sialic acid residue was linked to the hydroxyl group of tyrosine as an *O*-glycoside.

5. Mammals. Most of the earlier studies on sialic acids were on mammalian tissues, and there was some confusion over early sialic acid nomenclature: neuraminic acid from brain glycolipid (Klenk, 1941); hemataminic acid from horse erythrocytes (Yamakawa and Suzuki, 1952); sialic acid from salivary gland (Blix *et al.*, 1952); lactaminic acid from cow colostrum (Kuhn *et al.*, 1954); sero-lactaminic acid from horse serum (Yamakawa and Suzuki, 1955); and gynaminic acid from human milk (Zilliken *et al.*, 1956). "Neuraminic acid" is now used specifically for aminodeoxynonulosonic acid (5-amino-3,5-dideoxy-D-glycero-D-galactononulosonic acid) while "sialic acid" is the group name for the acylated derivatives (Blix *et al.*, 1957) which occur in nature, as outlined in Chapter 1.

Sialic acids are ubiquitous in tissues of all mammalian species so far studied.

a. Free Sialic Acids. Free *N*-acetylneuraminic acid is found in human cerebrospinal fluid (Uzman and Rumley, 1956; Bogoch, 1958; William and Roboz, 1958; Saifer and Siegel, 1959; Papadopoulos and Hess, 1960; Jakoby and Warren, 1961). Both free *N*-acetyl- and *N*-glycolylneuraminic acids are found in hog gastric tissue (Atterfelt *et al.*,

1958). The seminal vesicles of the Chinese hamster secrete a small but significant amount of free sialic acid (Fouquet, 1971), while an exceptionally high amount of free sialic acid of more than one species is observed in the golden hamster (Fouquet, 1972). An exceptionally high level of free sialic acid also occurs in a unique melituria, in man, sialuria, where the urine sialic acid content of the patient ranges as high as 11–36 g/liter or 5.8–7.2 g/day (Dupont *et al.*, 1967; Fontaine *et al.*, 1967; 1968; Montreuil *et al.*, 1967; 1968). In most normal mammalian tissues, however, the free sialic acid level is low. In rat brain, free N-acetylneuraminic acid is only about 3% of the total sialic acid content (Benedetta *et al.*, 1969). Normal human urine contains some free N,O-diacetylneuraminic acid (Carrion *et al.*, 1970).

 b. Heterosaccharides. In all sialo compounds, sialic acids are ketosidically bound to either D-galactose (in positions 3 or 6) or to N-acetyl-D-galactosamine (in position 6) or to another sialic acid residue (in position 8) (Tuppy and Gottschalk, 1972; Gottschalk and Drzeniek, 1972).

 The sialoheterosaccharides may be grouped into three classes: (i) containing lactose or lactosamine, (ii) containing galactosyl galactosamine, and (iii) containing uridine diphosphate (UDP) as part of the molecule.

 Sialic acids are linked to lactose, a uniquely mammalian compound, to give several species of sialyllactose: N-acetyl-, N-glycolyl-, and N,O-diacetylneuramin-(2→3)-lactoses, which are important components of milk during early lactation (Kuhn and Brossmer, 1956; Kuhn and Wiegandt, 1963; Barra *et al.*, 1969). Sialyllactose isomers with an N-acetylneuraminic acid residue linked to C_6 of the galactose residue are also found in milk and colostrum of several mammalian species (Kuhn, 1958; Kuhn and Brossmer, 1958). Also of wide occurrence is N-acetylneuramin(2→6)lactosamine (Kuhn and Brossmer, 1958). A sialyllactose derivative sulfated at the galactose residue was reported in rat mammary gland during the first to sixth days after delivery. It is absent in the mammary gland after 18 days of lactation (Carubelli *et al.*, 1961).

 Recently a new species of sialyllactose was found in the milk of the primitive Australian egg-laying mammal, the echidna monotreme, *Tachyglossus aculeatus*. The sialic acid residue was neither N-acetyl-, nor N-glycolylneuraminic acid. It was not released by *Clostridium perfringens* neuraminidase, but was susceptible to influenza A2 virus neuraminidase. Treatment with alkali gave N-acetylneuraminlactose which was then susceptible to bacterial neuraminidase. Periodate oxidation yielded 1 mole of formic acid per mole of sialic acid. This species of sialyllactose, which has not been reported in other mammalian species was concluded to be N-acetyl-4-O-acetyl-neuramin-($2\rightarrow3$)-lactose. The milk also contained a small amount of N-acetylneuraminlactose (Messer, 1974).

Sialyllactoses are also found in rat urine, with N-acetylneuramin-($2\rightarrow3$)-lactose present in greatest amount (Maury, 1971; Huttunen et al., 1972). The sialyllactoses and sialyllactosamines are also normal constituents of human urine and may originate from the mammary glands as well as from extramammary tissues (Huttunen and Miettinen, 1965; Huttunen et al., 1972).

Two species of sialic-acid-containing galactosyl-galactosamines have been found in soluble urinary oligosaccharides of the human. One of these is of the structure NANA-($\alpha2\rightarrow8$)-NANA-($\alpha2\rightarrow3$)-gal-($\beta1\rightarrow3$)-galNAc; (Cornillot and Bourrillon, 1964; Huttunen and Miettinen, 1965); the other species was not obtained in homogeneous form and was tentatively identified as NANA-($\alpha2\rightarrow3$)-gal-($\beta1\rightarrow3$)-galNAc (Huttunen and Miettinen, 1965).

The nucleotide-oligosaccharides belong to an interesting class of compounds. In goat colostrum, the following species containing sialic acids have been identified (Jourdian et al., 1961; Jourdian and Roseman, 1963):

    (i) NANA-($2\rightarrow?$)-Gal-($\beta1\rightarrow4$)-GlcNAc-UDP
    (ii) NGNA-($2\rightarrow?$)-Gal-($\beta1\rightarrow4$)-GlcNAc-UDP
    (iii) NANA-($2\rightarrow?$)-Gal-($\beta1\rightarrow6$)-GlcNAc-UDP
    (iv) NGNA-($2\rightarrow?$)-Gal-($\beta1\rightarrow6$)-GlcNAc-UDP

Species (i) and (iii) and a variant species of structure NANA-($2\rightarrow?$)-gal-($\beta1\rightarrow4$)-glcNAc-UDP have been synthesized *in vitro* with a partially purified sialyltransferase from goat milk and colostrum (Jourdian and Distler, 1973). The sialic acid residue is most likely linked to a galactose residue by an ($\alpha2\rightarrow6$) bond (Bartholomew et al., 1973). Other nucleotide-oligosaccharides without sialic acid have also been found in human milk and colostrum and hen oviduct (reviewed by Warren, 1972). Despite their natural occurrence, the biological roles of these nucleotide-oligosaccharides are obscure. One of the possibilities is that they may be intermediates in the biosynthesis of oligosaccharides, glycoproteins, and glycolipids as well as proteoglycans (Jourdian and Distler, 1973).

c. Glycoproteins. In most animal tissues, bound sialic acids are mainly associated with proteins. N-acetylneuraminic acid (cf. Chapter 1), was first isolated in crystalline form from bovine submaxillary mucin (Klenk and Faillard, 1954), and has since been obtained from pseudomyxomatous gel (Odin, 1955a), human serum protein and meconium (Odin, 1955b), human cervical mucus, hog seminal gel, and ovomucin (Odin, 1955c).

Glycoproteins with MN antigenic activities can be isolated from human erythrocyte membrane and contain 25–27.8% sialic acid (Marchesi *et al.*, 1972; Winzler, 1972). The sialic acid residues are presumably required for antigenicity of the glycoproteins, since neuraminidase treatment of erythrocytes (Springer and Ansell, 1958; Klenk and Uhlenbruck, 1960) and amidation of the antigenic glycoproteins (Ebert *et al.*, 1972) abolished MN antigenicity. However, briefer exposure to neuraminidase resulted in the abolition of only M antigenicity, making the N antigen appear instead (Yokoyama and Trams, 1962; Huprikar and Springer, 1970). Thus it appears that the N antigen, which bears sialic acid itself, may be a precursor of M antigen and that the additional sialic acid residues are required to confer the correct configuration of the protein moiety of N antigen to give it M antigenicity.

Sialic acid also has a role in a separate P system of blood groups (Ebert *et al.*, 1971, 1972; Merz and Roelcke, 1971; Roelcke *et al.*, 1971).

The histocompatibility antigens contain sialic acid residues although they may not contribute to the antigenic properties (Yamane and Nathenson, 1970).

Sialic acid is also found associated with erythrocyte antigens in other mammals. Spooner and Maddy (1971), for example, have described a sialoglycoprotein antigen in the ox.

Three sialoglycopeptides have been isolated from trypsin and pronase digests of human erythrocyte stroma, and their saccharide structures have been elucidated. All these saccharide chains are branched and some of them have sialic acid as their terminal group. Their structures are different from saccharide sequences found in glycopeptides and glycolipids of M, N, A, B, and H antigenicities (Kornfeld and Kornfeld, 1970; Winzler, 1972).

Papain-solubilized HL-A antigens from human lymphoblastoid culture cells showed substantial charge heterogeneity in isoelectric focusing gels. This charge heterogeneity was due to the sialic acid content, which varied up to three residues on each antigen molecule. Neuraminidase treatment did not alter specificity and specific activity of the purified antigen (Parham *et al.*, 1974).

Although A, B, and H antigeneity of human erythrocytes have been found associated with a series of neutral glycolipids (Watkins, 1972), recent studies have shown that they also can be associated with glycopeptides or glycoproteins in erythrocytes (Gardas and Koscielak, 1971; Liott *et al.*, 1972; Koscielak, 1973). In fact in secretors of these antigens, the glycolipid content was demonstrated to be too low to account for all the antigenic sites. A class of molecules low in protein and lipid content but high in carbohydrates was subsequently isolated by Koscielak (1973). These substances were not only extremely antigenic as A, B, and H antigens, but also contained small amounts of sialic acid (1–3%). It was proposed that the sialylated species are derivatives from the blood group glycolipids or their precursors and may account for the loss of the blood group antigens on leukocytes in some leukemias (Koscielak, 1973).

Glycoproteins behaving as infectious mononucleosis heterophile antigens were isolated from goat and horse erythrocyte membrane. The glycoprotein from goat erythrocytes inhibited agglutination of sheep erythrocytes by serum from a patient with infectious mononucleosis. Chemical analysis revealed an equimolar ratio of sialic acid, galactose, and hexosamine (Fletcher and Lo, 1974). The glycoprotein from horse erythrocytes, which had a slightly larger molecular weight than the one studied from goat erythrocytes, consisted of 54% carbohydrate with sialic acid, hexose, and hexosamine in the ratio of 2:1:1 (Fujita and Cleve, 1975).

Sialic acid is found in all classes of immunoglobulins in man, rabbit, horse, pig, guinea pig, and sheep. Human IgG has 0–5 sialic acid residues per molecule, while other immunoglobulin classes have a higher content, from 3 to 19 sialic acid residues per molecule. The carbohydrates on immunoglobulins may contribute to their resistance to proteolysis and antigenicity, but appear not to be involved in antibody activities or complement fixation reactions (reviewed by Clamp and Johnson, 1972).

Sialic acids are constituents of many serum glycoproteins, and are mainly responsible for their electrophoretic microheterogeneity (reviewed by Montgomery, 1972). They also determine the survival in circulation, and thus are related to the biological roles of orosomucoid, fetuin, ceruloplasmin, haptoglobin, $\alpha 2$-macroglobulin, thyroglobulin, lactoferrin, human chorionic gonadotropin, and follicle-stimulating hormone (Morell et al., 1968; van den Hamer et al., 1970; Hickman et al., 1970; Morell et al., 1968, 1971; Ross et al., 1972). See Chapter 9.

Recent studies on the survival of glycoproteins in circulation have shown that catabolism of asialoglycoproteins begins with absorption to specific receptors on liver cells. A rabbit liver binding protein specific for asialoglycoproteins has been isolated. It is a glycoprotein containing 10% carbohydrate, with sialic acid, galactose, mannose, and glucosamine in the molar ratio of 1:1:2:2. The sialic acid residues were terminal and were required for binding (Hudgin et al., 1974). The binding protein had no detectable transferase activity for sialic acid, galactose, glucosamine, or fucose, which indicates that initiation of catabolism of glycoproteins is by mechanisms other than glycosylation (Hudgin and Aswell, 1974).

Biologically active follicle-stimulating hormone, human chorionic gonadotropin, and erythropoietin are partially or completely inactivated by neuraminidase treatment (Papkoff, 1966), which suggests that the sialic acid moiety of these hormones contributes to the native active structural conformation required for interaction with the hormone receptors. Sheep and human follicle-stimulating hormones, with sialic acid residues chemically shortened by one or two carbon atoms in the side chain, exhibit biological activity *in vivo* which is greater than 50% of that of the native hormones (Suttajit *et al.*, 1971).

The human pituitary luteinizing hormone has been found to be a sialoglycoprotein, with α and β subunits differing in their sialic acid content (Shownkeen *et al.*, 1973).

Highly purified human kininogen, molecular weight 50,000 daltons, was found to have 8.6 sialic acid residues/molecule (Londesborough and Hamberg, 1975).

The list of enzymes containing sialic acid as part of the molecules has grown rapidly. Prothrombin and some of its derivatives contain sialic acid. Using sialyl-[^{3}H]prothrombin, Butkowski *et al.* (1974) showed that this glycoprotein was converted to intermediates 1 and 3 with thrombin and to intermediates 1, 3, and 4 with factor Xa. Intermediates 1 and 3 contained sialic acid, and in the factor Xa catalyzed reaction, one-third of the prothrombin sialic acid went into intermediate 1, while the rest was found on intermediate 3. Intermediate 1 was then converted into intermediate 2 to form α-thrombin.

Two forms of rabbit plasminogen have been separated by affinity chromatography. Each isoenzyme had five subforms which possessed distinct isoelectric points. It was further observed that the survival time in circulation of the individual subforms depended upon their sialic acid content (Siefring and Castellino, 1974).

Phosphatase (Stepan and Ferwerda, 1973), acid phosphatase (Campbell *et al.*, 1973), and alkaline phosphatase (Ramadose *et al.*, 1974) were all found to exist in three forms, and in each case only one of the isoenzymes was susceptible to neuraminidase. In the case of alkaline phosphatase neuraminidase treatment of the sialic-acid-containing isoenzyme did not alter it with respect to reaction rate, affinity for substrates, or response to activators and inhibitors. The sialic acid residues thus seemed not to be essential to the catalytic properties of the isoenzyme, although it differed from the other two nonsialo species in thermal stability, metal ion effect, and affinity toward some substrates (Ramadoss *et al.*, 1974). Nucleotide pyrophosphatase on the hepatocyte external surface was identified as a sialoglycoprotein with a molecular weight of 130,000 daltons (Evans, 1974). Pyrophosphatase activity of confluent nervous system cells in tissue culture (N18 mouse neuroblasts and NN newborn hamster astroblasts) was dependent on the presence of sialic acid on the cell surface. Removal of a fraction of total cell surface sialic acid by *Clostridium perfringens* neuraminidase increased the activity of pyrophosphatase without causing a loss of cell intactness or adhesion (Stefanovic *et al.*, 1975).

In some enzymes removal of most of the carbohydrates does not impair enzyme function. Examples of this type of glycoprotein are thrombin (Skaug and Christensen, 1971), glycoamylase (Pazur *et al.*, 1970), and chloroperoxidase (Lee and Hager, 1970). Catalytic properties of serum cholinesterase, α-glutamyltranspeptidase, enterokinase, and serum atropinesterase (Papkoff, 1966) are also not greatly affected by neuraminidase. In some other enzymes, the sialic acid moiety appears to be involved in the catalytic functions. Neuraminidase removal of sialic acid from arylsulfatase (Goldstone *et al.*, 1971) and α-galactosidases (Ho *et al.*, 1972; Mapes and Sweeley, 1973) shifts the pH optimum of these enzymes.

Enzymatic removal of sialic acid from the A form of N-acetyl-β-D-hexosaminidase produces a species with an electrophoretic mobility similar to the B form (Robinson and Stirling, 1968). While both hexosaminidase A and B catalyze *in vitro* the degradation of lipids accumulated in Tay–Sachs disease (trihexosylceramide and kidney globoside), only enzyme A degrades the main storage compound, the Tay–Sachs ganglioside, to a significant degree (Sandhoff, 1970). These observations, however, may not agree with a more recent finding that highly purified hexosaminidase A, upon heating alone, appears to undergo transformation and acquires the electrophoretic and ion-exchange properties of hexosaminidase B, which suggests that the inherent difference between the two enzyme forms may not be entirely due to their sialic acid content. Similar kinetic parameters were observed with both enzymes when either fluorogenic compounds or natural lipids were used as substrates. Furthermore, hexosaminidase B was as effective as hexosaminidase A in catalyzing the hydrolysis of the Tay–Sachs ganglioside (Tallman and Brady, 1973).

A new isoenzyme of hexosaminidase, called the C form, has been described. It differs from the A and B forms in having a greater molecular weight, a more acidic pH optimum, and a higher net charge (Hooghwinkel *et al.*, 1972; Poenaru and Dreyfus, 1973, Poenaru *et al.*, 1973; Braidman *et al.*, 1974). So far, differences between hexosaminidases A and B have not been clarified (Tallman et al., 1974; Tallman, 1974). In addition to previously proposed differences in sialic acid content (Robinson and Stirling, 1968) and molecular conformation (Tallman and Brady, 1973), differences in disulfide linkages were also suggested (Carmody and Rattazzi, 1974). Results from recent genetic analyses on the expression and linkage relationships in man–mouse somatic cell hybrids, however, have clearly indicated that the A and B isoenzymes were coded by two different and unlinked genes (Gilbert et al., 1975; Lalley et al., 1974). Furthermore, the expression of the gene coding for hexosaminidase A did not necessarily require the presence of the gene coding for hexosaminidase B, which indicated that the former may not be a conversion product of the latter (Gilbert et al., 1975). It is apparent that any electrophoretic data must be interpreted with caution, and that a definitive understanding of the interrelationship among the isoenzymes must wait for a detailed structural analysis of these isoenzymes. Detailed examination of the differences between similar enzyme forms and an understanding of the mechanisms of their interconversion may prove to be important in elucidating the molecular pathology of many genetic diseases. See Chapter 6.

The antiviral substance, interferon, induced by poly I:C from rabbit kidney cells in culture, has been shown to be a glycoprotein containing sialic acid. Treatment with neuraminidase from *Vibrio cholera* converts the glycoprotein from heterogeneity to homogeneity in charge as revealed by isoelectric focusing. The terminal oligosaccharide sequence is sialic acid→galactose (Dorner et al., 1973).

There is an appreciable content of sialic acids in brain tissues, about one-third of which is associated with glycoproteins, the rest with gangliosides (di Benedetta et al., 1969).

Glycoproteins isolated from bovine, porcine, ovine, and human bile all contain more than one species of sialic acid (Cabezas and Ramos, 1972).

Sialic acids are common in urinary glycoproteins in man (Tamm and Horsfall, 1950), rabbit (Cornelius et al., 1965; Neuberger and Ratcliffe, 1972), cattle, horses, and sheep (Cornelius et al., 1963; Mia and Cornelius, 1966; reviewed by Bourrillon, 1972).

Glycopeptides rich in sialic acids (16–49%) and galactose (20.7–25%) and of molecular weight of about 4000 daltons have been studied in nondialyzable urinary fractions from normal human (Carrion et al., 1969a,b; Goussault and Bourrillon, 1970). While the amino acid pattern is similar to that described for glycoproteins secreted by glandular and epithelial cells, the sialic acid in these glycopeptides is more easily split off by mild acid or neuraminidase than from the plasma glycoproteins (Cornillot and Bourrillon, 1964).

With the exception of the golden hamster (Fouquet, 1972), sialic acids in semen are chiefly in the bound form (Mann, 1964). The bulbourethral glands of the boar produce a glycoprotein containing 25% N-acetylneuraminic acid. This glycoprotein, by interacting with basic proteins secreted by the seminal vesicles, swells and gives the rigid elastic gel of boar semen (Hartree, 1962; Boursnell et al., 1970). In ram spermatozoa, the sialic acids appear to be localized in the acrosomes, where they are components of two glycoproteins (Hartree and Srivastava, 1965). A bovine cervical glycoprotein contains sialic acid with an O-acetyl group (Neuberg and Ratcliffe, 1972).

d. *Gangliosides.* The other major class of biomolecules bearing sialic acids is that of the glycolipids—gangliosides and related compounds. (Refer to Table I for chemical structures and nomenclatures of the gangliosides found in vertebrate brain tissue.)

Interest in the sialoglycolipids has recently been accelerated by evidence that the gangliosides may play important roles in cell functioning. Possible involvement of glycosyltransferase mediated metabolism of glycolipids and glycoproteins on cell surfaces in specific intercellular interactions is well documented (Deman et al., 1974; Hakomori, 1970a,b, 1974; Hakomori and Murakami, 1968; Hakomori et al., 1968; 1972; Oppenheimer et al., 1969; Roseman, 1970, 1974a,b,c; Roth and White, 1972; Roth et al., 1971). Gangliosides may also be involved in the regulation of cell growth (Hollenberg et al., 1974). Ganglioside G_{M1a} acts as a receptor for cholera toxin in intestinal epithelial cells (and other tissues as well) and so is coupled to the adenyl cyclase system (Bennett et al., 1975; Cuatrecasas, 1974a,b,c,d; Hewlett et al., 1974; van Heyningen, S., 1974; van Heyningen, W. E., 1974; Peterson 1974; Staerk et al., 1974). Gangliosides G_{D1b} and G_{T1b} may act as receptors for the neurotoxin from tetanus in nervous tissues (van Heyningen, W. E., 1974). Ganglioside G_{D3} may act as a receptor for serotonin (van Heyningen, W. E., 1974). The θ-antigen of mouse thymocytes appears to be a glycolipid, possibly G_{D1b} (Esselman and Miller, 1974a,b; Vitetta et al., 1973). Even the hemolytic action of intact as well as heated cells of *Crytococcus laurentii* may be mediated through an interaction with glycolipids on the erythrocyte cell membrane. Preincubation of the bacteria with gangliosides blocked the hemolytic action in this order of effectiveness: $G_{M2} > G_{M1a} > G_{D1a} > G_{T1}$ (Ali et al., 1975). In nervous tissues ganglioside metabolism can be more responsive to nervous activities than previously supposed. The effects of penicillin induced convulsion (Balasubramanian et al., 1973), light (Dreyfus et al., 1974), different levels of sensory and visual stimulation (Maccioni et al., 1971; 1974) and electroconvulsive shock (ECS) on ganglioside metabolism have been studied. With ECS, a significant increase in free sialic acid content of cerebral

TABLE I. Gangliosides of Vertebrate Brain

		Code systems					
Chemical structure[a]	Generic term[b] [trivial name]	Svennerholm[c]	Svennerholm modified[d]	Wiegandt[e]	Wiegandt modified[f]	Korey et al.[g]
Gal(1 → 1)Cer 3 (↑) α2 NANA	Monosialosylgalactosylcer-amide	—	G_{M4}	$G_{gal}1$	$G_{gal}1$	G_7
Gal(β1 → 4)Glc(β1 → 1)Cer 3 (↑) α2 NANA	Monosialosyllactosylcer-amide [Hematoside]	G_{M3}	G_{M3}	$G_{lact}1$	$G_{lact}1$	G_6
Gal(β1 → 4)Glc(β1 → 1)Cer 3 (↑) α2 NANA(8 ← 2α)NANA	Disialosyllactosylcer-amide	G_{D3}	G_{D3}	$G_{Lact}2$	$G_{Lact}2$	G_{3A}
GalNAc(β1 → 4)Gal(β1 → 4)Glc(β1 → 1)Cer 3 (↑) α2 NANA	Monosialosyl-N-triglycosyl-ceramide [Tay–Sachs ganglioside]	G_{M2}	G_{M2}	$G_{GNTrII}1$	$G_{Gtri}1$	G_5
GalNAc(β → 4)Gal(β1 → 4)Glc(β1 → 1)Cer 3 (↑) α2 NANA(8 ← 2α)NANA	Disialosyl-N-triglycosylcer-amide	G_{D2}	G_{D2}	$G_{GNTrII}2$	$G_{Gtri}2$	G_{2A}

Structure	Name					
Gal(β1 → 3)GalNAc(β1 → 4)Gal(β1 → 4)Glc(β1 → 1)Cer 3 (←) α2 NANA	Monosialosyl-N-tetraglycosylceramide [Monosialoganglioside]	—	G_M^h	—	—	—
Gal(β1 → 3)GalNAc(β1 → 4)Gal(β1 → 4)Glc(β1 → 1)Cer 3 (←) α2 NANA	Monosialosyl-N-tetraglycosylceramide [Monosialoganglioside]	G_{M1}	G_{M1a}^i	$G_{GNT}1$	$G_{Glet}1$	G_4
Gal(β1 → 3)GalNAc(β1 → 4)Gal(β1 → 4)Glc(β1 → 1)Cer 3 (←) α2 NANA(8 ← 2α)NANA	Disialosyl-N-tetraglycosylceramide [Disialoganglioside]	G_{D1a}	G_{D1a}	$G_{GNT}2a$	$G_{Glet}2a$	G_3
Gal(β1 → 3)GalNAc(β1 → 4)Gal(β1 → 4)Glc(β1 → 1)Cer 3 (←) α2 NANA	Disialosyl-N-tetraglycosylceramide [Disialoganglioside]	G_{D1b}	G_{D1b}	$G_{GNT}2b$	$G_{Glet}2b$	G_2
Gal(β1 → 3)GalNAc(β1 → 4)Gal(β1 → 4)Glc(β1 → 1)Cer 3 (←) α2 NANA(8 ← 2α)NANA	Trisialosyl-N-tetraglycosylceramide [Trisialoganglioside]	G_{T1a}	G_{T1a}	—	—	—
Gal(β1 → 3)GalNAc(β1 → 4)Gal(β1 → 4)Glc(β1 → 1)Cer 3 (←) α2 NANA(8 ← 2α)NANA	Trisialosyl-N-tetraglycosylceramide [Trisialoganglioside]	G_{T1b}	G_{T1b}	$G_{GNT}3$	$G_{Glet}3a$	G_1

(continued)

TABLE I. (Continued)

Chemical structure[a]	Generic term[b] [trivial name]	Svennerholm[c]	Svennerholm modified[d]	Wiegandt[e]	Wiegandt modified[f]	Korey et al.[g]
Gal($\beta 1 \to 3$)GalNAc($\beta 1 \to 4$)Gal($\beta 1 \to 4$)Glc($\beta 1 \to 1$)Cer $\begin{pmatrix}3\\\leftarrow\\\alpha 2\end{pmatrix}$ NANA(8 $\leftarrow 2\alpha$)NANA (8 $\leftarrow 2\alpha$)NANA	Trisialosyl-N-tetraglycosyl-ceramide [Trisialoganglioside]	—	G_{T1c}[j]	—	G_{Gtet}3b	—
Gal($\beta 1 \to 3$)GalNAc($\beta 1 \to 4$)Gal($\beta 1 \to 4$)Glc($\beta 1 \to 1$)Cer $\begin{pmatrix}3\\\leftarrow\\\alpha 2\end{pmatrix}$ NANA(8 $\leftarrow 2\alpha$)NANA NANA(8 $\leftarrow 2\alpha$)NANA	Tetrasialosyl-N-tetraglyco-sylceramide [Tetrasialoganglioside; Polysialoganglioside]	G_{Q1}	G_{Q1b}	$G_{GNT}4$	G_{Gtet}4b	(G_0)
Gal($\beta 1 \to 3$)GalNAc($\beta 1 \to 4$)Gal($\beta 1 \to 4$)Glc($\beta 1 \to 1$)Cer $\begin{pmatrix}3\\\leftarrow\\\alpha 2\end{pmatrix}$ NANA NANA(8 $\leftarrow 2\alpha$)NANA (8 $\leftarrow 2\alpha$)NANA	Tetrasialosyl-N-tetraglyco-sylceramide [Tetrasialoganglioside; Polysialoganglioside]	—	G_{Q1c}	—	$G_{Gtet}4$	(G_0)
Gal($\beta 1 \to 3$)GalNAc($\beta 1 \to 4$)Gal($\beta 1 \to 4$)Glc($\beta 1 \to 1$)Cer $\begin{pmatrix}3\\\leftarrow\\\alpha 2\end{pmatrix}$ NANA(8 $\leftarrow 2\alpha$)NANA NANA(8 $\leftarrow 2\alpha$)NANA (8 $\leftarrow 2\alpha$)NANA	Pentasialosyl-N-tetraglyco-sylceramide [Pentasialoganglioside; Polysialoganglioside]	—	G_{P1c}	—	$G_{Gtet}5$	—

[a] The structure of ceramide, abbreviated cer, N-acylsphingosine, is variable: The acyl group is mainly stearic acid, while the sphingoside is mainly a mixture of C_{18} and C_{20} long chain bases. The hexoses are in the D-configuration. The glycosides are in the form of pyranosides.
[b] Generic terms assigned according to the rules suggested by the IUPAC-IUB Commission on Biochemical Nomenclature (Svennerholm, 1970a; 1970b).
[c] In this code system, G stands for ganglioside; M, D, T, and Q for mono-, di-, tri-, and tetrasialyl groups; and index 1 for the major carbohydrate chain with four monosaccharide residues, 2 for the carbohydrate chain lacking the terminal galactose, and 3 for the chain lacking the terminal galactose and N-acetylgalactosamine. The letters a and b indicate isomers with sialic acid residues at different positions. This code system applies only to the brain gangliosides. Nonnervous tissues contain, in addition to gangliosides, ganglioside-like glycolipids with various different carbohydrate backbones (Svennerholm, 1963; 1964; 1970a; 1970b).
[d] The modification of the Svennerholm code system proposed here by these authors confers a numerical value to the index letters indicating the number of sialic acids attached to the inner galactose residue. Thus a, b, c, d, and e denote 1, 2, 3, 4, and 5 sialic acid residues at that position; the absence of a letter in the index denotes the absence of a sialic acid residue on the inner galactose. According to this scheme, the number of sialic acid residues on the terminal galactose is easily deduced as the difference of the total number of sialic acid residues on the whole ganglioside molecule and the number of sialic acid residues on the inner galactose indicated by the index letter. Thus, the G_{M1} of the original Svennerholm code system is now designated G_{M1a}, while the isomer with the sialic acid residue attached to the terminal galactose is G_{M1}. A similar notation is assigned to the polysialogangliosides whose structures have been elucidated more recently, and P denotes pentasialoganglioside. The modification as proposed above requires only minor changes of the original Svennerholm code system and yet allows systematic coding of all potentially existing ganglioside species. This is made possible by the evidence so far, indicates that sialic acid is only linked to galactose by the $(\alpha 2 \rightarrow 3)$ linkage [except the extraneural species (II) discussed in Section II, F, 5, d] (Wiegandt, 1973) and to another sialic acid by $(\alpha 2 \rightarrow 8)$ linkage, and that there is no branching of the sialic acid chain. The only exception in the carbohydrate skeleton sequence is in G_{M4}, the monosialosylgalactosylceramide, with galactose (instead of glucose) linked to ceramide.

Further elaborations on the original Svennerholm code system, as may be necessary in extraneural ganglioside-like glycolipids, are still possible: the prefix GlcNAc- can be used for the glucosamine-containing gangliosides (e.g., GlcNAc-G_{M1} in Li et al., 1973; and Wiegandt, 1973); the nature of the sialic acid residues can be indicated by adding NANA or NGNA (N-Glycolylneuraminic acid) to the index number (e.g., G_{M3NANA} and G_{M8NGNA}); and the non-sialic-acid-containing (poly)glycosylceramide skeletons that are obtained naturally, chemically, or enzymatically, can be indicated by adding the prefix A- (for asialo-) to the corresponding parent ganglioside species (e.g., A-G_{M1a}, A-G_{M2}, A-G_{M3}, and A-G_{M4}).
[e] In this earlier code of Wiegandt, G stands for ganglioside; the indexes Lact, GNTri1, and GNT designate carbohydrate chains with 2, 3, and 4 residues, respectively. The numericals indicate the number of sialic acids, with letters a and b indicating isomers. G_{gal} 1 has been used to stand for the monosialosylgalactosylceramide (Wiegandt, 1966; 1968).
[f] In this more recent code system of Wiegandt, Lact, Gtri, and Gtet designate carbohydrate chains of 2, 3, and 4 residues. The nature of the sialic acid residues (as may be necessary in extraneural gangliosides) could be indicated at the end of the code, e.g., G_{Lact}1neuNA and G_{Lact}1neuNGl stand for hematosides with N-acetylneuraminic acid and N-glycolylneuraminic acid, respectively (Wiegandt, and Bucking, 1970; Wiegandt, 1971; Ishizuka and Wiegandt, 1972).
[g] This code system, when first proposed, refers to the patterns of brain ganglioside species resolved on silica gel thin layers with bands numbered as G_0–G_6 in order of increasing mobility. The solvent used was n-propanol-water (7:3, v/v). The monosialosylgalactosylceramide, NANA($\alpha 2 \rightarrow 3$)Gal($\beta 1 \rightarrow 1$)Cer, moves faster than G_6 on thin layer and has been labeled as G_7. The disialosyllactosylceramide and disialosyl-N-triglycosylceramide move faster than G_3 and G_2 and are referred to as G_{3A} and G_{2A}, respectively (Korey and Gonatas, 1963; Suzuki, 1974; Yu et al. 1974).
[h] This novel ganglioside was synthesized in vitro and was named as G_{M1b} by Yip (1973). Its natural occurrence, however, has not yet been reported.
[i] Similarly named by Yip (1973).
[j] Similarly named by Mestrallet et al. (1974).

homogenate was observed concomitant with a decrease in both the gangliosidic and proteinaceous sialic acid of the synaptosomal fraction but not of the mitochondrial and microsomal fractions (Dunn, 1975).

The gangliosides are found in all vertebrate nervous tissues; the concentration is highest in mammals and birds (reviewed by Martensson, 1969; Svennerholm, 1970; Wiegandt, 1971). Although the total sialic acid content of gangliosides in mammals is relatively constant from species to species (Schengrund and Garrigan, 1969), the population of the constituent gangliosides varies. The brain tissues of dog and sheep contain a monosialo-, a disialo-, and a trisialoganglioside; cat brain has an additional trisialoganglioside, and pig brain has one monosialo- and two disialogangliosides (Schengrund and Garrigan, 1969). The gangliosides have also been analyzed in brain tissues of some other mammals (Klenk and Gielen, 1961; Wherrett and Cumings, 1963; Prokhorova *et al.*, 1965; Tettamanti *et al.*, 1965, 1972; Suzuki, 1965). In general, four main gangliosides with one to three sialic acid residues in their molecules constitute 80–90% of the total sialic acid content; the polysialogangliosides, containing four to five sialic acid residues, are found in minute amounts or are nondetectable (different from fish and amphibia; Tettamanti *et al.*, 1965; Avrova, 1971).

The gangliosides are constituents of the outer membrane of nerve endings (Whittaker, 1966; Lapetina *et al.*, 1967; Wiegandt, 1967), and presumably are essential in the normal functioning of the neuronal membranes (Lehninger, 1968; Tettamanti, 1968; Bondareff and Sjostrand, 1969; Schengrund and Rosenberg, 1972; reviewed by Johnston and Roots, 1972). The glial cells (Derry and Wolfe, 1967; Fewster and Mead, 1968) and the myelin structures (Cumings *et al.*, 1968; Suzuki *et al.*, 1967) were once believed to contain relatively little or no gangliosides. However, more recent studies by Norton and Poduslo (1971) and by Hamberger and Svennerholm (1971) revealed that gangliosides are present in both neuronal and glial cell preparations. In fact a greater concentration of gangliosides was found in the glial cell preparation, possibly due to a greater loss of neuronal plasma membrane as well as a larger membrane to volume ratio of the glial cells.

The question of the distribution of gangliosides between neurons and glial cells has been clarified to an appreciable extent by studying the biosynthetic capabilities of the two cell types. It was shown in an earlier investigation that the incorporation of DL-[3-^{14}C]serine and N-acetyl-[4-^{14}C]neuraminic acid into gangliosides was greater in the neuronal fraction than in the gial cell fraction (Jones *et al.*, 1972). More recently, the neuronal cells have been found to possess significantly higher activities of CMP–N-acetylneuraminic acid synthetase and sialyltransferase than the glial cells (Gielen and Hinzen, 1974; Van den Eijnden, 1973). Even in these studies, the biosynthetic capability of neuronal cells may be underestimated due to extensive loss of cellular processes which very likely are active metabolically. The above findings, however, have been substantiated by studies with cultured cell strains which revealed that higher gangliosides are synthesized by neuronal cells only. While G_{M2}, G_{M1a}, and G_{D1a} have been isolated from a C1300 mouse neuroblastoma and several derived cell strains, these gangliosides were virtually absent from mouse glial tumors or glioma cells, which were shown to exhibit little or no UDP-GalNAc:G_{M3} N-acetylgalactosaminyltransferase activity (Stoolmiller *et al.*, 1975; Robert *et al.*, 1975).

Gangliosides are not limited to the nervous tissues. They are widely distributed in various tissues, but the extraneural gangliosides are usually of shorter carbohydrate chain length and, according to species, may bear N-glycolylneuraminic acid instead of N-acetylneuraminic acid. Hematoside appears to be the main ganglioside of the extraneural tissues (reviewed by Martensson, 1969; Svennerholm, 1970; Wiegandt, 1971).

A variant of hematoside was isolated from equine erythrocyte membranes. The sialic acid residue is N-glycolyl-O-acetylneuraminic acid instead of the commoner N-acetylneuraminic acid and N-glycolylneuraminic acid. The position of the acetyl group has not been ascertained (Hakomori and Saito, 1969).

Several other variants of basic ganglioside structures have been reported. Glucosamine-containing gangliosides have been detected since the very early studies (reviewed by Martensson, 1969). The structures of four of these gangliosides have been elucidated recently (Li *et al.*, 1973; Wiegandt, 1973, 1974).

$$\text{Gal}(\beta1\rightarrow4)\text{GlcNAc}(\beta1\rightarrow3)\text{Gal}(\beta1\rightarrow4)\text{Glc}(\beta1\rightarrow1)\text{Cer}$$
$$\begin{pmatrix}3\\\uparrow\\\alpha2\end{pmatrix}$$
$$\text{NANA}$$

(I)

$$\text{Gal}(\beta1\rightarrow4)\text{GlcNAc}(\beta1\rightarrow3)\text{Gal}(\beta1\rightarrow4)\text{Glc}(\beta1\rightarrow1)\text{Cer}$$
$$\begin{pmatrix}6\\\uparrow\\\alpha2\end{pmatrix}$$
$$\text{NANA}$$

(II)

$$\text{Gal}(\beta1\to4)\text{GlcNAc}(\beta1\to3)\text{Gal}(\beta1\to4)\text{GlcNAc}(\beta1\to3)\text{Gal}(\beta1\to4)\text{Glc}(\beta1\to1)\text{Cer}$$
$$\begin{pmatrix}3\\\uparrow\\\alpha2\end{pmatrix}$$
NANA (III)

$$\text{Gal}(\beta1\to3)\text{GlcNAc}(\beta1\to4)\text{Gal}(\beta1\to4)\text{Glc}(\beta1\to1)\text{Cer}$$
$$\begin{pmatrix}2\\\uparrow\\\alpha1\end{pmatrix}\qquad\begin{pmatrix}3\\\uparrow\\\alpha2\end{pmatrix}$$
Fuc   NANA
            (IV)

Ganglioside (I) is present in both neural and extraneural tissues in man. It is the major ganglioside of peripheral nerve (Li et al., 1973). In contrast to G_{M1a} but similar to G_{M1} (Yip, 1973; nomenclature according to modified Svennerholm system as listed in Table I), ganglioside (I) is susceptible to neuraminidase treatment (Svennerholm et al., 1972; Li et al., 1973). Ganglioside (II), a structural isomer of (I), is found in human and bovine spleen and kidney (Wiegandt, 1973). It is unique among gangliosides in its NANA ($\alpha2\to6$) Gal linkage. Ganglioside (III), with its more complex carbohydrate structure, is found in human spleen (Wiegandt, 1974). Ganglioside (IV) is found in bovine liver and is the first ganglioside species shown to possess fucose as part of the oligosaccharide chain (Wiegandt, 1973). Interestingly, this ganglioside exhibits blood group H(O) antigen activity and inhibits the agglutination of H(O) erythrocytes by *Ulex europaeus* lectin (Wiegandt, 1973).

Other gangliosides of yet unknown structures include one containing both galactosamine and glucosamine from the human spleen (Wagner and Weicker, 1966) and one containing glucose, galactose, and sialic acid in the proportion 1:2:1 in dog intestine (Vance et al., 1966).

e. Ungulic Acids. The existence of these compounds is in doubt. They have been claimed to be another class of sialic-acid-containing glycolipids called "ungulic acids" because of their presence in hooves of horse and cattle. They reportedly contain ceramide, galactose, N-acetylgalactosamine, N-acetylneuraminic acid, and sulfate. They are present in human epidermis, hair, nail, and kidney, and constitute about half of the total lipids of bovine and horse hooves (Leikola et al., 1969). Evidence has been presented that they are fortuitous mixtures of gangliosides and sterol sulfate.

f. Forms of Sialic Acids in Mammals. While sialic acids in the invertebrates are mainly of the N-glycolyl type, those of the vertebrates have in addition the N-acetyl type and its O-acetyl derivatives (Warren, 1963; Tuppy and Gottschalk, 1972; cf. Chapter 1).

An early work reported that the crystalline sialic acids isolated from bovine, ovine, porcine, and equine submaxillary mucin contained 2 acetyl, 1 acetyl, 1 glycolyl, and 2 acetyl groups, respectively (Blix *et al.*, 1955, 1956). Later studies identified the presence in bovine submaxillary mucin of N-acetyl-8-O-acetyl-, N-acetyl-7,8-di-O-acetyl-, and N-acetyl-tri-O-acetylneuraminic acid, in addition to a small amount of N-acetyl- and N-glycolylneuraminic acids. A sixth species was detected, but it was unstable, possibly O-acetylated at the C_7 position (Blix and Lindberg, 1960). The major sialic acid of equine submaxillary mucin is N-acetyl-4-O-acetylneuraminic acid (Blix and Lindberg, 1960) and that of the ovine submaxillary mucin is N-acetylneuraminic acid (Graham and Gottschalk, 1960).

The simultaneous occurrence of more than one sialic acid species was also demonstrated in calf serum (Faillard and Cabezas, 1963), colostrum (Cabezas *et al.*, 1966), and bile (Cabezas, 1966); human urine (Bourrillon *et al.*, 1961; Carrion *et al.*, 1969b) and bile (Cabezas and Ramos, 1972); rat serum (Cabezas *et al.*, 1968a); and goat tissues (Cabezas *et al.*, 1968b). The relative proportions of the individual species of sialic acids vary with the organisms and the tissues concerned as well as the physiological states of the organisms.

In glycolipids heterogeneity due to sialic acids also exists. Besides the major N-acetylneuraminic acid, N-glycolylneuraminic acid has been found in beef brain gangliosides (Tettamanti, 1964). More recently a new sialic acid was isolated from an equine erythrocyte membrane glycolipid having a carbohydrate structure similar to that of hematoside, but with a higher migration rate on thin layer chromatography (Hakomori and Saito, 1969). This sialic acid residue, which was not releasable by bacterial neuraminidase, was identified as N-glycolyl-O-acetylneuraminic acid, with the O-acetyl group either on the hydroxyl group of glycolic acid or on the C_4-hydroxy group of neuraminic acid.

The main sialic acid species in humans is N-acetylneuraminic acid (Gottschalk, 1960). Low levels of N-glycolylneuraminic acid were suggested in human serum (Martensson *et al.*, 1958) and mingin, a urinary trypsin inhibitor (Astrup and Nissen, 1964). The N-glycolyl compound was identified in human chorionic gonadotropin (Got *et al.*, 1960), in a nondialyzable urinary fraction (Bourrillon, 1961) and in erythrocyte stroma (Eylar *et al.*, 1962b). Apparently the level of N-glycolylneuraminic acid in human tissues is quite low and in certain cases may not represent *de novo* synthesis by the human tissue cells. As mentioned earlier, cf. Chapter 1, the existence of N-glycolylneuraminic acid in HeLa S3 cells in culture (Carubelli and Griffin, 1967; 1968) has been shown by Hof and Faillard (1973) to arise from the calf serum added to the growth medium and is not due to the cancerous nature of the HeLa cells. When the cells are grown with human serum as supplement, only N-acetylneuraminic acid can be found.

Another species of sialic acid found in human tissue is N,O-diacetylneuraminic acid, possibly the N-acetyl-4-O-acetylneuraminic acid, in urine (Carrion et al., 1969b; 1970) and bile (Cabezas and Ramos, 1972). In the urine, the free form of the diacetyl derivative was detected.

Improved methods for detection of N-acyl and O-acyl groups by two-dimensional thin-layer chromatography and by gas–liquid chromatography have established the structures of three new sialic acids: (a) N-glycolyl-8-O-acetylneuraminic acid (in glycoproteins of bovine submandibular glands); (b) N-acetyl-4-O-glycolylneuraminic acid (in glycoprotein of equine submandibular glands, blood plasma, and erythrocyte membrane); (c) N-acetyl-7 or 8-O-glycolylneuraminic acid (in glycoprotein from bovine submandibular glands) (Buscher et al., 1974). The O-acetylated N-glycolylneuraminic acid, first detected in horse serum by Pepper (1968) and in horse erythrocyte membranes by Hakomori and Sato (1969), was characterized as N-glycolyl-4-O-acetylneuraminic acid; it was also found in submandibular glands of horse (Buscher et al., 1974).

Biosynthesis of the various sialic acids and their activation to CMP-derivatives have been extensively studied in bovine, porcine, and equine submaxillary glands (Kean and Roseman, 1966; Schoop and Faillard, 1967; Schoop et al., 1969; Schauer, 1970a,b,c,d; Schauer and Wember, 1971; Schauer, 1972; Schauer et al., 1972). See Chapter 4.

III. THE EVOLUTION OF SIALIC ACIDS

The possible relationship between the occurrence of sialic acids in nature and their evolution has been discussed by Warren (1963). Sialic acids appear relatively late in the course of evolution. They are absent in primitive invertebrates and are characteristic only of the higher invertebrates and the chordates (except the urochordates) (Warren, 1963). Their absence in the primitive multicellular invertebrates suggests that sialic acids were not prerequisite for the evolution of the metazoans from a unicellular ancestor (Warren, 1963). Furthermore, the apparent absence of sialic acids in the plant kingdom suggests that the sialic acids were acquired after the divergence of plants from the animals. If this is true, the appearance of sialic acids in bacteria and viruses might be secondarily acquired, through their contact or association with the higher animals. This speculation is based on the observations that sialic acids are present only in the pathogenic bacteria (Irani and Ganapathi, 1962; Luppi and Cavazzini, 1966). The enveloped viruses which contain sialic acid in their membrane proteins have mammals, birds, and insects as their hosts (Casals and Reeves, 1965). While the bacteria apparently have acquired adequate enzyme systems for the metabolism of sialic acids, the sialylation of the membrane protein in viruses may be partly, if not entirely, dependent on the infected cells. In Sindbis virus, a difference in sialic acid content was found depending upon whether the virus was grown in chick or in hamster cells (Strass et al., 1970; Burge and Strass, 1970). In this context, the acquisition of sialic acids by the viruses may be considered an evolutionary consequence of ontogenic parasitism.

Despite their wide occurrence in nature, the biological roles of the macromolecule-linked carbohydrates in general and sialic acids in particular are far from clear. A unified concept on this subject is tentatively not feasible. This is not unexpected. The existence of a large number of sialylated biomolecules implies that the biological roles of sialic acids depend on their being linked glycosidically to another sugar and not in the free form. It is possible that a mutational appearance of sialic acids was selected and subsequently survived in the glycosidically linked form. The biologically significant evolution of sialic acids thus corresponds to the appearance of the enzyme sialyltransferase(s), cf. Chapter 4. But functionally speaking, the carbohydrate moiety may have evolved to modify previously existing macromolecules to make them more efficient in their original or modified functions. Coupled with the possibility of the lack of substrate specificity of the sialyltransferase(s), generalization on the biological roles of sialic acids may be very difficult.

Evidence available so far has suggested some possible roles for glycosylation processes in general. The carbohydrate moieties of complex carbohydrates may be involved in (i) the regulation of intracellular enzyme activity: the incorporation of sialic acid residue(s) onto a substrate molecule may signal the completion of a finished cell product (Winterburn and Phelps, 1972) and may modify the catalytic properties of some enzymes (refer to Section II,F, 5,c; also Brodbeck et al., 1973). (ii) The control of extracellular sialo compounds: sialic acids may facilitate the secretion of sialylated molecules (Eylar, 1966) and may determine the extracellular fates of those compounds (Winterburn and Phelps, 1972). (iii) Membrane function: sialic acids may form part of membrane components that transmit extracellular stimuli to the intracellular environment (Roseman, 1974); they may also be responsible for intercellular membrane interactions (reviewed by Curtis, 1973; Hakomori et al., 1974; Roseman, 1974).

There still exists a wide gap in our understanding of the multitudinous and diversified classes of sialo compounds. Only when that gap is narrowed may we be in a position to evaluate with some confidence the biological roles which sialic acids may contribute. Even then there is the possibility that diversity may override uniformity, and our search toward simple generalization may not be realized.

Acknowledgment

This work was supported in part by grant NS-05104 from the National Institutes of Health.

IV. REFERENCES

Aaronson, A., and Lessie, T., 1960, Nonulosaminic acid (sialic acid) in protists, *Nature* **186**:719.

Acton, R. T., Weinheimer, P. F., and Niedermeir, W., 1973, The carbohydrate composition of invertebrate hemagglutinin subunits isolated from the lobster *Panulirus argus* and the oyster *Crassostrea virginica, Comp. Biochem. Physiol* **44B**:185.

Ali, B., Raizada, M. K., and Ankel, H., 1975, Inhibition of hemolytic activity Crytococcus laurentii cells by gangliosides, *Fed. Proc.* **34**:646.

Astrup, T., and Nissen, U., 1964, Urinary trypsin inhibitor (mingin) transformation into a new trypsin inhibitor by acid hydrolysis or by sialidase, *Nature* **203**:255.

Atterfelt, P., Blohme, I., Norby, A., and Svennerholm, L., 1958, The sialic acids of hog gastric mucosa, *Acta Chem. Scand.* **12**:359.

Avrova, N. F., 1968, The content and comparative characteristics of brain gangliosides of vertebrates, *Zh. Evol. Biochim. Fiziol.* **4**:128.

Avrova, N. F., 1971, Brain ganglioside patterns of vertebrates, *J. Neurochem.* **18**:667.

Balasubramanian, A. S., Taori, G. M., Mokashi, S., and Bachhawat, B. K., 1973, Role of sialo-compounds in the metabolic processes of brain, *in:* Proceedings of a Symposium on Control Mechanisms in Cellular Processes, pp. 545–56, Bhabha At. Res. Cent., Bombay, India.

Barra, H. S., Cumar, F. A., and Caputto, R., 1969, The synthesis of neuramin lactose by preparations of rat mammary gland and its relation to the synthesis of lactose, *J. Biol. Chem.* **244**:6233.

Barry, G. T., 1958, Colominic acid, a polymer of N-acetylneuraminic acid, *J. Exp. Med.* **107**:507.

Barry, G. T., 1959, Detection of sialic acid in various *Escherichia coli* strains and in other species of bacteria, *Nature* **183**:117.

Barry, G. T., 1965, Un nouveau mucopolyoside des enterobacteries, *Bull. Soc. Chim. Biol.* **47**:529.

Barry, G. T., and Goebel, W. F., 1957, Colominic acid, a substance of bacteria origin related to sialic acid, *Nature* **179**:206.

Barry, G. T., Tsai, T., and Chen, F. P., 1960, Chemical and serological relationships of certain bacterial polysaccharides containing sialic acid, *Nature* **185**:597.

Barry, G. T., Abbot, V., and Tsai, T., 1962, Relationship of colominic acid (poly N-acetylneuraminic acid) to bacteria which contain neuraminic acid, *J. Gen. Microbiol* **29**:335.

Barry, G. T., Hamm, J. D., and Graham, M. G., 1963, Evaluation of colorimetric methods in the estimation of sialic acid in bacteria, *Nature* **200**:806.

Bartholomew, B. A., Jourdian, G. W., and Roseman, S., 1973, The sialic acids. XV. Transfer of sialic acid to glycoproteins by a sialyltransferase from colostrum, *J. Biol. Chem.* **248**:5751.

Bennett, V., O'Keefe, E., and Cuatrecasas, P., 1975, Mechanism of action of cholera toxin and the mobile receptor theory of hormone receptor-adenylate cyclase interactions, *Proc. Natl. Acad. Sci. U.S.* **72**:33.

Blix, G., and Lindberg, E., 1960, The sialic acids of bovine and equine submaxillary mucins, *Acta Chem. Scand.* **14**:1809.

Blix, G., Svennerholm, L., and Werner, I., 1952, The isolation of chondrosamine from gangliosides and from submaxillary mucin, *Acta Chem. Scand.* **6**:358.

Blix, G., Lindberg, E., Odin, L., and Werner, I., 1955, Sialic acids, *Nature* **175**:340.

Blix, G., Lindberg, E., Odin, L., and Werner, I., 1956, Sialic acids, *Acta Soc. Med. Upsalien* **61**:1.

Blix, F. G., Gottschalk, A., and Klenk, E., 1957, Proposed nomenclature in the field of neuraminic and sialic acids, *Nature* **179**:1088.

Bogoch, S., 1958, Cerebrospinal fluid neuraminic acid deficiency in schizophrenia, *Arch. Neurol. Psychiat.* **80**:221.

Bohm, P., and Baumeister, L., 1956, Uber das Vorkommen Neuramininsaure haltiger Glycoproteide in Korperflussigkeiten, *Z. Physiol. Chem.* **305**:42.

Bondareff, W., and Sjostrand, J., 1969, Cytochemistry of synaptosomes, *Exp. Neurol.* **24**:450.

Bourrillon, R., 1972, Urinary glycoproteins, glycopeptides and related heterosaccharides, in: *Glycoproteins: Their Composition, Structure and Function* (A. Gottschalk, ed.), 2nd ed., pp. 909–925, Elsevier Publishing Co., Amsterdam.

Bourrillon, R., Got, R., and Michon, J., 1961, Urinary glycoproteins. II. Study of a nondialyzable fraction soluble in 65% ethanol, *Clin. Chim. Acta* **6**:91.

Boursnell, J. C., Hartree, E. F., and Briggs, P. A., 1970, Studies of the bulbo-urethral (Cowper's)-gland mucin and seminal gel of the boar, *Biochem. J.* **117**:981.

Braidman, I., Carroll, M., Dance, N., and Robinson, D., 1974, Separation and properties of human brain hexosaminidase C, *Biochem. J.* **143**:295.

Brodbeck, U., Gentinetta, R., and Lundin, S. J., 1973, Multiple forms of a cholinesterase from body muscle of plaice (*Pleuronectes platessa*) and possible role of sialic acid in cholinesterase reaction specificity, *Acta Chem. Scand.* **27**:561.

Burge, B. W., and Strass, J. H., 1970, Glycopeptides of the membrane glycoprotein of Sindbis virus, *J. Mol. Biol* **47**:449.

Buscher, H. P., Casals-Stenzel, J. and Schauer, R., 1974, New sialic acids. Identification of N-glycolyl-O-acetylneuraminic acids and N-acetyl-O-glycolyl-neuraminic acids by improved methods for detection of N-acyl-O-acyl groups and by gas–liquid chromatography, *Eur. J. Biochem.* **50**:57.

Butkowski, R. J., Bajaj, S. P., and Mann, K. G., 1974, The preparation and activation of sialyl-^{3}H prothrombin, *J. Biol. Chem.* **249**:6562.

Cabezas, J. A., 1966, Acides N-glycolyl et acetylneuraminiques dans la bile de veau, *Bull. Soc. Chim. Biol* **48**:381.

Cabezas, J. A., 1968, Sialic acids in various biological materials, *An. Real Acad. Farm.* **34**:155.

Cabezas, J. A., 1973, The type of naturally occurring sialic acids, *Rev. Esp. Fisiol.* **29**:307.
Cabezas, J. A., and Feo, F., 1969, Sialic acids. XI. On the thiobarbituric acid positive reaction in several materials from the vegetal kingdom, *Rev. Esp. Fisiol.* **25**:153.
Cabezas, J. A., and Frois, M. D., 1966, Neuraminic acid. VI. Acetylneuraminic acids in lamprey liver and eggs, and in eggs from two teleostei species, *Rev. Esp. Fisiol.* **22**:147.
Cabezas, J. A., and Ramos, M., 1972, The type and content of the sialic acid of bile from several animal sources, *Carbohyd. Res.* **24**:486.
Cabezas, J. A., Trigueros, E., and Vazquez-Porto, J., 1966, Sialic acids. V. Content in *N*-glycolyl-neuraminic acid of the glycoproteins from cow colostrum, *Rev. Esp. Fisiol.* **22**:15.
Cabezas, J. A., Carrion, A., Gomez-Gonzalez, M. C., and Ramos, M., 1968*a*, Neuraminic acids. VIII. Identification of *N*-acetyl- and *N*-glycolylneuraminic acids in rat serum, and study of their absorption by the rat intestine, *Rev. Esp. Fisiol.* **24**:99.
Cabezas, J. A., Frois, M. D., and Vazquez-Porto, J., 1968*b*, Sialic acids. IX. Nature and concentration of the acylneuraminic acids isolated from various goat products, *Rev. Esp. Fisiol.* **24**:133.
Campbell, H. D., Dudman, N. P. B., and Ferner, B., 1973, Multiple forms of acid phosphatase in pig liver, *FEBS Lett.* **31**:123.
Carmody, P. J., and Rattazzi, M. C., 1974, Conversion of human hexosaminidase A to hexosaminidase B by crude *Vibrio cholerae* neuraminidase preparations: Merthiolate is the active factors, *Biochim. Biophys. Acta* **371**:117.
Carnillot, P., and Bourrillon, R., 1964, Characterisation d'un oligosaccharide riche en acide sialique dans l'urine humaine normale, *Prot. Biol. Fluids, Proc. Colloq.* **12**:320.
Carrion, A., Bourrillon, R., and Cabezas, J. A., 1969*a*, Preparation and properties of a sialoglycopeptide rich in sialic acid in normal human urine, *Clin. Chim. Acta.* **24**:351.
Carrion, A., Bourrillon, R., and Cabezas, J., 1969*b*, *N*-Acetyl- and *N,O*-diacetylneuraminic acids in a sialoglycopeptide from normal human urine, *Clin. Chim. Acta* **26**:481.
Carrion, A., Bourrillon, R., and Cabezas, J. A., 1970, Sialic acids. XII. On the nature of the acetylneuraminic acids from human normal urine, *Rev. Esp. Fisiol.* **26**:171.
Carubelli, R., and Griffin, M. J., 1967, Sialic acids in HeLa cells: Effects of hydrocortisone, *Science* **157**:693.
Carubelli, R., and Griffin, M. J., 1968, On the presence of *N*-glycolylneuraminic acid in HeLa cells, *Biochim. Biophys. Acta* **170**:446.
Carubelli, R., Ryan, L. C., Trucco, R. E., and Raputto, R., 1961, Neuraminlactose sulfate, a new compound isolated from the mammary gland of rats, *J. Biol. Chem.* **236**:2381.
Casals, J., and Reeves, W. C., 1965, The arborviruses, in: *Viral and Rickettsial Infection of Man* (F. H. Horsfall and I. Tamm, eds.), 4th ed., pp. 580–582, Lippincott, Philadelphia.
Clamp, J. R., and Johnson, I., 1972, Immunoglobulins, in: *Glycoprotein: Their Composition, Structure, and Function* (A. Gottschalk, ed.), 2nd ed., pp. 612–652, Elsevier Publishing Co., Amsterdam.
Compans, R. W., 1971, Location of the glycoprotein in the membrane of Sindbis virus, *Nat. New Biol.* **229**:114.
Cook, G. M. W., and Stoddart, R. W., 1973, *Surface Carbohydrates of the Eukaryotic Cell*, Academic Press, London.
Cook, G. M. W., Heard, D. H., and Seaman, G. V. F., 1960, A sialo-mucopeptide liberated by trypsin from the human erythrocyte, *Nature* **188**:1011.

Cornelius, C. E., Pangborn, J., and Heckly, R. J., 1963, Isolation and characterization of a urinary mucoprotein from ovine urine, *Arch. Biochem. Biophys.* **101**:403.
Cornelius, C. E., Mia, A. S., and Rosenfeld, S., 1965, Ruminant urolithiasis. III. Studies on the origin of Tamm–Horsfall urinary mucoprotein and its presence in ovine calculous matrix, *Invest. Urol.* **2**:453.
Correll, D. L., 1964, Sialic-acid-containing glycopeptide from Chlorella, *Science* **145**:588.
Cuatrecasas, P., 1973a, Interaction of *Vibrio cholerae* enterotoxin with cell membranes, *Biochem.* **12**:3547.
Cuatrecasas, P., 1973b, Gangliosides and membrane receptors for cholera toxin, *Biochemistry* **12**:3558.
Cuatrecasas, P., 1973c, Cholera toxin-fat cell interaction and the mechanism of activation of the lipolytic response, *Biochemistry* **12**:3567.
Cuatrecasas, P., 1973d, Vibrio cholerae choleragenoid. Mechanism of inhibition of cholera toxin action, *Biochemistry* **12**:3577.
Cumings, J. N., Thompson, E. J., and Goodwin, H., 1968, Sphingolipids and phospholipids in microsomes and myelin from normal and pathological brains, *J. Neurochem.* **15**:243.
Curtis, A. S. G., 1973, Cell adhesion, *Prog. Biophys. Mol. Biol.* **27**:317.
Dain, J. A., and Yip, M. C. M., 1969, Regulatory mechanisms in ganglioside biosynthesis: a possible role in Tay–Sachs disease, *Metabolismo* **5**:129.
Deman, J. J., Bruyneel, E. A., and Mareel, M. M., 1974, Mechanism of intercellular adhesion. Effects of neuraminidere, calcium and trypsin on the aggregation of suspended Hela cells, *J. Cell Biol.* **60**:641.
DeMartinez, H. R., and Olavarria, J. M., 1973, Sialic acids of toad oviduct mucoprotein, *Biochim. Biophys. Acta* **320**:295.
Derry, D. M., and Wolfe, L. S., 1967, Gangliosides in isolated neurons and glial cells, *Science* **158**:1450.
DeWitt, C. W., 1958, Biological and biochemical comparison of *Escherichia coli* endotoxins, *Bact. Proc.* **1958**:75, abstract M61.
DeWitt, C. W., and Rowe, J. A., 1959, N,O-Diacetylneuraminic acid and N-acetylneuraminic acid in *Escherichia coli*, *Nature* **184**:381.
DeWitt, C. W., and Rowe, J. A., 1961, Sialic acids ($N,7$-diacetylneuraminic acid and N-acetylneuraminic acid) in *Escherichia coli*. I. Isolation and identification, *J. Bacteriol.* **82**:838.
di Benedetta, C., Brunngraber, E. G., Whitney, G., Brown, B. D., and Aro, A., 1969, Compositional patterns of sialofucohexosaminoglycans derived from rat brain glycoproteins, *Arch. Biochem. Biophys.* **131**:404.
Dorner, F., Scriba, M., and Weil, R., 1973, Interferon: Evidence for its glycoprotein nature, *Proc. Natl. Acad. Sci. U.S.* **70**:1981.
Dreyfus, H., Urban, P. F., Boseh, P., Edel-Harth, S., Rebel, G., and Mandel, P., 1974, Effect of light on gangliosides from calf retina and photoreceptors, *J. Neurochem.* **22**:1073.
Dunn, A. J., 1975, Decreased sialic acid content of isolated synaptosomes following ECS, in: Sixth Meeting of the American Society of Neurochemistry, Abstract No. 272, Mexico City.
Dupont, A., Farriaux, J. P., Biserte, G., Montreuil, J., and Fontaine, G., 1967, Particularites de l'ultra structure hepatique chez un enfant presentant une sialurie (acid N-acetyl nueraminique), *Lille Med.* **12**:654.
Dzulynska, J., Potemkowska, E. A., Walkowiak, H., and Fabijanska, I., 1969, Studies on

serum glycoproteins in the fowl during the growing and laying periods, *Bull. Acad. Pol. Sci.* **17**:523.

Ebert, W., Metz, J., Weicker, H., and Roelcke, D., 1971, Ficin catalyzed fragmentation of glycoproteins from erythrocyte membranes, *Z. Physiol. Chem.* **352**:1309.

Ebert, W., Metz, J., and Roelcke, D., 1972, Modifications of N-acetylneuraminic acid and their influence on the antigen activity of erythrocyte glycoproteins, *Eur. J. Biochem.* **27**:470.

Eldredge, N. F., Read, G., and Cutting, W., 1963, Sialic acids in the brain and tissues of various animals: analytical and physiological data, *Med. Exp.* **8**:265.

Esselman, W. J., and Miller, H. C., 1974a, The ganglioside nature of θ antigens, *Fed. Proc.* **31**:771.

Esselman, W. J., and Miller, H. C., 1974b, Brain and thymus lipid inhibition of antibrain-associated θ-cytotoxicity, *J. Exp. Med.* **139**:445.

Etchison, J. R., and Holland, J. J., 1974, Carbohydrate composition of the membrane glycoprotein of vesicular stomatitis virus grown in four mammalian cell lines, *Proc. Natl. Acad. Sci. U.S.* **71**:4011.

Evans, W. H., 1974, Nucleotide pyrophosphatase, a sialoglycoprotein located on the hepatocyte surface, *Nature* **250**:381.

Eylar, E. H., 1966, On the biological role of glycoproteins, *J. Theoret. Biol.* **10**:89.

Eylar, E. H., Doolittle, R. F., and Madoff, M. A., 1962a, Sialic acid from blood cells of the lamprey eel, *Nature* **193**:1183.

Eylar, E. H., Madoff, M. A., Brody, O. V., and Oncley, J. L., 1962b, The contribution of sialic acid to the surface charge of the erythrocyte, *J. Biol. Chem.* **237**:1992.

Faillard, H., 1969, Unpublished observation, cited in Cabezas and Feo, 1969.

Faillard, H., and Cabezas, J. A., 1963, Isolierung von N-acetyl- und N-glykolyl-neuraminsaure aus Kalber-und Huhnerserum, *Z. Physiol. Chem.* **333**:266.

Faillard, H., and Schauer, R., 1972, Glycoproteins as lubricants, protective agents, carriers, structural proteins and as participants in other functions, *in: Glycoproteins: Their Composition, Structure and Function*, (A. Gottschalk, ed.), 2nd ed., pp. 1246–1267, Elsevier Publishing Co., Amsterdam.

Feeney, R. E., Anderson, J. S., Azari, P. R., Bennett, N., and Rhodes, M. B., 1960, The comparative biochemistry of avian egg white proteins, *J. Biol. Chem.* **235**:2307.

Fewster, M. E., and Mead, J. F., 1968, Lipid composition of glial cells isolated from bovine white matter, *J. Neurochem.* **15**:1041.

Fletcher, M. A., and Lo, T. M., 1974, Isolation of characterization of a glycoprotein from goat erythrocyte membrane, *Fed. Proc.* **33**:777.

Fontaine, G., Gaudier, B., Biserte, G., Montreuil, J., Dupont, A., Farriaux, J. P., Strecker, G., Spik, G., Puvion, E., Puvion-Dutilluel, M., and Sezille, G., 1967, Elimination urinaire permanente d'acide sialique libre chez un enfant de trois ans atteint de troubles cliniques divers, *Pediatrie* **22**:705.

Fontaine, G., Biserte, G., Montreuil, J., Dupont, A., Farriaux, J. P., Strecker, G., Spik, G., Puvion, E., Puvion-Dutilluel, F., Sezille, G., and Picque, M. T., 1968, La sialurie: un trouble metabolique original, *Helvet. Paed. Acta.*, Suppl. XVII **23**:3.

Fouquet, J. P., 1971, Secretion of free glucose and related carbohydrates in male accessory organs of rodents, *Comp. Biochem. Physiol.* **40**:305.

Fouquet, J. P., 1972, Free sialic acids in the seminal vesicle secretion of the golden hamster, *J. Reprod. Fert.* **28**:273.

Frits, O., Orskov, I., Jann, B., and Jann, K., 1971, Immunoelectrophoretic patterns of extracts from all *Escherichia coli* O & K antigen test strains correlation with pathogenicity, *Acta. Pathol. Microbiol. Scand. Sec. B. Microbiol. Immunol.* **79**:142.

Fujita, S., and Cleve, H., 1975, Isolation and partial characterization of the major glycoproteins of horse and swine erythrocyte membranes, *Fed. Proc.* **34**:1033.

Gardas, A., and Koscielak, J., 1971, A, B, and H blood-group specificities in glycoprotein and glycolipid fraction of human erythrocyte membrane, *Vox Sang.* **20**:137.

Garrigan, O. W., Chargaff, E., 1963, Studies on the mucolipids and the cerebrosides of chicken brain during embryonic development, *Biochim. Biophys. Acta* **70**:452.

Gielen, W., 1968a, Neuraminsaure in Pflanzen? Die-2-keto-3-deoxyaldonsauren in Pflanzen und die synthese der 3-desoxy-D-glycero-β-D-galakto-nonulosonsaure, *Z. Naturforsch* **23b**:1598.

Gielen, W., 1968b, Ganglioside function. Distribution of the serotonin receptor, *Z. Naturforsch* **23b**:117.

Gielen, W., and Hinzen, D. H., 1974, Acetylneuraminate cytidyltransferase and sialyltransferase of neuronal and glial cells isolated from rat cerebral cortex, *Z. Physiol. Chem.* **355**:895.

Gilbert, F., Kucherlapati, R., Creagan, R. P., Murnane, M. J., Darlington, G. J., and Ruddle, F. H., 1975, Tay–Sachs and Sandhoffs diseases: The assignment of genes for hexosamindase A and B to individual human chromosomes, *Proc. Natl. Acad. Sci. U.S.* **72**:263.

Goldstone, A., Konecny, P., and Koenig, H., 1971, Lysosomal hydrolases: conversion of acidic to basic forms by neuraminidase, *FEBS Lett.* **13**:68.

Got, R., Bourrillon, R., and Michon, J., 1960, The glucidic constituents of human chorionic gonadotropin, *Bull. Soc. Chim. Biol.* **42**:41.

Gottschalk, A., 1960, *The Chemistry and Biology of Sialic Acids and Related Substances*, p. 30–35, Cambridge University Press, London.

Gottschalk, A. (ed.), 1972, *Glycoproteins: Their Composition, Structure and Function*, 2nd ed., Elsevier Publishing Co., Amsterdam.

Gottschalk, A., and Drzeniek, R., 1972, Neuraminidase as a tool in structural analysis, *in: Glycoproteins: Their Composition, Structure and Function*, (A. Gottschalk, ed.), 2nd ed., pp. 381–402, Elsevier Publishing Co., Amsterdam.

Goussault, Y., and Bourrillon, R., 1970, Chemical characterization of two urinary sialic acid-rich glycopeptides, *Biochem. Biophys. Res. Commun.* **40**:1404.

Graham, E. R. B., and Gottschalk, A., 1960, Studies on mucoproteins. I. The structure of the prosthetic group of ovine submaxillary gland mucoprotein, *Biochim. Biophys. Acta* **38**:513.

Hakomori, S., 1970a, Physiological variation of glycolipids in cultured cells and change of the pattern in malignant-transformed cells, *in:* 160th Meeting of the American Chemical Society, Abstract No. 33.

Hakomori, S., 1970b, Cell density-dependent changes of glycolipid concentrations in fibroblasts and loss of this response in virus-transformed cells, *Proc. Natl. Acad. Sci. U.S.* **67**:1741.

Hakomori, S., 1971, Glycolipid changes associated with malignant transformation *in: The Dynamic Structure of Cell Membranes* (H. Fisher, ed.) pp. 65–96, Springer-Verlag, Berlin.

Hakomori, S., and Murakami, W. T., 1968, Glycolipids of hamster fibroblasts and derived malignant-transformed cell lines, *Proc. Natl. Acad. Sci. U.S.* **59**:254.

Hakomori, S., and Saito, T., 1969, Isolation and characterization of a glycosphingolipid having a new sialic acid, *Biochemistry* **8**:5082.

Hakomori, S., Teather, C., and Andrew, H., 1968, Organizational difference of cell surface "hematoside" in normal and virally transformed cells, *Biochem. Biophys. Res. Commun.* **33**:563.

Hakomori, S., Kijimoto, S., and Siddiqui, S., 1972, Glycolipids of normal and transformed cells: A difference in structure and dynamic behavior, *in: Membrane Research* (C. F. Fox, ed.), pp. 253–280, Academic Press, New York.

Hakomori, S., Gahmberg, C. G., and Laine, R. A., 1974, Growth behavior of transformed cells as related to the surface structures of glycolipids and glycoproteins, *in:* Sixth Annual Miami Winter Symposia, Dept. of Biochem., Univ. of Miami, pp. 49–52.

Hamberger, A., and Svennerholm, L., 1971, Composition of gangliosides and phospholipids of neuronal and glial cell enriched fractions, *J. Neurochem.* **18:**1821.

Hartree, E. F., 1962, Sialic acid in the bulbo-urethral glands of the boar, *Nature* **196:**483.

Hartree, E. F., and Srivastava, P. N., 1965, Chemical composition of the acrosomes of ram spermatozoa, *J. Reprod. Fert.* **9:**47.

Hewlett, E. L., Guerrant, R. L., Evans, D. J., and Greenough, W. B., 1974, Toxins of *Vibrio cholerae* and *Escherichia coli* stimulate adenyl cyclase in rat fat cells, *Nature* **249:**371.

Hickman, J., Ashwell, G., Morell, A. G., Van den Hamer, C. J. A., and Scheinberg, I. H., 1970, Physical and chemical studies on ceruloplasmin. VIII. Preparation of N-acetylneuraminic acid-1-^{14}C-labeled ceruloplasmin, *J. Biol. Chem.* **245:**759.

Hirschberg, C. B., and Robbins, P. W., 1974, The glycolipids and phospholipids of Sindbis virus and their relation to the lipids of the host cell plasma membrane, *Virology* **61:**602.

Ho, M. W., Beutler, S., Tennant, L., and O'Brien, J. S., 1972, Fabry's disease: Evidence for a physically altered α-galactosidase, *Am. J. Human Genet.* **24:**256.

Hof, L., and Faillard, H., 1973, The serum dependence of the occurrence of N-glycolylneuraminic acid in HeLa cells, *BBA* **297:**561.

Hollenberg, M. D., Fishman, P. H., Bennett, V., and Cuatrecasas, P., 1974, Cholera toxin and cell growth: Role of membrane gangliosides, *Proc. Natl. Acad. Sci. U.S.* **71:**4224.

Hooghwinkel, G. L. M., Veltkamp, W. A., Overdijk, B., and Lisman, J. J. W., 1972, Electrophoretic separation of β-N-acetylhexosaminidases of human and bovine brain and liver and of Tay–Sachs Brain, *Z. Physiol. Chem.* **353:**839.

Hotta, K., Hamazaki, H., Kurokawa, M., and Isaka, S., 1970a, Isolation and properties of a new type of sialopolysaccharide protein complex from the jelly coat of sea urchin eggs, *J. Biol. Chem.* **245:**5434.

Hotta, K., Kurokawa, M., and Isaka, S., 1970b, Isolation and identification of two sialic acids from the jelly coat of sea urchin eggs, *J. Biol. Chem.* **245:**6307.

Hotta, K., Kurokawa, M., and Isaka, S., 1973, A novel sialic acid and fucose-containing disaccharide isolated from the jelly coat of sea urchin eggs, *J. Biol. Chem.* **248:**629.

Hudgin, R. L., and Ashwell, G., 1974, Studies on the role of glycosyltransferases in the hepatic binding of asialoglycoproteins, *J. Biol. Chem.* **249:**7369.

Hudgin, R. L., Pricer, W. E., Jr., Ashwell, G., Stockert, R. J., and Morell, A. G., 1974, The isolation and properties of a rabbit liver binding protein specific for asialoglycoproteins, *J. Biol. Chem.* **249:**5536.

Hughes, R. C., 1973, Glycoproteins as components of cellular membranes, *Prog. Biophys. Mol. Biol.* **26:**189.

Huprikar, S., and Springer, G. F., 1970, Structural aspects of human blood-group M and N specificity, *in: Blood and Tissue Antigens* (E. Aminoff, ed.) pp. 327–335, Academic Press, New York.

Huttunen, J. K., and Miettinen, T. A., 1965, Isolation of neuraminlactose from human male urine, *Acta Chem. Scand.* **19:**1486.

Huttunen, J. K., Maury, P., and Miettinen, T. A., 1972, Increased urinary excretion of neuraminic-acid-containing oligosaccharide after myocardial infraction, *J. Mol. Cell. Cardiol.* **4:**59.

Immers, J., 1968, N-Acetyl-neuraminic acid in the sea urchin jelly coat, *Acta Chem. Scand.* **22:**2204.

Irani, R. J., and Ganapathi, K., 1962, Occurrence of sialic acid in some gram-positive and gram-negative pathogenic bacteria, *Nature* **195:**1227.

Isemura, M., Zahn, R. K., and Schmid, K., 1973, A new neuraminic acid derivative and three types of glycopeptides isolated from the Cuvierian tubules of the sea cucumber *Holothuria forskali*, *Biochem. J.* **131:**509.

Ishizuka, I., and Wiegandt, H., 1972, An isomer of trisialoganglioside and the structure of tetra- and pentasialogangliosides from fish brain, *Biochim. Biophys. Acta* **260:**279.

Ishizuka, I., Kloppenburg, M., and Weigandt, H., 1970, Characterization of gangliosides from fish brain, *Biochim. Biophys. Acta* **210:**299.

Isono, Y., and Nagai, Y., 1966, Biochemistry of glycolipids of sea urchin gametes, *Jap. J. Exp. Med.* **36:**461.

Jakoby, R. K., and Warren, L., 1961, Identification and quantitation of N-acetylneuraminic acid in human cerebrospinal fluid, *Neurology* **11:**232.

Johnston, P. V., and Roots, B. I., 1972, *Nerve Membranes: A Study of the Biological and Chemical Aspects of Neuron–Glial Relationships*, pp. 202–209, Pergamon Press, Oxford.

Jones, J. P., Ramsey, R. B., Aexel, R. T., and Nicholas, H. J., 1972, Lipid biosynthesis in neuron-enriched and glial-enriched fractions of rat brain; ganglioside biosynthesis, *Life Sci.* **11:**309.

Jourdian, G. W., and Distler, J. J., 1973, Formation *in vitro* of uridine-5'-(oligosaccharide)-1-pyrophosphates, *J. Biol. Chem.* **248:**6781.

Jourdian, G. W., and Roseman, S., 1963, Intermediary metabolism of the sialic acids, *Ann. NY Acad. Sci.* **106:**202.

Jourdian, G. W., Shimizu, F., and Roseman, S., 1961, Isolation of Nucleotide-oligosaccharides containing sialic acid, *Fed. Proc.* **20:**161.

Kathan, R. H., and Weeks, I. D., 1969, Structural studies of collocalia mucoid. I. Carbohydrate and amino acid composition, *Arch. Biochem. Biophys.* **134:**572.

Kauffmann, F., Jann, B., Kruger, L., Luderitz, O., and Westphal, O., 1962, Zur Immunchemie der O-antigene von Enterobacteriaceae. VIII. Analyse der Zuckerbausteine von Polysacchariden, Weiterer Salmonella-und Arizona-O-Gruppen, *Parasitenk, Abt. I. Orig.* **186:**509.

Kean, E. L., and Roseman, S., 1966, The sialic acids. X. Purification and properties of cytidine 5'-monophosphosialic acid synthetase, *J. Biol. Chem.* **241:**5643.

Kedzierska, B., Mikulaszek, E., and Pogonowska-Goldhar, J., 1968, Immunochemical studies on Salmonella serotype 48. III. N-Acetyl-neuraminic acid as an immunodominant sugar in *S. dahlem*, *Bull. Acad. Pol. Sci., Ser. Sci. Biol.* **16:**673.

Kennan, T. W., and Berridge, L., 1973, Identification of gangliosides as constituents of egg yolk, *J. Food Sci.* **38:**43.

Klenk, E., 1941, Neuraminic acid, the cleavage product of a new brain lipoid, *Z. Physiol. Chem.* **268:**50.

Klenk, E., and Faillard, H., 1954, The carbohydrate groups of mucoproteins, *Z. Physiol. Chem.* **298:**230.

Klenk, E., and Gielen, W., 1961, A chromatographically pure human brain ganglioside containing hexosamine, *Z. Physiol. Chem.* **326:**158.

Klenk, E., and Uhlenbruck, G., 1960, Neuraminic acid mucoids from the stroma of human erythrocytes, a contribution to the chemistry of agglutinogens, *Z. Physiol. Chem.* **319**:151.

Kochetkov, N. K., Zhukova, I. G., and Smirnova, G. P., 1968, Separation of sphingolipids containing sialic acids from gonads of *Strongylocentrotus intermedius, Dokl. Akad. Nauk SSSR, Ota Biokh.* **180**:999.

Kochetkov, N. K., Zhukova, I. G., Smirnova, G. P., and Glukhoded, I. S., 1973, Isolation and characterization of a sialoglycolipid from the sea urchin *Strongylocentrotus intermedius, Biochim. Biophys. Acta* **326**:74.

Korey, S. R., and Gonatas, J., 1963, Separation of human brain gangliosides, *Life Sci.* **2**:296.

Kornfeld, R., and Kornfeld, S., 1970, The structure of a phytohemagglutinin receptor site from human erythrocytes, *J. Biol. Chem.* **245**:2536.

Koscielak, J., Gardas, A., Pacuszka, T., and Piasek, A., 1973, A, B, and H active blood-group substances of human erythrocyte membrane, in: *Membrane-Mediated Information*, Vol. 1, Biochemical Functions (P. W. Kent, ed.), pp. 95–103, American Elsevier Publishing Co., New York.

Krantz, M. J., Lee, Y. C., and Hung, P. P., 1974, Carbohydrate groups in the major glycoproteins of Rous sarcoma virus, *Nature* **248**:684.

Krysteva, M. A., Mancheva, I. N., and Dobrev, I. D., 1973, Tyrosine environments of chicken ovomucoid. Environment of the most exposed tyrosine, *Eur. J. Biochem.* **40**:155.

Kuhn, R., 1958, The oligosaccharides of milk, *Bull. Soc. Chim. Biol.* **40**:297.

Kuhn, R., and Brossmer, R., 1956, *O*-Acetyllactaminic acid-lactose from cow colostrum and its cleavability by the influenza virus, *Chem. Ber.* **89**:2013.

Kuhn, R., and Brossmer, R., 1958, The constitution of lactaminic acid lactose: α-ketosidase effect of viruses of the influenza group, *Angew. Chem.* **70**:25.

Kuhn, R., and Wiegandt, H., 1963, Die Konstitution der Ganglio-*N*-tetraose und des Gangliosids G_1, *Chem. Ber.* **96**:866.

Kuhn, R., Brossmer, R., and Schulz, W., 1954, The prosthetic group of the mucoproteins of cow colostrum, *Chem. Ber.* **87**:123.

Kundig, F. D., Aminoff, D., and Roseman, S., 1971, The sialic acids. XII. Synthesis of colominic acid by a sialyltransferese from *Escherichia coli* K-235, *J. Biol. Chem.* **246**:2543.

Lalley, P. A., Rattazzi, M. C., and Shows, T. B., 1974, Human β-D-*N*-acetylhexosaminidases A and B: Expression and linkage relationships in somatic cell hybrids, *Proc. Natl. Acad. Sci. U.S.* **71**:1569.

Lee, T., and Hager, L. P., 1970, Carbohydrate Stripping from glycoproteins with retention of biological activity, *Fed. Proc.* **29**:599.

Lehninger, A. L., 1968, The neuronel membrane, *Proc. Natl. Acad. Sci. U.S.* **60**:1069.

Leikola, E., Nieminen, E., and Teppo, A. M., 1969, New sialic acid-containing sulfolipid: "ungulic acid," *J. Lipid Res.* **10**:440.

Lemoine, A.-M., 1974, Action du cortisol chezl'anguille hypophysectomisèe, en eau de mer. Effet sur l'acide *N*-acètyl-neuraminique branchial, *C. R. Soc. Biol. (Paris)* **168**:402.

Lemoine, A.-M., and Olivereau, M., 1973, Action of prolactin in intact and hypophysectomized eels. IX. Effect on *N*-acetyl-neuraminic acid content of the skin, in sea water, *Acta Zool.* **54**:223.

Lemoine, A.-M., and Olivereau, M., 1974, Effect of cortisol on *N*-acetyl-neuraminic acid of the skin in intact and hypophysectomized eels in fresh water and sea water, *Acta Zool.* **55**:255.

Li, Y.-T., Mansson, J-E., Vanier, M. T., and Svennerholm, L., 1973, Structure of the major glucosamine-containing ganglioside of human tissues, *J. Biol. Chem.* **248:**2634.

Licht, P., and Papkoff, H., 1974, Phylogenetic survey of the neuraminidase sensitivity of reptilian gonadotropin, *Gen. Comp. Endocrinol.* **23:**415.

Liotta, I., Ouintilliani, M., Ouintiliani, L., Buzzonetti, A., and Guiliani, E., 1972, Extraction and partial purification of blood group substances A, B, and H from erythrocyte stroma, *Vox Sang.* **22:**171.

Londesborough, J. C., and Hamberg, U., 1975, The sialic acid content and isoelectric point of human kininogen, *Biochem. J.* **145:**401.

Lopes, R. A., and Valeri, V., 1972, Morphological and histochemical study of mucosubstances of the lacrimal and Harder's glands of the lizard *Ameiva ameiva* (Lacertihia, Teiidae), *Cieno. Cult. (Sao Paulo)* **24:**1163.

Lopes, R. A., Valeri, V., Campos, G. M., Lopes, O. V. P., and DeFaria, R. M., 1973, Histochemical study of mucopolysaccharides of the cephalic glands of *Micrurus corallinus*, *Ann. Histochim.* **18:**131.

Lopes, R. A., DeOliveira, C., Campos, M. N. M., Campos, S. M., and Birman, E. G., 1974, Morphological and histochemical study of the cephalic glands of *Bothrops jararaca*, *Acta Zool.* **55:**17.

Luppi, A., and Cavazzini, G., 1966, Sialic acid determination in some Corynebacteriaceae, *Nuovi Ann. Ig. Microbiol.* **17:**183.

Maccioni, A. H. R., Gimenez, M. S., and Caputto, R., 1971, The labeling of the gangliosidic fraction from brains of rats exposed to different levels of stimulation after injection of 6-^{3}H glucosamine, *J. Neurochem.* **18:**2363.

Maccioni, A. H. R., Gimenez, M. S., Caputto, B. L., and Caputto, R., 1974, Labeling of the gangliosidic fraction from brains of chickens exposed to different levels of stimulation after injection of 6-^{3}H glucosamine, *Brain Res.* **73:**503.

Mann, T., 1964, *The Biochemistry of Semen and of the Male Reproductive Tract*, Methuen, London.

Mapes, C. A., and Sweeley, C. C., 1973, Interconversion of the A and B forms of ceramide trihexosidase from human plasma, *Arch. Biochem. Biophys.* **158:**297.

Marchesi, V. T., Tillack, T. W., Jackson, R. L., Segrest, J. P., and Scott, R. E., 1972, Chemical characterization and surface orientation of the major glycoprotein of the human erythrocyte membrane, *Proc. Natl. Acad. Sci. U.S.* **69:**1445.

Marin, B. J., and Philpott, C. W., 1974, Biochemical nature of the cell periphery of the salt gland secretory cells of fresh and salt water adapted mallard ducks, *Cell Tissue Res.* **150:**193.

Marshall, R. D., 1972, Glycoproteins, *Ann. Rev. Biochem.* **41:**673.

Martensson, E., 1969, Glycosphingolipids of animal tissue, *Prog. Chem. Fats Lipids* **10:**365.

Martensson, E., Raal, A., and Svennerholm, L., 1958, Sialic acids in blood serum, *Biochim. Biophys. Acta* **30:**124.

Maury, P., 1971, Identification of N,O-diacetyl-, N-acetyl-, and N-glycolyl neuraminyl-(2-3)-lactose in rat urine, *Biochim. Biophys. Acta* **252:**472.

Mayer, F. C., Dam, R., and Pazur, J. H., 1964, Occurrence of sialic acids in plant seeds, *Arch. Biochem. Biophys.* **108:**356.

McCluer, R. H., and Agranoff, B. W., 1972, Studies on gangliosides of goldfish brain, *J. Neurochem.* **19:**2307.

McGuire, E. J., and Binkley, S. B., 1964, The structure and chemistry of colominic acid, *Biochem.* **3:**247.

Mehrishi, J. N., 1972, Molecular aspects of the mammalian cell surface, *Prog. Biophys. Mol. Biol.* **25**:1.
Merz, W., and Roelcke, D., 1971, Biochemische Differenzierung der P_{r1}-P_{r2}-determinicerenden vonder. MN-determinierenden N-Acetyl-Neuraminsaure durch Acelylurungsversuche mit erythrocyten-glykoproteinen, *Eur. J. Biochem.* **23**:30.
Messer, M., 1974, Identification of N-acetyl-4-O-acetylneuraminyl-lactose in Echidna milk, *Biochem. J.* **139**:415.
Mestrallet, M. G., Cumar, F. A., and Caputto, R., 1974, On the pathway of biosynthesis of trisialogangliosides. *Biochem. Biophys. Res. Commun.* **59**:1.
Mia, A. S., and Cornelius, C. E., 1966, Biosynthesis and turnover of Tamm and Horsfall mucoprotein in sheep, *Invest. Urol.* **3**:334.
Montgomery, R., 1972, Heterogeneity of the carbohydrate groups of glycoproteins, *in: Glycoproteins: Their Composition, Structure and Function* (Gottschalk, ed.), 2nd ed., pp. 518–528, Elsevier Publishing Co., Amsterdam.
Montreuil, J., Biserte, G., Strecker, G., Spik, G., Fontaine, G., and Farriaux, J. P., 1967, Description d'un nouveau type de meliturie: La sialurie, *C.R. Acad. Sci., Paris, Ser. D.* **265**:97.
Montreuil, J., Biserte, G., Strecker, G., Spik, G., Fontaine, G., and Farriaux, J-P., 1968, Description d'un nouveau type de meliturie: la sialurie, *Clin. Chim. Acta* **21**:61.
Morell, A. G., Irvine, R. A., Sternlieb, I., Scheinberg, I. H., and Ashwell, G., 1968, Physical and chemical studies on cerulplasmin. V. Metabolic studies on sialic-acid-free ceruloplasmin *in vivo, J. Biol. Chem.* **243**:155.
Morell, A. G., Gregoriadis, G., Scheinberg, I. H., Hickman, J., and Ashwell, G., 1971, The role of sialic acid in determining the survival of glycoproteins in the circulation, *J. Biol. Chem.* **246**:1461.
Neuberger, A., and Ratcliffe, W. A., 1972, The acid and enzymic hydrolysis of O-acetylated sialic acid residues from rabbit Tamm–Horsfall glycoproteins, *Biochem. J.* **129**:683.
Noren, R., and Svennerholm, L., 1970, The gangliosides of fish, crab, lobster and octopus, cited in Svennerholm, 1970a.
Norton, W. T., and Poduslo, S. E., 1971, Neuronal perikarya and astroglia of rat brain: chemical composition during myelination, *J. Lipid Res.* **12**:84.
Odin, L., 1955a, Sialic acid in pseudomyxomatous gels, *Acta Chem. Scand.* **9**:714.
Odin, L., 1955b, Sialic acid in human serum protein and in meconium, *Acta Chem. Scand.* **9**:862.
Odin, L., 1955c, Sialic acid in human cervical mucus, in hog seminal gel and in ovomucin, *Acta Chem. Scand.* **9**:1235.
Olivereau, M., and Lemoine, A.-M., 1972, Effects of external salinity changes on N-acetyl-neuraminic acid content of the skin in eels, *J. Comp. Physiol.* **79**:411.
Onodera, K., Hirano, S., and Hayashi, H., 1966, Sialic acid and related substances. IV. Sialic acid content of some biological materials, *Agr. Biol. Chem. (Tokyo)* **30**:1170.
Oppenheimer, S. B., Edidin, M., Orr, C. W., and Roseman, S., 1969, An L-glutamine requirement for intercellular adhesion, *Proc. Natl. Acad. Sci. U.S.* **63**:1395.
Papadopoulos, N. M., and Hess, W. C., 1960, Determination of neuraminic (sialic) acid, glucose, and fructose in spinal fluid, *Arch. Biochem. and Biophys.* **88**:167.
Papkoff, H., 1966, Glycoproteins with biological activity *in: Glycoproteins: Their Composition, Structure, and Function,* (A. Gottschalk, ed.), 1st ed., pp. 532–557, Elsevier Publishing Co., New York.

Parham, P., Humphreys, R. E., Turner, M. J., and Stromiger, J. L., 1974, Heterogeneity of HL-A antigen preparations is due to variable sialic acid content, *Proc. Natl. Acad. Sci. U.S.* **71:**3998.

Pazur, J. H., Knull, K. R., and Simpson, D. L., 1970, Glycoenzyme: A note on the role of the carbohydrate moieties, *Biochim. Biophys. Res. Commun.* **40:**110.

Pepper, D. S., 1968, The sialic acids of horse serum with special reference to their virus-inhibitory properties, *Biochim. Biophys. Acta* **156:**317.

Peterson, J. W., 1974, Tissue-binding properties of the cholera toxin, *Inf. Immunol.* **10:**157.

Poenaru, L., and Dreyfus, J. C., 1973, Electrophoretic study of hexosaminidases, *Clin. Chim. Acta* **43:**439.

Poenaru, L., Weber, A., Vibert, M., and Dreyfus, J. C., 1973, Hexosaminidase C. Immunological distinction from hexosaminidase A., *Biomedicine* **19:**538.

Prokhorova, M. I., Penjeva, T. I., Romanova, L. A., and Tumanova, S., 1965, Gangliosides of brain, *Ukr. Biokhim. Zh.* **37:**778.

Pustzai, A., 1964, Hexosamines in the seeds of high plants, *Nature* **201:**1328.

Ramadoss, C. S., Selvam, R., Shannugasundaram, K. R., and Shamugasundarum, E. R. B., 1974, Rabbit kidney alkaline phosphatase. Role of sialic acid in the heterogeneity, *Experientia* **30:**982.

Robert, J., Freysz, L., Sensenbrenner, M., Mandel, P., and Rebel, G., 1975, Gangliosides of glial cells. A comparative study of normal astroblasts in tissue culture and glial cells isolated on sucrose-ficoll gradients, *FEBS Lett.* **50:**144.

Robinson, D., and Stirling, J. L., 1968, N-Acetyl-β-glucosaminidase in human spleen, *Biochem. J.* **107:**321.

Roelcke, D., Ebert, W., Metz, J., and Weicker, H., 1971, I, MN, and $P_f/P_{\bar{f}}$ activity of human erythrocyte glycoprotein fractions obtained by ficin treatment, *Vox Sang.* **21:**352.

Roseman, S., 1970, The synthesis of complex carbohydrates by multiglycosyltransferase systems and their potential function in intercellular adhesion, *Chem. Phys. Lipids* **5:**270.

Roseman, S., 1974a, Complex carbohydrates and intercellular adhesion, Sixth Annual Miami Winter Symposia, pp. 1–4, Dept. of Biochem., Univ. of Miami.

Roseman, S., 1974b, The biosynthesis of cell-surface components and their potential role in intercellular adhesion, *in: The Neurosciences: Third Study Program* (F. O. Schmitt and F. G. Worden, eds.), pp. 795–804, MIT Press, Cambridge, Mass.

Roseman, S., 1974c, The biosynthesis of complex carbohydrates and their potential role in intercellular adhesion, *in: The Cell Surface in Development* (A. A. Moscona, ed.), pp. 255–272, John Wiley, New York.

Ross, G. T., Van Hall, E. V., Vaitukaitis, J. L., Braunstein, G. D., and Rayford, P. L., 1972, Sialic acid and the immunologic and biologic activity of gonadotropins, *in: Gonadotropins* (B. B. Saxena, C. G. Beling, and H. M. Gandy, eds.), pp. 417–423, Wiley-Interscience, New York.

Roth, S. and White, D., 1972, Intercellular contact and cell-surface galactosyl transferase activity, *Proc. Natl. Acad. Sci. U.S.* **69:**485.

Roth, S., McGuire, E. J., and Roseman, S., 1971, Evidence for cell-surface glycosyltransferases: Their potential role in cellular recognition, *J. Cell Biol.* **51:**536.

Sachtleben, P., Gsell, R., and Mehrishi, J. N., 1972, Cited in Mehrishi, J. N., 1972.

Saifer, A., and Siegel, H. A., 1959, The photometric determination of the sialic (N-acetylneuraminic) acid distribution in cerebrospinal fluid, *J. Lab. Clin. Med.* **53:**474.

Sakamoto, W., Nishikaze, O., and Sakakibara, E., 1974, Isolation of an inhibitor of glucuronidase from human saliva, *J. Biochem. (Tokyo)* **75:**675.
Sandhoff, K., 1970, The hydrolysis of Tay–Sachs ganglioside (TSG) by human N-acetyl-β-hexosaminidase A., *FEBS Lett.* **11:**342.
Schauer, R., 1970a, Biosynthesis of N-acetyl-O-neuraminic acid. I. Incorporation of ^{14}C acetate into slices of the submaxillary salivary glands of ox and horse, *Z. Physiol. Chem.* **351:**595.
Schauer, R., 1970b, Biosynthesis of N-acetyl-O-acetylneuraminic acid. II. Substrate and intracellular localization of the bovine, acetyl-coenzyme A: N-acetyl-neuraminate-7- and 8-O-acetyl transferase, *Z. Physiol. Chem.* **351:**749.
Schauer, R., 1970c, Biosynthesis of N-glycoloylneuraminic acid by an ascorbate or NADPH-dependent, N-acetyl hydroxylating "N-acetylneuraminate: oxygen oxidoreductase" in homogenates of porcine submaxillary gland, *Z. Physiol. Chem.* **351:**783.
Schauer, R., 1970d, Studies on the subcellular site of the biosynthesis of N-glycoloylneuraminic acid in porcine submaxillary gland, *Z. Physiol. Chem.* **351:**1353.
Schauer, R., 1972, Biosynthesis of glycoprotein precursors and the mechanism of their assembly, *Biochem. J.* **128:**112.
Schauer, R., 1973, Chemistry and biology of the acylneuraminic acids, *Angew. Chem. Intern. Ed.* **12:**127.
Schauer, R., and Wember, M., 1971, Hydroxylation and O-acetylation of N-acetylneuraminic acid bound to glycoproteins of isolated subcellular membranes from porcine and bovine submaxillary gland, *Z. Physiol. Chem.* **352:**1282.
Schauer, R., Wember, M., and doAmaral, C. F., 1972, Synthesis of CMP-glycosides of radioactive N-acetyl-, N-glycolyl-, N-acetyl-7-O-acetyl- and N-acetyl-8-O-acetylneuraminic acids by CMP-sialate synthetase from bovine submaxillary glands, *Z. Physiol. Chem.* **353:**883.
Schengrund, C. L., and Garrigan, O. W., 1969, A comparative study of gangliosides from brains of various species, *Lipids* **4:**488.
Schengrund, C. L., and Rosenberg, A., 1972, Neuronal sialidase action in isolated nerve endings, *in:* Third Annual Meeting of the American Society of Neurochemistry, Vol. 3, No. 1, pp. 118.
Schloemer, R. H., and Wagner, R. R., 1974, Sialoglycoprotein of vesicular stomatitis virus. Role of the neuraminic acid in infection, *J. Virol.* **14:**270.
Schoop, H. J., and Faillard, H., 1967, Biosynthesis of the glycolyl group of N-glycolylneuraminic acid, *Z. Physiol. Chem.* **348:**1518.
Schoop, H. J., Schauer, R., and Faillard, H., 1969, Biosynthesis of N-glycolylneuraminic acid. Oxidative conversion of N-acetylneuraminic acid to N-glycolylneuraminic acid, *Z. Physiol. Chem.* **350:**155.
Seed, T. M., Aikawa, M., Sterling, C., and Rabbege, J., 1974, Surface properties of extracellular malaria parasites: morphological and cytochemical study, *Inf. Immunity* **9:**750.
Sefton, B., and Keegstra, K., 1974, Glycoproteins of Sindbis virus: preliminary characterization of oligosaccharides, *J. Virol.* **14:**522.
Seiter, C., 1970, The structural Analysis of Brain Gangliosides from Goldfish (*Carassius Auratus*): The presence of N, 8-O-diacetylneuraminic acid in the polysialogangliosides, Ph.D. Thesis, Ohio State University, Columbus, Ohio.
Seyama, Y., and Yamakawa, T., 1974, Multiple components of β-N-acetylhexosaminidase from equine kidney. Their action on glycolipids and allied oligosaccharide, *J. Biochem.* **75:**495.

Shownkeen, R. C., Thomas, M. B., and Hartree, A. S., 1973, Sialic acid and tryptophan content of subunits of human pituitary luteinizing hormone, *J. Endocrinol.* **59**:201.
Siefring, G. E., Jr., Castellino, F. J., 1974, The role of sialic acid in the determination of distinct properties of the isoenzymes of rabbit plasminogen, *J. Biol. Chem.* **249**:7742.
Skaug, K., and Christensen, T. B., 1971, The significance of the carbohydrate constituents of bovine thrombin for the clotting activity, *Biochim. Biophys. Acta* **230**:627.
Spooner, R. L., and Maddy, A. H., 1971, The isolation of ox red cell membrane antigens: antigen associated with sialoprotein, *Immunology* **21**:809.
Springer, G. F., and Ansell, N. J., 1958, Inactivation of human erythrocyte agglutinogens M and N by influenza viruses and receptor-destroying enzyme, *Proc. Natl. Acad. Sci. U.S.* **44**:182.
Springer, G. F., and Huprikar, S. V., 1972, On the biochemical and genetic basis of the human blood group MN specificities, *Hematologia* **6**:81.
Staerk, J., Ronneberger, H. J., Wiegandt, H., and Ziegler, W., 1974, Interaction of ganglioside G_{tet} and its derivation with choleragen, *Eur. J. Biochem.* **48**:103.
Stefanovic, V., Mandel, P., and Rosenberg, A., 1975, Ectopyrophosphatase activity of nervous system cells in tissue culture and its enhancement by removal of cell surface sialic acid, *in:* Sixth Meeting of the American Society of Neurochemistry, Abstract No. 49, Mexico City.
Stepan, J., and Ferwerda, W., 1973, Phosphatase. IX. Differences in sialic acid content of rat liver alkaline phosphatase isoenzymes, *Experientia* **29**:948.
Stoolmiller, A. C., Dawsen, G., and Schachner, M., 1975, Comparison of glycosphingolipid metabolism in mouse glial tumors and cultured cell strains of neural origin, *Fed. Proc.* **34**:634.
Strass, J. H., Burge, B. W., and Darnell, J. E., 1970, Carbohydrate content of the membrane protein of Sindbis virus, *J. Mol. Biol.* **47**:437.
Sullivan, C. W., and Volcani, B. E., 1974, Isolation and characterization of plasma and smooth membrane of the marine diatom *Nitzschia alba*, *Arch. Biochem. Biophys.* **163**:29.
Suttajit, M., Reichert, L. E., Jr., Winzler, R. J., 1971, Effect of modification of N-acetylneuraminic acid on the biological activity of human and ovine follicle-stimulating hormone, *J. Biol. Chem.* **246**:3405.
Suzuki, K., 1964, A simple and accurate micromethod for quantitative determination of ganglioside patterns, *Life Sci.* **3**:1227.
Suzuki, K., 1965, The pattern of mammalian brain gangliosides. III. Regional and developmental difference, *J. Neurochem.* **12**:969.
Suzuki, K., Poduslo, S. E., and Norton, W. T., 1967, Gangliosides in the myelin fraction of developing rats, *Biochim. Biophys. Acta* **144**:375.
Svennerholm, L., 1963, Chromatographic separation of human brain gangliosides, *J. Neurochem.* **10**:613.
Svennerholm, L., 1964, The gangliosides, *J. Lipid Res.* **5**:145.
Svennerholm, L., 1970a, Gangliosides, *in: Handbook of Neurochemistry*, Vol. 3 (A. Lajtha, ed.), pp. 425–452, Plenum Press, New York.
Svennerholm, L., 1970b, Ganglioside metabolism, *in: Comprehensive Biochemistry* (M. Florkin and E. H. Stotz, eds.), Vol. 18, pp. 201–228, Elsevier Publishing Co., Amsterdam.
Svennerholm, L., Bruce, A., Mansson, J. E., Rynmark, B. M., and Vanier, M. T., 1972, Sphingolipids of human skeletal muscle, *Biochim. Biophys. Acta* **280**:626.
Tallman, J. F., 1974, Hexosaminidases and ganglioside catabolism in the G_M-gangliosidoses, *Chem. Phys. Lipids* **13**:292.

Tallman, J. F., and Brady, R. O., 1973, Properties of purified hexosaminidases A and B and their relationship to Tay–Sachs disease, *in:* Ninth International Congress of Biochemistry, Stockholm, Abstract Book p. 396, Abstract 9a6.

Tallman, J. F., Brady, R. O., Quirk, J. M., Villalba, M., and Jal, A. E., 1974, Isolation and relationship of human hexosaminidases, *J. Biol. Chem.* **249:**3489.

Tamm, I., and Horsfall, F. L., Jr., 1950, Characterization and separation of an inhibitor of viral hemagglutination present in urine, *Proc. So. Exp. Biol. Med.* **74:**108.

Tettamanti, G., 1968, I glicolipidi del tessuto nervoso e le loro proprieta biologiche, *Riv. Istoch. Norm. Pat.* **14:**5.

Tettamanti, G., and Zambotti, V., 1971, Nature and properties of brain neuraminidase, *in: Glycolipids, Glycoproteins, and Mucopolysaccharides of the Nervous System: Chemical and Metabolic Correlations,* pp. 12, Summary NB-4, Satellite Symposium of the XXV International Congress of Physiological Sciences, Milano.

Tettamanti, G., Bertona, L., Berra, B., and Zambotti, V., 1964, On the evidence of glycolylneuraminic acid in beef brain ganglioside, *Ital. J. Biochem.* **13:**315.

Tettamanti, N. G., Bertoana, L., Gualandi, V., and Zambotti, V., 1965, Sulla distribuzione dei gangliosidi del sistema nervoso in varie specie animali, *Instituto Lombardo (Rend. Sc.)* **B99:**173.

Tettamanti, G., Venerando, B., Preti, A., Lombardo, A., and Zambotti, V., 1972, Brain neuraminidases, *in: Glycolipids, Glycoproteins, and Mucopolysaccharides of the Nervous System* (V. Zambotti, G. Tettamanti, and M. Arrigoni, eds.), pp. 161–181, Plenum Press, New York.

Troy, F. A., Vijay, I., and Tesche, N., 1973, Lipid-bound *N*-acetylneuraminic acid (N-AN) as a precursor in the biosynthesis of sialyl polymers in *E. coli, Fed. Proc.* **32:**481.

Tuppy, H., and Gottschalk, A., 1972, The structure of sialic acids and their quantitation, *in: Glycoproteins: Their Composition, Structure and Functions* (A. Gottschalk, ed.), 2nd ed., pp. 403–449, Elsevier Publishing Co., Amsterdam.

Turumi, K., and Saito, Y., 1953, A chemical study of the loach mucin, *Tohoku J. Exp. Med.* **58:**247.

Uzman, L. L., and Rumley, M. K., 1956, Neuraminic acid as a constituent of human cerebrospinal fluid, *Proc. Soc. Exp. Biol. Med.* **93:**497.

Van den Eijnden, D. H., 1973, The subcellular localization of cytidine 5'-monophospho-*N*-acetylneuraminic acid synthetase in calf brain, *J. Neurochem.* **21:**949.

van den Hamer, C. J. A., Morell, A. G., Scheinberg, I. H., Hickman, J., and Ashwell, G., 1970, Physical and chemical studies on ceruloplasmin. IX. The role of Galactosyl residues in the clearance of ceruloplasmin from the circulation, *J. Biol. Chem.* **245:**4397.

Van Heyningen, S., 1974, Cholera toxin: interaction of subunits with ganglioside G_M, *Science* **183:**656.

Van Heyningen, W. E., 1974, Gangliosides as membrane receptors for tetanous toxin, cholera toxin and serotonin, *Nature* **249:**415.

Vance, W. R., Shook, C. P., and McKibbin, J. M., 1966, The glycolipids of dog intestine, *Biochem.* **5:**435.

Vaskovasky, V. E., Kostetsky, E. Y., Svetashev, V. I., Zhukova, I. G., and Smirnova, G. P., 1970, Glycolipids of marine invertebrates, *Comp. Biochem. Physiol.* **34:**163.

Vitetta, E. S., Boyse, E. A., and Uhr, J. W., 1973, Isolation and characterization of a molecular complex containing Thy-1 antigen from the surface of murine thymocytes and T cells, *Eur. J. Immunol.* **3:**446.

Wagner, A., and Weicker, H., 1966, Ganglioside-like substances from human spleen, *Z. Klin. Chem.* **4:**73.

Warren, L., 1959, The thiobarbitunic acid assay of sialic acid, *J. Biol. Chem.* **234**:1971.
Warren, L., 1960, Unbound sialic acids in fish eggs, *Biochim. Biophys. Acta* **44**:347.
Warren, L., 1963, The distribution of sialic acids in nature, *Comp. Biochem. Physiol.* **10**:153.
Warren, L., 1972, The biosynthesis and metabolism of amino sugars and amino sugar-containing heterosaccharides, *in: Glycoproteins: Their Composition, Structure and Function* (A. Gottschalk, ed.), pp. 1097–1126, Elsevier Publishing Co., Amsterdam.
Warren, L., and Hathaway, R. R., 1960, Lipid-soluble sialic acid containing material in Arbacia eggs, *Biol. Bull.* **119**:354.
Watkins, W. M., 1972, Blood-group specific substances, *in: Glycoproteins: Their Composition, Structure and Function,* (A. Gottschalk, ed.), 2nd ed., pp. 830–891, Elsevier Publishing Co., Amsterdam.
Watson, R. G., Marinetti, G. V., and Scherp, H. W., 1958, The specific hapten of group C (group II α) meningococcus II. Chemical nature, *J. Immunol.* **81**:337.
Weiss, L., 1973, Neuraminidase, sialic acids, and cell interactions, *J. Nat. Canc. Inst.* **50**:3.
Wessler, E., and Werner, I., 1957, On the chemical composition of some mucous substances of fish, *Acta Chem. Scand.* **11**:1240.
Wherrett, J. R., and Cumings, J. N., 1963, Detection and resolution of gangliosides in lipid extracts by thin-layer chromatography, *Biochem. J.* **86**:378.
Whittaker, V. P., 1966, Some properties of synaptic membranes isolated from the central nervous system, *Ann N.Y. Acad. Sci.* **137**:982.
Wiegandt, H., 1966, Gangliosides, *Erg. Physiol. Biol. Chem. Exp. Pharm.* **57**:190.
Wiegandt, H., 1967, The subcellular localization of gangliosides in the brain, *J. Neurochem.* **14**:671.
Wiegandt, H., 1968, The structure and functions of gangliosides, *Angew. Chem. Intern. Ed.* **7**:87.
Wiegandt, H., 1971, Glycosphingolipids, *Adv. Lipid Res.* **9**:249.
Wiegandt, H., 1973, Gangliosides of extraneural organs, *Z. Physiol. Chem.* **354**:1049.
Wiegandt, H., 1974, Monosialolactoisohexaosylceramide, a ganglioside from human spleen, *Eur. J. Biochem.* **45**:367.
Wiegandt, H., and Bucking, H. W., 1970, Carbohydrate components of extraneuronal gangliosides from bovine and human spleen, and bovine kidney, *Eur. J. Biochem.* **15**:287.
William, M., and Roboz, E., 1958, Neuraminic acid in cerebrospinal fluid, *Fed. Proc.* **17**:335.
Winterburn, P. J., and Phelps, C. F., 1972, The significance of glycosylated proteins, *Nature* **236**:147.
Winzler, R. J., 1972, Glycoproteins of plasma membranes chemistry and function, *in: Glycoproteins: Their Composition, Structure and Function* (A. Gottschalk, ed.), 2nd ed., pp. 1268–1293, Elsevier Publishing Co., Amsterdam.
Winzler, R. J., and Bocci, V., 1972, Turnover of plasma glycoproteins, *in: Glycoproteins: Their Composition, Structure and Function* (A. Gottschalk, ed.), 2nd ed., pp. 1228–1245, Elsevier Publishing Co., Amsterdam.
Yamakawa, T., and Suzuki, S., 1952, The chemistry of the lipids of posthemolytic residue or stroma of erythrocytes. II. The structure of hemataminic acid, *J. Biochem. (Tokyo)* **39**:175.
Yamakawa, T., and Suzuki, S., 1955, Presence of serolactaminic acid and glucosamine as constituents of serum mucoproteins, *J. Biochem (Tokyo)* **42**:727.

Yamane, K., and Nathenson, S. G., 1970, Biochemical similarity of papain-solubilized H-2^d alloantigens from tumor cells and from normal cells, *Biochem* **9**:4743.

Yiamouyiannis, J. A., and Dain, J. A., 1968, The appearance of ganglioside during embryological development of the frog, *J. Neurochem.* **15**:673.

Yip, M.C.-M., 1973, A novel monosialoganglioside synthesized by a rat brain cytidine-5'-monophosphate-*N*-acetylneuraminic acid: galactosyl-*N*-acetylgalactosaminyl-galactosyl-glucosylceramide sialyltransferase, *Biochem. Biophys. Res. Commun.* **53**:737.

Yokoyama, M., and Trams, E. G., 1962, Effect of enzymes on blood group antigens, *Nature* **194**:1048.

Yu, R. K., Ledeen, R. W., and Eng, L. F., 1974, Ganglioside abnormalities in multiple sclerosis, *J. Neurochem.* **23**:169.

Zilliken, F., Braun, G. A., and Gyorgi, P., 1956, "Gynaminic Acid" and other naturally occurring forms of *N*-acetylneuraminic acid, *Arch. Biochem. Biophys.* **63**:394.

Chapter 3

The Distribution of Sialic Acids Within the Eukaryotic Cell

Leonard Warren

I. INTRODUCTION

In this chapter the distribution of sialic acid within the cell will be discussed. It is fairly safe to assume that essentially all of the sialic acid of the cell is in a bound form incorporated in glycoproteins and glycolipids and that these are to a large extent components of the cell membranes. In glandular secretory tissues large amounts of macromolecules containing sialic acid may be present in granular from awaiting elimination from the cell—a rather special situation that will not be considered further.

In this discussion we are, in fact, using sialic acid as a marker for membrane glycoproteins and glycolipids. In these structures sialic acids are, with very few exceptions, in terminal positions at the nonreducing ends of oligosaccharides which in turn are attached to proteins or lipids. We should keep in mind that the sialic acid content of a particular organelle may be high not only because there are a large number of sialic-acid-bearing molecules present, but also because the amount of sialic acid per macromolecule is high. Considerations such as this and ignorance of the number of existing populations of macromolecules in a given membrane make evaluation of function and relevance particularly fuzzy. The complexity grows when dynamic aspects such as turnover are introduced.

In order to determine the distribution of sialic acids within cells

LEONARD WARREN • Wistar Institute of Anatomy and Biology, Philadelphia, Pennsylvania 19104.

they must be broken up and fractionated. Assays for sialic acid are then carried out on each fraction. The assumption that each fraction is "pure" must be guarded. As will be discussed, there is little question that most of the sialic acid of the cell is in the surface membrane. Thus, a small degree of contamination of nuclear, mitochondrial, or endoplasmic reticular preparations by plasma membrane will clearly affect the sialic acid values of internal fractions. Further, during purification it is possible that sialic-acid-containing glycoproteins and glycolipids redistribute, to some extent, among the different fractions. In spite of these traps the general conclusions that can be drawn regarding distribution of sialic acids within the cell are essentially correct.

II. EXTRACELLULAR SIALIC ACIDS

Sialic-acid-containing macromolecules produced within the cell are frequently excreted and found as normal constituents of a number of extracellular fluids such as saliva (Blix, 1936), gastric juice, bronchial and nasal secretions, gall bladder fluid, cervical mucins (Werner, 1953), and in the semen of various animal species: boar (Odin, 1955), human (Svennerholm, 1958; Warren, 1959). It is found in urine (Bourillon, 1972), blood serum, and cerebrospinal fluid (see Gottschalk, 1960; 1973). Further information can be found in the books written (1960) and edited (1972) by Gottschalk. The most notable sites in which sialic acid is found in free form are gastric juice (Atterfelt et al., 1958) urine (Warren, 1960b), and trout eggs (Warren, 1960a), although a small amount may be found in cerebrospinal fluid (Uzman and Rumley, 1956; see Gottschalk, 1960). Since there appears to be free sialic acid in the urine but not detectable free sialic acid in the blood, it is probable that sialic acid has a very low renal threshold value. Perhaps unbound sialic acid is produced during normal kidney function, from tubular membrane or from reabsorbed serum glycoprotein. Just as sialic-acid-free glycoproteins are adsorbed serumand taken up by liver cells (Pricer and Ashwell, 1971), so the removal of sialic acid might be necessary for reabsorption of glycoproteins by the kidney tubule.

It is not surprising that free sialic acid can be found in human feces since there is present considerable amounts of sialidase (neuraminidase), probably of bacterial origin (unpublished).

Sialic acids have also been found in a component of a heterosaccharide-protein complex in the jelly coat of the sea urchin (Immers 1968; Hotta et al; 1970a,b; 1973). A form of sialic acid not described previously and differing from N-glycolyneuraminic acid was found

(Hotta et al., 1970a,b). (See Chapters 1 and 2.) Partial hydrolysis of the sialic-acid-containing material yielded an unusual disaccharide whose structure is probably fucopyranosyl-(1→4)-N,O-glycolylneuraminic acid (Hotta et al., 1973).

III. DISTRIBUTION WITHIN THE CELL

A. The Plasma Membrane

There have been very few complete, quantitative surveys on where sialic acid is found within the cell. The results of the most complete studies are shown in Table I. It can be seen that most of the sialic acid of the cell is found in its surface structure—approximately 65–70% of the total. It is assumed that the "microsome" fraction of rate liver (Table I) consists largely of surface membrane. Since the surface membrane probably constitutes no more than 5% of the total mass of the cell, the actual concentration of sialic acid per mg of membrane protein is correspondingly high (see Table V). The surface membrane is the chief site not only of sialic acid but of other sugars found in glycoproteins and glycolipids as well, (Table II). Although glycoproteins are concentrated at the cell surface, it is questionable whether the same can be said for

TABLE I. The Sialic Acid Present in the Various Organelles of Cells

	Fraction	Percent	Reference
Ehrlich ascites carcinoma	Whole cells	100	Wallach and Eylar, 1961
	Microsomes	69.5	
	Mitochondria	17.3	
	Soluble	13.6	
	Nuclei	3.2	
L cell (Mouse fibroblast)	Whole cell	100	Glick et al., 1971
	Surface membrane	66	
	Lysozomes	16	
	Soluble	8	
	Microsomes (ER)	7	
	Mitochondria	4	
	Nucleus	<0.6	
Rat brain cortex	Microsomes	53.0[a]	Lapetina et al., 1967
	Mitochondria	41.7	
	Nuclei	3.2	
	Supernatant	2.1	

[a] Sialic acid found in gangliosides.

TABLE II. The Carbohydrate Content of the L Cell and Its Surface Membrane[a]

	Whole cell μmole $\times$ 10^{-10}	Surface membranes[b] μmole $\times$ 10^{-10}	Percent of total in surface membrane
L-fucose	6.3 ± 3.8	2.3 ± 2.2	36
D-mannose	52.8 ± 12.0	8.5 ± 2.5	16
D-galactose	32.8 ± 4.6	21.6 ± 2.9	66
D-hexosamine	58.3 ± 17.3	25.9 ± 5.5	44
Sialic acid	8.9 ± 1.7	6.5 ± 1.4	73
Cells $N = 16$	Surface membrane $N = 7$		

[a] Glick et al., 1970.
[b] Isolated by the zinc ion method of Warren and Glick, 1969.

glycolipids, at least in the mouse fibroblast (Weinstein et al., 1970). Histochemists have, by a variety of staining procedures, visualized bound sialic acids on the surface membrane as well as on some intracellular membranes of several tissues (Gasic and Berwick, 1962; Gasic and Gasic, 1963; Bennett, 1963; Rambourg and Leblond, 1967).

The data of Tables III and IV support the statement that sialic acid (and glycoprotein) is concentrated at the cell surface. Treatment of most living cells with neuraminidase cleaves 50–80% of their sialic acids leaving the cells intact and viable. The erythrocyte, which has no internal membrane systems, bears all of its sialic acid at its surface, from which it can be entirely removed by neuraminidase (Table III). Since sialic acids are almost always in terminal positions on glycoprotein chains, they are the only cell components removed by neuraminidase treatment. The underlying sugars to which sialic acid is attached, usually D-galactose or N-acetyl-D-galactosamine, then become terminal. Such treatment changes various surface specificities and obviously considerably decreases charge at the cell surface. Treated cells do not migrate to the positive pole in an electrophoretic cell as rapidly as do untreated cells (Weiss, 1967).

The disposition of sialic acid within the surface may be studied by measuring the rate of electrophoretic migration of cells before and after treatment with neuraminidase. A most significant and interesting finding is that when mouse sarcoma cells (Cook, et al., 1963), liver or hepatoma cells (Emmelot and Bos, 1972) derived from a tissue, or the same cells grown as individuals in culture, are exposed to neuraminidase, appreciable amounts of sialic acid are released. Yet a decrease in electrophoretic mobility is seen only in the single cells grown in culture. There appears to be a fundamental difference in the conformation of glycoproteins in the surface of these cells growing in two different environments. In the cells derived directly from tissues the sialic acid is exposed sufficiently at the surface to be cleaved by a hydrolytic enzyme, yet it does not appear to contribute to the surface charge density.

It is surprising that when living L cells (mouse fibroblasts) grown in tissue culture are treated with neuraminidase, more than 50% of the total sialic acid of the cell is freed; however, the amount of sialic-acid-containing glycolipid does not decrease (Weinstein et al., 1970). Apparently the glycolipids (gangliosides and hematoside) are not susceptible to enzyme action or are not suitably exposed. L cell gangliosides are digestible by neuraminidase (Weinstein et al., 1970) when removed from the membrane by lipid solvents. Similar results have been found in a

TABLE III. Percentage of Total Sialic Acid of Intact Cell Removed by Treatment with V. cholerae Neuraminidase

Cell	Percent release	Reference
Ehrlich ascites (Swiss white mice)	73	Wallach and Eylar, 1961
Erythrocyte		Eylar et al., 1962
Human	100	
Calf	100	
Chicken	100	
Pig	>90	
Erythrocyte (mouse)	69	Miller et al., 1970
Polymorphonuclear leukocyte (guinea pig)	50	Noseworthy et al., 1972
Polymorphonuclear leukocytes (human)	54	Lichtman and Weed, 1970
Chronic lymphatic leukemia lymphocytes (human)	63	Lichtman and Weed, 1970
Cells in culture		
L cells (mouse fibroblast)	59.3	Kraemer, 1965
L5178Y (murine lymphoma)	65.3	
HeLa (human)	60.1	
C_{13} (Syrian hamster)	65	
Py183 (same, polyoma virus)	67	
CHO (Chinese hamster ovary)	68	
L cells (mouse fibroblast)	45–60	Glick et al., 1970
BHK_{21} (Syrian hamster)	56	Ohta et al., 1968
PyY (Syrian hamster, transformed)	75	
3T3 (mouse)	66	
3T3 (Py (mouse, transformed)	49	
TA_3 (mouse ascites tumor)	80	Codington et al., 1970

TABLE IV. Percentage of Total Sialic Acid of Cell Released as Glycopeptide by Trypsin

Type of cell	Notes	Percent release	Reference
HeLa (human)	Confluent cells on glass treated for 10 min 37°C with 0.25% solution of trypsin	47	Shen and Ginsburg, 1968
HeLa	In suspension culture	55	Shen and Ginsburg, 1968
BHK_{21}/C_{13} (baby hamster kidney)	Log phase, on glass surface	24.0	Buck et al., 1970
C_{13}/B_4 virus transformed		23.0	Buck et al., 1970
C_{13}/SR_7 virus-transformed		20.4	Buck et al., 1970
Erythrocytes (or stroma) human		48	Eylar and Madoff, 1962
Erythrocytes (human)		33–50	Winzler et al., 1967
Erythrocytes (mouse)		89	Miller et al., 1963
Ascites tumor (mouse)		71–74	Codington et al., 1970

variety of cells (Barton and Rosenberg, 1973) although the arrangement of gangliosides in the surface membrane of cat erythrocytes may be different (Wintzer and Uhlenbruck, 1967), Here, some types of sialic-acid-containing glycolipids of the surface membrane are susceptible to neuraminidase action while others are not.

Although it is assumed that neuraminidase does not penetrate the cell but removes sialic acid from the surface alone, there is some evidence both direct (Nordling and Mayhew, 1966) and indirect (Wallach and Eylar, 1970; Glick et al., 1971) that some enzyme does, in fact, get in. Cleavage of interior sialic acids would obviously lead to an overestimation of the relative proportion of sialic acid at the cell surface.

The data in Table IV lend further support to the belief that most of the sialic acid of cells is at their surfaces. Trypsin, which is assumed to be unable to enter living cells, cleaves glycoproteins exposed at the cell surface and releases sialic-acid-containing glycopeptides. In many instances at least 50% of the total sialic acid of the cell is released.

Approximately one quarter of the total sialic acid of the L cell is bound to lipid in the form of hematoside, mono- and disialogangliosides. The remaining three-quarters of the sialic acid is bound to protein. Weinstein et al. (1970) found no significant preference of location of gangliosides and hematosides for the surface membranes of the L cell, while glycoproteins clearly concentrated in this organelle. On the other hand, Dod and Gray (1968a,b) and Renkonen et al. (1970) have reported that plasma membrane fractions are enriched in glycolipids.

There is a considerable literature on the distribution of sialic-acid-containing complex glycolipids (gangliosides) in brain and its components (Burton et al., 1965; Wolfe, 1961; Lapetina et al., 1967; Wiegandt, 1967; Wherrett and McIlwain, 1962; Eichberg et al., 1964; others). The gangliosides, whose function is essentially unknown, are concentrated in the grey matter.

The data of Lapetina et al. (1967) on rat brain cortex seen in Table I are rather similar to those of others listed above. In general, the microsomal and especially the mitochondrial fractions can be further fractionated to yield pinched off nerve endings or synaptosomes. These are essentially of plasma membrane origin and are especially rich in gangliosides and bound acetylcholine. Synaptosomes can be ruptured by osmotic shock to release synaptic vesicles that are virtually free of gangliosides.

The concentrations of lipid soluble (ganglioside) sialic acids of the cerebral cortex of guinea pig and ox brain are seen in Table V. Gangliosides are abundant in the nerve endings or in crude fractions containing them. They are low in nuclei and supernatant fractions. In these analyses the amount of sialic acid bound to glycoproteins, i.e., lipid-insoluble, was not determined, but it is likely that the sialic acid distribution pattern would not change much if sialoglycoproteins were considered, since they probably are a relatively minor component of brain from a quantitative point of view. Sialoglycoproteins have been found in crude mitochondrial fractions from dog brain (Brunngraber and Brown, 1963) as well as in the microsomal fraction of rat brain (Brunngraber and Brown 1964a,b). Sialoglycoproteins may be a component of the 5-hydroxytryptamine receptor in the central nervous system (Wesemann et al., 1971).

B. Endoplasmic Reticulum

The total amount of sialic acids in the smooth and rough endoplasmic reticulum (ER) are difficult to estimate accurately because of the inadequacy of current methods of isolation of these fractions, both in terms of purity and yield. Microsomal fractions are highly complex in their composition and consist of ER, surface membrane, and probably Golgi membranes and other organelles. It is difficult to estimate how much of the sialic acid actually belongs to endoplasmic reticulum. It is present, but the concentration is certainly less than in plasma membrane. Further, it can be seen in Table V (Ehrlich ascites tumor and

TABLE V. Concentration of Sialic Acids in Membranes of Various Cells[a]

Cell	Membrane system	Sialic acid, nmoles/mg protein	Reference
Ehrlich ascites (mouse)	Whole cells	4.0	Molnar, 1967
	16,000 × g sed.	2.2	
	"Rough" microsomes	6.1	
	"Smooth" microsomes (pl. memb.)	29.7	
	Supernatant fraction	2.1	
	Plasma membrane	28.0	Wallach and Kamat, 1966
	Endoplasmic reticulum[b]	16.0	
HeLa cells (human)	Homogenate	5.6	Bosmann et al., 1968
	Plasma membrane	14.6	
	Smooth internal membranes	10.0	
	Ribosome bearing membranes	6.5	
L cells (mouse)	Whole cells	3.3	Glick et al., 1971
	Plasma membrane	20.0	
	Lysosomes	11.0	
	Mitochondria	5.1 ± 0.2	
	Mitochondria (further purified)	2.8 ± 0.2	
	Microsomes	12.2	
	Soluble fraction	1.2	
	Nuclei	0.05	
Thymocytes (calf)	Homogenate	12.9 ± 1.5	Van Blitterswijk et al., 1973
	Microsomes	28.9 ± 3.8	
	Plasma membrane	65.8 ± 4.7	

Cell	Fraction	Value	Reference
BHK$_{21}$/C$_{13}$	Plasma membrane	34.5	
BHK-Py	Plasma membrane	13.1	
3T3 (mouse)	Whole cells	4.4^c	Sheinin et al., 1971
	Plasma membrane	11.8	
3T3-SV$_{40}$	Whole cells	4.6	Sheinin et al., 1971
	Plasma membrane	6.4	
3T3-Py	Whole cells	3.3	Sheinin et al., 1971
3T3	Mitochondria	3.3	Wu et al., 1969
3T3-SV	Mitochondria	0.8	
3T3-ST^{40}	Mitochondria	2.3	
Chick embryo Fibroblast (CEF)	Homogenate	21.0	Purdue et al., 1971a,b
	Surface membrane	106.0	
CEF-RBA (avian sarcoma virus)	Homogenate	18.0	Purdue et al., 1971a,b
	Surface membrane	79.0	
CEF-RAV 49 (avian leukosis virus)	Homogenate	26.0	Purdue et al., 1971a,b
	Surface membrane	114.0	
Chicken breast muscle	Homogenate	24.7 ± 1.5	Purdue et al., 1973
	Plasma membrane	72.0 ± 8.5	
Chicken wing tumor (avian sarcoma virus)	Homogenate	30.3 ± 1.3	
	Plasma membrane	83.2 ± 7.6	
Mouse liver	Plasma membrane	30 ± 1	Emmelot and Bos, 1972

(continued)

TABLE V. (Continued)

Cell	Membrane system	Sialic acid, nmoles/mg protein	Reference
Mouse hepatoma-147042	Plasma membrane	361	Emmelot and Bos, 1972
4189	Plasma membrane	27 ± 1	
143066	Plasma membrane	28 ± 0 ± 2	
Rat liver	Plasma membrane	33 ± 2	
Rat hepatoma 484	Plasma membrane	45 ± 2	
484 A	Plasma membrane	61 ± 1	
Rat liver cells	Homogenate	14.0	Purdue et al., 1971a,b
	Plasma membrane	71.0	
Rat liver cells-MSV transformed	Homogenate	14.0	
	Plasma membrane	65.0	
Rat ascites hepatoma	Plasma membrane	28.8	Shimizu and Funakoshi, 1970
Rat liver	Nuclear membrane	0.33 ± 0.47	Kashnig and Kasper, 1969
Rat liver	Mitochondria—whole	3.1	deBernard et al., 1971
	Mitochondria—inner membrane	1.8	
	Mitochondria—outer membrane	6.5	
	Microsomes	12.6	
	Homogenate	0.5	Patterson and Touster, 1962
	Microsomes	11.2	
	Mitochondria	3.1	
	Nuclei	not detected	
	Soluble (supernatant)	5.4	

Rabbit liver	Mitochondria	4.0–5.2	Yamashina et al., 1965
	Microsomes	14.0–22.5	
	Rough microsomes	3.9–10.3	
	Smooth microsomes	24.0–37.9	
Cerebral cortex (guinea pig)	Original dispersion	2.3[c]	Wherrett and McIlwain, 1962
	Microsomes	5.1	
	Mitochondria	1.2	
	Nuclei	1.4	
	Supernatant	0.5	
Cerebral cortex (ox brain)	Myelin	2.5[c]	Weigandt, 1967
	Microsomes	3.2	
	Mitochondria	2.6	
	Nerve endings	5.3	

[a] Wherever sialic acid values were calculated from the original data in μg/mg protein, the molecular weight of NANA (309) was used. If the data were presented as nmoles/mg N, the values were divided by 6.25 to convert to nmoles/mg protein.
[b] Maximum amount of sialic acid removed from isolated membrane by neuraminidase.
[c] Lipid-bound, nmoles ganglioside sialic acid/mg protein.

rabbit liver), that the sialic acid concentration is considerably higher in smooth ER than in rough ER. This conclusion is supported by studies on the long-term *in vivo* incorporation of ^{14}C and ^{3}H-labeled D-glucosamine and L-fucose into various subcellular fractions of tissue culture cells. It has been found that far more isotope is incorporated into smooth ER than into rough ER (Buck *et al.*, 1974).

C. Mitochondria

As with ER, estimation of the fraction of sialic acid of the cell present in mitochondria is made difficult because of problems of purity. While sialic-acid-rich surface membrane material can be eliminated from preparations of mitochondria, lysosomes which also are rich in sialic acids (Tables I, V) often contaminate mitochondrial preparations. The data of Table V clearly show that the concentration of sialic acid in mitochondria from a variety of cells is low. In L cells only 4% of the total sialic acid is present in mitochondria (Table I), while in Ehrlich ascites carcinoma 17)% and in rat liver 11% are in the mitochondria (Wallach and Eylar, 1961; Patterson and Touster, 1962). It is possible that the latter two values are high because of contamination due to inadequate fractionation techniques.

In the analysis of the L cell it was found that the more extensively purified the mitochondria were, the lower was their content of sialic acid. This could have been due to a loss of outer membrane or of soluble glycoprotein (de Bernard *et al.*, 1971) but probably was the result of elimination of lysosomal contamination, as measured by acid phosphatase and neuraminidase activities. The latter activity was very low and could have reduced the sialic acid level only slightly (Glick *et al.*, 1971).

There is some indirect evidence that glycoproteins and sialic acids are found in mitochondria. When cells are grown in the presence of radioactive carbohydrate precursors, radioactivity can be found in isolated mitochondria (Molnar, 1967; Meezan, 1969; Wu, 1969; Buck *et al.*, 1974). Further, it has been shown that preparations of mitochondria from rat liver (Bosmann and Martin, 1969) contain transferases that add various sugars to glycoproteins, although transfer of sialic acid was not measured. The question of cross-contamination of mitochondria is as relevant in these studies as it is in direct chemical analysis for sialic acids.

D. Nuclei

The nucleus is poorer in bound sialic acid than are the other organelles (Tables I and V). Patterson and Touster (1962) could not detect sialic acid in rat liver nuclei, while Kashnig and Kasper (1969) could find only 0.33–0.47 nmoles of sialic acid per milligram of purified nuclear membrane protein. The low sialic acid concentration is suprising in view of the observations of Kean (1970) that the nucleus is the sole site of formation of cytidine-5′-monophosphosialic acid, the activated substrate of sialyltransferases. Perhaps this unusual location for activation of sialic acid forms part of an elaborate controlling mechanism for glycoprotein and glycolipids synthesis within the cell.

There is approximately 6×10^{-10} μmoles of sialic acid in the surface membrane and 10^{-11} μmoles in the nucleus of the L cell. If one assumes that the L cell and its nucleus are spherical, of diameters of 15 and 5 μm, and of surface areas of 705 and 78.5 μm^2, respectively, one can calculate that there is at least ten times more sialic acid per unit area of cell surface than of nuclear surface. Since there is a double membrane surrounding the nucleus, the density of sialic acids on the nuclear membrane may be one twentieth that of the surface membrane. An assumption in these calculations about which nothing is known is that all of the sialic acid of the nucleus is in its membranes. Marcus *et al.* (1965) had previously come to the same conclusion upon comparing the number of sialic acid residues per square micrometer of HeLa cell versus nuclear surface. However, they pointed out that if the area of villi on the cell surface is taken into account, the density of sialic acid residues on the surface may be no greater than that on the smooth nuclear exterior.

They also showed that Newcastle disease virus was capable of attaching to receptors on the nuclear and exterior surface of HeLa cells. The receptors are glycoproteins terminating in an essential residue of sialic acid. When intact, cytoplasm-free nuclei were treated with neuraminidase the virus no longer attached. Nordling and Mayhew (1966) have provided further evidence for the presence of sialic acid on nuclei when they treated nuclei of tumor cells with neuraminidase and observed a decrease in electrophoretic mobility.

E. Other Fractions

The lysosomes of L cells contain considerable amounts of bound sialic acid, despite the presence of neuraminidase in these organelles. The high level might be expected because lysosomal membrane is derived in part from the plasma membrane (de Duve and Wattiaux, 1966). The phospholipids and cholesterol contents of lysosomes and plasma membrane of rat liver are similar. (Thinès-Sempoux, 1967). Histochemical studies of Rambourg (1969) indicate that the lysosomes are rich in carbohydrate. It has also been shown that radioactive D-glucosamine can be incorporated into the sialic acid of lysosomes (Buck et al., 1974).

The soluble fractions of cells, i.e., the supernatant solution remaining after centrifugation at high speeds, appear to be very poor in sialic acid (Table I). Essentially all cellular glycoproteins and gangliosides appear to be bound to particulate matter, i.e., membranes.

III. CONCLUSIONS

It seems quite probable that no membrane system of the cells is devoid of sialic acid. Generally most of the sialic acid is bound in glycoproteins, while in nerve tissue and brain it is largely a component of the glycolipids. The concentration of sialic acid generally is highest in the plasma membrane, lower in the smooth endoplasmic reticulum (and Golgi apparatus?), and lowest in the rough endoplasmic reticulum. This pattern of distribution is to be expected on the basis of what is known about the assembly of polypeptide chains on the ribosome and the further addition of sugars in the smooth ER and possibly in the surface membrane. (Pricer and Ashwell, 1971; Warren et al., 1972; Bosmann, 1972). The enrichment of the cell surface structure in sialic acids may be the consequence of a high affinity of the plasma membrane for glycoproteins whose biosynthesis is complete, as indicated by a high content of chain-terminating sialic acid residues. It is also possible that the plasma membrane may incorporate glycoproteins which are partially deficient in sialic acids. This "deficiency" would be overcome by sialyltransferases known to be present in the surface structure. (Pricer and Ashwell, 1971; Warren et al., 1972; Bosmann, 1972).

The overall levels of sialic acids in the various organelles can also be influenced by endogenous neuraminidase, but no information is available on the *in vivo* physiological activity or role of this enzyme. In recent work radioactive glycopeptides were derived from isolated preparations of surface membrane, mitochondria, smooth endoplasmic reticulum, nuclei, and lysosomes of hamster cells. The glycopeptides derived from each type of membrane were compared by double-label techniques on columns of Sephadex G-50. It was evident that overall glycopeptide patterns of all the membranes eluting from the columns were quite similar and that they changed coordinately when the hamster cells were transformed into malignant cells by oncogenic virus (Buck et al., 1974). Obviously there are, in fact, differences between the glycopeptides of the various membranes that are not resolved by columns of Sephadex G-50. However these data suggest that the levels of sialic acid in the various membranes differ because the level of glycoproteins bearing similar carbohydrate components varies, and not because the oligosaccharide chains are incomplete—lacking sialic acid. Sakiyama and Burge (1972) believe that the lower levels of sialic acid in DNA-oncogenic virus-transformed cells compared to controls is due to a reduced amount of glycoprotein per cell. Similarly, the differences in the sialic acid levels characteristic of various intracellular membranes may follow from their inherent capacity to bind glycoproteins that, possibly, are held in common by all membranes.

V. REFERENCES

Atterfelt, P., Blohme, I., Norrby, A., and Svennerholm, L., 1958, Sialic acids of hog gastric mucosa, *Acta Chem. Scand.* **12**:359–360.

Barton, N. W., and Rosenberg, A., 1973, Action of *Vibrio cholerae* neuraminidase (sialidase) upon the surface of intact cells and their isolated sialolipid components, *J. Biol. Chem.* **248**:7353–7358.

Bennett, H. S., 1963, Morphological aspects of extracellular polysaccharides, *J. Histochem. Cytochem.* **11**:15.

Blix, G., 1936, Über die Kohlenhydratgruppen des Submaxillarismucins, *Hope-Seyl. Z.* **240**:43–53.

Bosmann, H. B., 1972, Sialyl transferase activity in normal and RNA and DNA virus transformed cells utilizing desialyzed, trypsinized cell plasma membrane external surface glycoproteins as exogenous acceptors, *Biochem. Biophys. Res. Commun.* **49**:1256.

Bosmann, H. B., and Martin, S. S., 1969, Mitochondrial autonomy, incorporation of monosaccharides into glycoprotein by isolated mitochondria, *Science* **164**:190.

Bosmann, H. B., Hagopian, A., and Eylar, E. H., 1968, Cellular membranes: the isolation and characterization of the plasma and smooth membranes of HeLa Cells, *Arch. Biochem. Biophys.* **128**:51–69.

Bourrillon, R., 1972, Urinary Glycoproteins, Glycopeptides and Related Heterosaccharides *in: Glycoproteins* (Gottschalk, A., ed.), pp. 909–925 Elsevier Publishing Co., Amsterdam.

Brunngraber, E. G., and Brown, B. D., 1963, Preparation of sialomucopolysaccharides from brain mitochondrial fraction, *Biochim. Biophys. Acta* **69**:581–582.

Brunngraber, E. G., and Brown, B. D., 1964a, Heterogeneity of sialomucopolysaccharides prepared from whole rat brain, *Biochim. Biophys. Acta* **83**:357–360.

Brunngraber, E. G., and Brown, B. D., 1964b, Fractionation of brain macromolecules. II. Isolation of protein-linked sialomucopolysaccharides from subcellular, particulate fractions from rat brain, *J. Neurochem.* **11**:449–459.

Buck, C. A., Glick, M. C., and Warren, L., 1970, A comparative study of glycoproteins from the surface of control and Rous sarcoma virus transformed hamster cells, *Biochemistry* **9**:4567–4575.

Buck, C. A., Fuhrer, J. P., Soslau, G., and Warren, L., 1974, Membrane glycopeptides from subcellular fractions of control and virus-transformed cells, *J. Biol. Chem.* **249**:1541–1550.

Burton, R. M., Howard, R. E., Baer, S., and Balfour, Y. M., 1965, Gangliosides and acetylcholine of the central nervous system, *Biochim. Biophys. Acta* **84**:441–447.

Codington, J. F., Sanford, B. H., and Jeanloz, R. W., 1970, Glycoprotein coat of the TA3cell. I. Removal of carbohydrate and protein material from viable cells, *J. Natl. Cancer Inst.* **45**:637–647.

Cook, G. M. W., Seaman, G. V. F., and Weiss, L., 1963, Physicochemical differences between ascites and solid forms of sarcoma 37 cells, *Cancer Res.* **23**:1813–1818.

de Bernard, B., Prigliarello, M. C., Sandri, G., Sottocasa, G. L., and Vittur, F., 1971, Glycoprotein components, sialic acid and hexosamines, bound to the inner and outer mitochondrial membranes, *FEBS Lett.* **12**:125–128.

de Duve, C., and Wattiaux, R., 1966, Functions of lysosomes, *Ann. Rev. Physiol.* **28**:435–492.

Dod, B. J., and Gray, G. M., 1968a, The lipid composition of rat-liver plasma membrane, *Biochim. Biophys. Acta* **150**:397–404.

Dod, B. J., and Gray, G. M., 1968b, The localization of the neutral glycosphingolipids in rat liver cells, *Biochem. J.* **110**:50P.

Eichberg, J., Whittaker, V. P., and Dawson, R. M. C., 1964, Distribution of lipids in subcellular particular particles of guinea pig brain, *Biochem. J.* **92**:91–99.

Eylar, E. H., and Madoff, M. A., 1962, Isolation of a glycopeptide from the red cell membrane, *Fed. Proc.* **21**:402.

Eylar, E. H., Madoff, M. A., Brody, O. V., and Oncley, J. L., 1962, The contribution of sialic acid to the surface charge of the erythroycte, *J. Biol. Chem.* **237**:1992–2000.

Emmelot, P., and Bos, C. J., 1972, Studies on plasma membranes, VVII. On the chemical composition of plasma membranes prepared from rat and mouse liver and hepatomas, *J. Membrane Biol.* **9**:83–104.

Gasic, G., and Berwick, L., 1962, Hale stain for sialic acid containing mucins—Adaptation to electron microscopy, *J. Cell Biol.* **19**:223–228.

Gasic, G., and Gasic, T., 1963, Removal of PAS positive surface sugars in tumor cells, by glycosidases, *Proc. Soc. Exp. Biol. Med.* **114**:660–663.

Glick, M. C., Comstock, C., and Warren, L., 1970, Membranes of animal cells. VII. Carbohydrates of surface membranes and whole cells, *Biochim. Biophys. Acta* **219**:290–300.

Glick, M. C., Comstock, C. A., Cohen, M. A., and Warren, L., 1971, Membranes of animal cells. 8. Distribution of sialic acid, hexosamines and sialidase in the L cell, *Biochim. Biophys. Acta* **233**:247–257.

Gottschalk, A., 1960, *The Chemistry and Biology of Sialic Acids and Related Substances*, Cambridge University Press, London.

Gottschalk, A., 1972, *Glycoproteins*, Elsevier Publishing Co., Amsterdam.

Hotta, K., Hamazaki, H., Kurokawa, M., and Isaka, S., 1970a, Isolation and properties of a new type of sialopolysaccharide-protein complex from the jelly coat of sea urchin eggs, *J. Biol. Chem.* **245**:5434–5440.

Hotta, K., Kurokawa, M., and Isaka, S., 1970b, Isolation and identification of two sialic acids from the jelly coat of sea urchin eggs, *J. Biol. Chem.* **245**:6307–6311.

Hotta, K., Kurokawa, M., and Isaka, S., 1973, A novel sialic acid and fucose containing disaccharide isolated from the jelly coat of sea urchin eggs, *J. Biol. Chem.* **248**:629–631.

Immers, J., 1968, N-acetylneuraminic acid in the sea urchin jelly coat, *Acta Chem. Scand.* **22**:2046–2048.

Kashnig, D. M., and Kasper, D. M., 1969, Isolation, Morphology, and composition of the nuclear membrane from rat liver, *J. Biol. Chem.* **244**:3786–3792.

Kean, E. L., 1970, Nuclear cytidine 5'-monophosphosialic acid synthetase, *J. Biol. Chem.* **245**:2301–2308.

Kraemer, P. M., 1965, Sialic acid of mammalian cell lines, *J. Cell Physiol.* **67**:23–34.
Lapetina, E. G., Soto, E. F., and DeRobertis, E., 1967, Gangliosides and acetylcholinesterase in isolated membranes of the rat-brain cortex, *Biochim. Biophys. Acta* **135**:33–43.
Lichtman, M. A., and Weed, R. I., 1970, Electrophoretic mobility and N-acetylneuraminic acid content of human normal and leukemic lymphocytes and granulocytes, *Blood* **35**:12–22.
Makita, A., and Seyama, Y., 1971, Alterations of Forssman-antigenic reactivity and monosaccharide composition in plasma membrane from polyoma-transformed hamster cells, *Biochim. Biophys. Acta* **241**:403–411.
Marcus, P. I., Salb, J. M., and Schwartz. V. G., 1965, Nuclear surface N-acetylneuraminic acid terminating receptors for myxovirus attachment, *Nature* **208**:1122–1124.
Meezan, E., Wu, H. C., Black, P. H., and Robbins, P. W., 1969, Comparative studies on the carbohydrate-containing membrane components of normal and virus-transformed mouse fibroblasts. II. Separation of glycoproteins and glycopeptides by sephadex chromatography, *Biochem.* **8**:2518–2524.
Miller, A., Sullivan, J. F., and Katz, J. H., 1963, Sialic acid content of the erythrocyte and of an ascites tumor cell of the mouse, *Cancer Res.* **23**:485.
Molnar, J., 1967, Glycoproteins of Ehrlich ascites carcinoma cells. Incorporation of (^{14}C)glucosamine and (^{14}C)sialic acid into membrane proteins, *Biochemistry* **6**:3064–3075.
Nordling, S. E., and Mayhew, E., 1966, On the intracellular uptake of neuraminidase, *Exp. Cell Res.* **44**:552–562.
Noseworthy, J. R., Korchak, H., and Karnovsky, M. L., 1972, Phagocytosis and the sialic acid of the surface of polymorphonuclear leukocytes, *J. Cell Physiol.* **79**:91–96.
Odin, L., 1955, Sialic acid in human cervical mucus, in hog seminal gel, and in ovomucin, *Acta Chem. Scand.* **9**:1235–1237.
Ohta, N., Pardee, A. B., McAuslan, B. R., and Burger, M. M., 1968, Sialic acid contents and controls of normal and malignant cells, *Biochim. et Biophys. Acta* **158**:98–102.
Patterson, M. K., and Touster, O., 1962, Intracellular distribution of sialic acid and its relationship to membrane, *Biochim. Biophys. Acta* **56**:626–628.
Pricer, W. E., Jr., and Ashwell, G., 1971, The binding of desialylated glycoproteins by plasma membranes of rat liver, *J. Biol. Chem.* **69**:1838.
Purdue, J. F., Kletzien, R., and Miller, K., 1971*a*, The isolation and characterization of plasma membrane from cultured cells. I. The chemical composition of membrane isolated from uninfected and oncogenic RNA virus-converted chick embryo fibroblasts, *Biochim. Biophys. Acta* **249**:419–434.
Purdue, J. F., Kletzien, R., Miller, K., Pridmore, G., and Wray, V. L., 1971*b*, The isolation and characterization of plasma membrane from cultured cells. II. The chemical composition of membrane isolated from uninfected and oncogenic RNA virus-converted parenchyma-like cells, *Biochim. Biophys. Acta* **249**:435–453.
Purdue, J. F., Warner, D., and Miller, K., 1973, The isolation and characterization of plasma membrane from cultured cells. III. The chemical composition of plasma membrane isolated from chicken tumors initiated with virus-transformed cells, *Biochim. Biophys. Acta* **298**:817–826.
Rambourg, A., and Leblond, C. P., 1967, Electron microscope observations on the carbohydrate-rich cell coat present at the surface of cells in the rat, *J. Cell. Biol.* **32**:27–53.
Rambourg, A., 1969, Localisation ultrastructurale et nature du material colore au niveau de la surface cellulaire par le melange chromique-phosphotung-stique, *J. Microscopie* **8**:325–341.

Renkonnen, O., Gahmberg, C. G., Simons, K., and Kääriäinen, L., 1970, Enrichment of gangliosides in plasma membranes of hamster kidney fibroblasts, *Acta Chem. Scand.* **24:**733–735.

Sakiyama, H., and Burge, B. W., 1972, Comparative studies of the carbohydrate-containing components of 3T3 and Simian virus 40 transformed 3T3 mouse fibroblasts, *Biochemistry* **11:**1366–1376.

Sheinin, R., and Onodera, K., Yogeeswaran, G., and Murray, R. K., Studies of components of the surface of normal and virus-transformed mouse cells, *in: Lepetit Colloquia on Biology and Medicine. 2. The Biology of Oncogenic Viruses,* 1971, (L. G. Silvestri, ed.), pp. 274–285, North-Holland Publishing Co., Amsterdam.

Shimizu, S., and Funakoshi, T. 1970, Carbohydrate composition of the plasma membranes of rat ascites hepatoma, *Biochim. Biophys. Acta* **203:**167–169.

Stahl, W. L., and Trams, E. G., 1968, Synthesis of lipids by liver plasma membranes. Incorporation of acyl-coenzyme derivatives into membrane lipids *in vitro, Biochim. Biophys. Acta* **163:**459–471.

Svennerholm, L., 1958, Quantitative estimation of sialic acid. III. An anion exchange-resin method, *Acta Chem. Scand.* **12:**547–554.

Thines-Sempoux, D., 1967, Chemical similarities between the lysosome and plasma membranes, *Biochem. J,* **105:**20p–21p.

Uzman, L. L., and Rumley, M. K., 1956, Neuraminic acid as a constituent of human cerebrospinal fluid, *Proc. Soc. Exp. Biol. Med.* **93:**497–500.

Van Blitterswijk, J., Emmelot, P., and Feltcamp, C. A., 1973, Studies on plasma membranes. XIX. Isolation and characterization of a plasma membrane fraction from calf thymocytes, *Biochim. Biophys. Acta* **298:**577–592.

Wallach, D. F. H., and Eylar, E. J., 1961, Sialic acid in the cellular membranes of Ehrlich ascites-carcinoma cells, *Biochim. Biophys. Acta* **52:**594–596.

Wallach, D. F. H., and Kamat, V. B., 1966, The contribution of sialic acid to the surface change of fragments of plasma membrane and endoplasmic reticulum, *J. Cell Biol.* **30:**660–663.

Warren, L., 1959, Sialic acid in human semen and in the male genital tract, *J. Clin. Invest.* **38:**755–761.

Warren, L., 1960*a*, Unbound sialic acids in fish eggs, *Biochim. Biophys. Acta* **44:**347–351.

Warren, L., 1960*b*, Studies on sialic acids, *Fed. Proc.* **19:**147.

Warren, L., and Glick, M. C., 1969, Isolation of surface membranes of tissue culture cells *in: Fundamental Techniques in Virology, Part II, Preparation of subcellular fractions,* (Habel, K. and Salzman, N. P., eds.), p. 66–71, Academic Press, New York.

Warren, L., Fuhrer, J. P. and Buck, C. A., 1972, Surface glycoproteins of normal and transformed cells: A difference determined by sialic acid and a growth-dependent sialyl transferase, *Proc. Natl. Acad. Sci. U.S.* **69:**1838–1842.

Wherrett, J. R., and McIlwain, H., 1962, Gangliosides, phospholipids, protein and ribonucleic acid in subfractions of cerebral microsomal material, *Biochem. J.* **84:**232–237.

Weigandt, H., 1967, The subcellular localization of gangliosides in the brain, *J. Neurochem.* **14:**671–674.

Weinstein, D., Marsh, J., Glick, M. C., and Warren, L., 1969, Membranes of animal cells. IV. Lipids of the L cell and its surface membrane, *J. Biol. Chem.* **244:**4103–4111.

Weinstein, D., Marsh, J., Glick, M. C., and Warren, L., 1970, Membranes of animal cells. VI. The glycolipids of the L cell and its surface membrane, *J. Biol. Chem.* **245:**3928–3937.

Weiss, L., 1967, *The Cell Periphery, Metastasis and Other Contact Phenomena*, pp. 267–270, John Wiley, New York.
Werner, I., 1953, Studies on glycoproteins from mucous epithelium and epithelial secretions, *Acta Soc. Med. Upsalien* **58**:1–55.
Wesemann, W., Henkel, R., and Marx, R., 1971, Receptors of neurotransmitters. V. Sialic acid distribution and characterization of the 5-hydroxytryptamine receptor in synaptic structures, *Biochem. Pharm.* **20**:1961–1966.
Wintzer, G., and Uhlenbruck, G., 1967, Topochemische Anordnung von Gangliosiden in der Erythrozyten-membran, *Immunitatsforsch. Allerg.* **133**:60–67.
Winzler, R. J., Harris, E. D., Pekas, D. J., Johnson, C. A., and Weber, P., 1967, Studies on glycopeptides released by trypsin from intact human erythrocytes, *Biochemistry* **6**:2195–2201.
Wolfe, L. S., 1961, The distribution of gangliosides in subcellular fractions of guinea-pig cerebral cortex, *Biochem. J.* **79**:348–355.
Woodin, A. M., and Wieneki, A. A., 1966, Composition and properties of a cell-membrane fraction from the polymorphonuclear leucocyte, *Biochem. J.* **99**:493–500.
Wu, H. C., Meezan, E., Black, P. H., and Robbins, P. W., 1969, Comparative studies on the carbohydrate-containing membrane components of normal and virus-transformed mouse fibroblasts. I. Glucosamine-labeling patterns in 3T3, spontaneously transformed 3T3, and SV-40-transformed 3T3 cells, *Biochemistry* **8**:2509–2517.
Yamashina, I., Izumi, K., Okawa, H., and Furuya, E., 1965, Hexosamine and sialic acid in mitochondria and microsomes from rabbit liver, *J. Biochem.* **58**:538–542.

Chapter 4
Anabolic Reactions Involving Sialic Acids

Edward John McGuire

I. INTRODUCTION: PERSPECTIVE AND DIRECTIONS

The sialic acids are a family of ubiquitous aminosugars found in some bacteria and invertebrates, and in all vertebrate tissues. They are normally found as nonreducing termini of complex heteropolymers such as glycoproteins, glycopeptide hormones, and glycolipids. The chemistry and structural features of these compounds are described in Chapter 1 of this volume.

The discovery and subsequent isolation of the sialic acids was an outgrowth and an intersection of several lines of research. Workers attempting to elucidate the chemical makeup of A, B, O(H) blood group substances turned from the surface of the red blood cell to mucinous secretions (mucoproteins), following the discovery of Lehrs (1930) and Putkonen (1930) that these viscous materials possessed blood group activity. While the sialic acids are not involved in the structure of the A, B, O(H) substances themselves (Watkins, 1966), they make up a significant percentage of the mucins and confer the property of high viscosity on these materials.

Working in a different area, Hirst (1941) as well as McClelland and Hare (1941) found that chicken red blood cells could be agglutinated by influenza virus. Burnet (1948) later found that mucins (mucoproteins) could inhibit this process and subsequently found that an enzyme from *Vibrio cholerae* could destroy this inhibitory activity. This enzyme was called RDE (receptor destroying enzyme) because it could also destroy the ability of red blood cells to absorb influenza virus.

EDWARD JOHN MCGUIRE • National Jewish Hospital and Research Center, Denver, Colorado 80206.

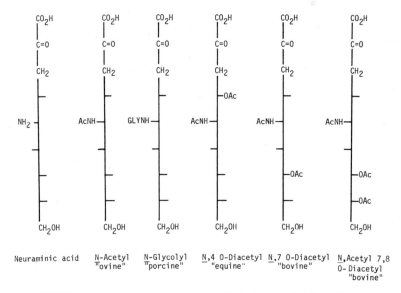

FIGURE 1. Schematic representation of the family of sialic acids.

Blix (1936), working from a chemical point of view, had isolated and crystallized a substance from bovine submaxillary mucin which he subsequently named sialic acid. Klenk (1941) isolated a similar material from glycolipids obtained from human brain.

These several lines of research were drawn together by the finding of Gottschalk and Lind (1949) that the enzyme RDE released a low-molecular-weight material from red blood cells or the mucoprotein inhibitors of viral hemagglutination, which had the properties of an acid-labile aminosugar. Through the efforts of Blix, Klenk, Gottschalk, Kuhn, and Roseman this material was finally identified as a sialic acid and a definitive structure was assigned and confirmed by subsequent synthesis. It was further shown that the sialic acids are a family of closely related neuraminic acids (see Figure 1 and Chapter 1). For a more detailed historical review, see Gottschalk (1972).

The recent rise in interest in the sialic acids and other carbohydrates is primarily due to evidence that these compounds confer important biological properties to macromolecules. The biological activity of the sialic acids greatly facilitated and provided the impetus for their subsequent chemical isolation and identification. It is against this background of a rising interest in the biological properties of the sialic acids and other carbohydrates that the author has written this chapter.

The intermediary metabolism of carbohydrates from an energy-yielding viewpoint has been studied and reviewed repeatedly. The emphasis in this chapter will be on the anabolism of sialic acids, which yields complex heteropolymers which possess *biological* activities.

A discussion of the anabolism of the sialic acids may be conveniently divided into three areas. The first area is the metabolic reactions leading to the basic nine-carbon chain of neuraminic acid. This involves the enzymatic interconversions leading from glucose, the primary precursor molecule, to N-acetylneuraminic acid and its derivatives (see Figure 1):

$$\text{Glucose} \to \to \to \to \text{NAN-9-PO}_4 \to \text{NAN} \to \text{Sialic Acids}$$

The second area is the activation of N-acetylneuraminic acid to its nucleotide derivative, CMP-NAN:

$$\text{NAN} + \text{CTP} \rightleftharpoons \text{CMP-NAN} + \text{PPi}$$

The third area of discussion is the incorporation of the activated sialic acids (by ketosidic linkage) into various oligosaccharides, glycoproteins, glycolipids, mucins, and connective tissue ground substances. It is as an integral part of such compounds that the sialic acids play their functional role in biological systems:

$$\text{CMP-NAN} + \text{Acceptor} \to \text{NAN-Acceptor} + \text{CMP}$$

In each of these areas, the anabolism of sialic acid will be emphasized. However, it will also be necessary to discuss the metabolism of other sugars (e.g., hexosamines) in order to place these discussions into an overall biochemical context.

II. BIOSYNTHESIS OF THE SIALIC ACIDS

A. Glucose to Sialic Acid

The steps leading to the biosynthesis of the sialic acids were elucidated primarily in the laboratories of Roseman (1962a) and Warren (1966). The enzymatic steps involved are shown in Figure 2. Early *in vivo* experiments (Becker and Day, 1953; Topper and Lipton, 1953; Roseman *et al.*, 1953; Rieder and Buchanan, 1958) demonstrated that ^{14}C-glucose is incorporated intact and without rearrangement into ^{14}C-glucosamine. It was subsequently shown that D-glucosamine is a precursor of sialic acid; hence, glucose is the principal metabolic precursor of sialic acid. Further, it was found by Spiro (1959) that in the rat, the liver is the primary site of synthesis of glucosamine from glucose.

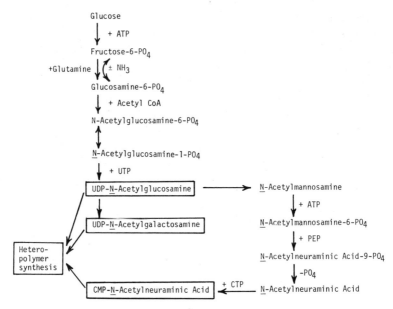

FIGURE 2. Major anabolic reactions leading to the synthesis of sialic acid and sialic-acid-containing heteropolymers.

1. Amination.

Fructose-6-PO$_4$ + L-Glutamine $\longrightarrow$
$\qquad$ Glucosamine-6-PO$_4$ + Glutamic Acid

The key enzyme which commits the six-carbon skeleton of glucose and its metabolic derivatives to aminosugar biosynthesis is L-glutamine: D-fructose-6-phosphate aminotransferase (glucosamine-6-phosphate synthetase). This enzyme siphons from the glycolytic pathway the necessary precursors for the hexosamine and sialic acid requirements of the cell. Lowther and Rogers (1953, 1955, 1956) found that the amino group of glucosamine was derived from the amide group of L-glutamine. Utilizing cell-free extracts of *Streptococcus haemolyticus*, ^{15}N-NH$_3$ was incorporated into glucosamine in the presence of unlabeled glutamate and glucose. Leloir and Cardini (1953), using cell-free extracts of *Neurospora crassa*, converted hexose phosphate and glutamine to a hexosamine. The presence of phosphoglucose isomerase in the extract, which interconverts glucose- and fructose-6-phosphates, made it impossible to determine which hexose phosphate was the true precursor. Blumenthal *et al.* (1955), also using *Neurospora* extracts, partially purified the enzyme and demonstrated that fructose-6-phosphate is the precursor.

Utilizing rat liver extracts, Pogell and Gryder (1957; 1969) demonstrated enzymatic activity which utilized glucose-6-phosphate and fructose-6-phosphate plus glutamine and yielded glucosamine-6-phosphate. Glucose-6-phosphate appeared to be the more efficient precursor and also served to stabilize an otherwise labile enzyme. Upon purifying some 25-fold, the enzyme still showed low levels of phosphoglucose isomerase activity, but utilized both glucose- and fructose-6-phosphates as precursors, apparently preferring glucose-6-phosphate. Glucose-6-phosphate plus glutamine was utilized in a rabbit epiphyseal cartilage extract to synthesize glucosamine (Castellani and Zambotti, 1956), while rat tumor homogenates were shown to utilize glucose, fructose, and mannose-6-phosphates to synthesize glucosamine-6-phosphate (Kizer and McCoy, 1959).

Roseman *et al.* (1958) purified the enzyme from microbial and mammalian extracts and demonstrated that fructose-6-phosphate, and not glucose-6-phosphate, was the required substrate. Ghosh *et al.* (1969) purified the enzyme from *Neurospora crassa, E. coli,* and rat liver until it was free from phosphoglucose isomerase and glutaminase. Using this more purified enzyme preparation, fructose-6-phosphate was shown to be the required substrate, and a 1:1 stoichiometry was demonstrated for glutamine and fructose-6-phosphate.

The apparent discrepancy between the findings of Pogell and Roseman have never adequately been resolved, although Winterburn and Phelps (1970, 1971*a,b,c*) have verified that fructose-6-phosphate is the true substrate for the rat liver enzyme. It is still possible that besides these, other transamidase enzymes exist.

An alternate enzymatic route for the biosynthesis of glucosamine-6-phosphate is the reversal of a reaction catalyzed by glucosamine-6-phosphate deaminase.

2. Deamination.

$$\text{Glucosamine-6-PO}_4 \rightleftharpoons \text{Fructose-6-PO}_4 + \text{NH}_3$$

Leloir and Cardini (1956) first found an enzyme in pig kidney cortex which catalyzed the synthesis of glucosamine-6-phosphate. The equilibrium of the enzyme reaction is strongly in favor of deamination. However, Comb and Roseman (1958*a*) showed that by coupling to a second enzyme (acetylase) the reaction can be pulled in the direction of synthesis. This enzyme (deaminase) apparently is ubiquitous and is stimulated by the presence of N-acetylglucosamine-6-phosphate.

There are two pathways for the synthesis of glucosamine-6-phosphate, a key intermediate since it is involved in the first step in the metabolism of the aminosugars. It appears that the reaction utilizing glutamine is anabolic, while that utilizing NH_3 is catabolic or degradative. Clark and Pasternak (1962) showed this to be true in bacteria, where growth on aminosugars represses the transamidase but induces the deaminase.

3. Acetylation.

Glucosamine-6-PO_4 + Acetyl-CoA ⟶
N-Acetylglucosamine-6-PO_4 + CoA

Chou and Soudak (1952) first demonstrated the acetylation of glucosamine by acetyl-CoA using pigeon liver extracts. The reaction is:

Glucosamine + Acetyl-CoA ⟶ N-Acetylglucosamine + CoA

It was later suggested that this enzymatic reaction was carried out by a *non*specific aromatic amine acetylase (Tabor *et al.*, 1953). Brown (1955), using a partially purified yeast preparation, demonstrated a specific enzyme which acetylated the phosphorylated hexosamine. The enzyme would not acetylate free glucosamine, did not require metal ions, and the equilibrium of the reaction was strongly in favor of acetylation. Davidson *et al.* (1959) purified the enzyme 254-fold from *Neurospora crassa* and demonstrated its specificity for glucosamine-6-phosphate. It did not acetylate free glucosamine, galactosamine, aromatic amines, or amino acids. Similar activity was demonstrated in crude extracts of *Penicillium,* Group A streptococci, rabbit kidney, liver and muscle, dog kidney, and human liver. This widespread enzyme, catalyzing an essentially unidirectional reaction, is important in converting the available glucosamine-6-phosphate to the metabolically more stable N-acetyl derivative.

Of interest is the finding that glycolyl-CoA is not utilized by the acetylase from hog tissues, nor is glycolic acid incorporated into acylated hexosamines or N-glycolylneuraminic acid using sheep colonic scrapings (Allen and Kent, 1968). A number of metabolic alterations (N-glycolation, O-acetylation) take place at the free-sugar level and possibly at the nucleotide-sugar or sugar-polymer level.

Ultimately the hexosamines and sialic acids are activated (converted to nucleoside mono- or diphosphate derivatives) and polymerized. Before this activation can be carried out by the appropriate pyrophosphorylase, the phosphate group must be transferred from the C-6 to C-1 position of the acylhexosamines. This reaction is catalyzed by a class of enzymes known as mutases.

4. Mutase.

$$N\text{-Acetylglucosamine-6-PO}_4 \underset{}{\overset{Mg^{2+}}{\rightleftharpoons}} N\text{-Acetylglucosamine-1-PO}_4$$

Brown (1953) demonstrated that phosphoglucomutase, which converts glucose-6- to glucose-1-phosphate, could also act on glucosamine-6-phosphate, although 100 times more slowly. The enzyme, which utilized the acyl derivative of glucosamine as its substrate, was described by Leloir and Cardini (1956) and Reissig (1956). This enzyme has been purified from *Neurospora* but is still not totally freed from phosphoglucomutase. At equilibrium, 84% N-acetylglucosamine-6-phosphate and 14% N-acetylglucosamine-1-phosphate are found. However, by coupling with the next reactions in the pathway, catalyzed by pyrophosphorylase and inorganic pyrophosphatase, the reaction is pulled in the direction of acylhexosamine "activation." Carlson (1966) has also purified the acylglucosamine mutase from pig submaxillary glands. A tissue survey has shown the enzymes to appear ubiquitously.

5. Pyrophosphorylase.

$$\text{Uridine-}\overset{*}{P}\text{-}\overset{*}{P}\text{-}\overset{*}{P} + \text{Acetylglucosamine-1-PO}_4 \rightleftharpoons$$
$$\text{Uridine-}\overset{*}{P}\text{-P-}N\text{-Acetylglucosamine} + \overset{*}{P}\overset{*}{P}i$$

The enzyme(s) which activates N-acetylglucosamine-1-phosphate is representative of a class of enzymes first described by Kornberg (1950) and Schrecker and Kornberg (1950). These enzymes catalyzed the synthesis or breakdown of unsymmetrical phosphoanhydrides:

$$\text{Nucleoside-}\overset{*}{P}\text{-}\overset{*}{P}\text{-}\overset{*}{P} + \text{Sugar-1-PO}_1 \rightleftharpoons \text{Nucleoside-}\overset{*}{P}\text{-P-Sugar} + \overset{*}{P}\overset{*}{P}i$$

The specific reactions first observed in yeast and liver extracts were

$$\text{ATP} + \text{NMN} \underset{}{\overset{Mg^{2+}}{\rightleftharpoons}} \text{NAD} + \text{PPi}$$

$$\text{ATP} + \text{FMN} \underset{}{\overset{Mg^{2+}}{\rightleftharpoons}} \text{FAD} + \text{PPi}$$

The reaction has an equilibrium constant of approximately 0.5 and may be measured in either direction. Many other pyrophosphorylases have been discovered, and a complete review is beyond the scope of this undertaking. However, it seems that the individual enzymes are relatively specific for the nucleoside triphosphate utilized (uridine, adenine, guanine, cytidine, etc.). For example, there are separate and distinct enzymes responsible for catalyzing the synthesis of UDP-glucose, ADP-glucose, CDP-glucose, TDP-glucose, and GDP-glucose, as well as separate enzymes for the synthesis of GDP-mannose and GDP-glucose (Hansen *et al.*, 1966a,b). The diversity and specificity of the

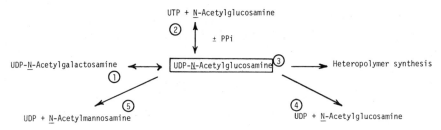

FIGURE 3. Metabolic conversions of UDP-*N*-acetylglucosamine.

pyrophosphorylases offers opportunity for cells to regulate how much and which sugars to activate.

Uridine diphosphate-*N*-acetylglucosamine (UDP-GlcNAc) was first found in yeast (Cabib *et al.*, 1953) and in liver extracts (Smith *et al.*, 1953; Hurlbert and Potter, 1954). Smith *et al.*, (1953) also demonstrated that rat liver nuclei could degrade UDP-GlcNAc by a reversal of the pyrophosphorylase reaction.

Glaser and Brown (1955) first achieved the synthesis of UDP-GlcNAc using extracts of yeast and Rous sarcoma tumors. Coupled with the chemical synthesis of α-D-glucosamine-1-phosphate and *N*-acetyl-α-D-glucosamine-1-phosphate achieved by Maley *et al.* (1956) and Maley and Lardy (1956), the synthesis of radioactively labeled UDP-GlcNAc became possible. The synthesis of radioactively labeled nucleotide sugars was greatly facilitated by the introduction of a chemical procedure for the synthesis of unsymmetrical phosphoanhydrides (Roseman *et al.*, 1961):

$$\text{Sugar-1-PO}_4 + \text{Nucleotide-O} - \overset{\overset{O^-}{|}}{\underset{\underset{O}{\downarrow}}{P}} - N \bigcirc \xrightarrow{\text{pyridine}}$$

$$\text{Nucleotide} - O - \overset{\overset{O}{\uparrow}}{\underset{\underset{\underline{O}}{|}}{P}} - O - \overset{\overset{O}{\uparrow}}{\underset{\underset{\underline{O}}{|}}{P}} - \text{Sugar} + H - N \bigcirc$$

Much of the progress in elucidating the metabolism of the sialic acids came as a result of the increased sensitivity of assays using

Anabolic Reactions Involving Sialic Acids 131

radioactive precursors whose synthesis was made possible by the previously mentioned chemical procedure.

UDP-GlcNAc is another key intermediate in the anabolism of aminosugars because it can undergo multiple reactions. It can itself be utilized in polymer biosynthesis, interconverted to other activated aminosugars (i.e., UDP-GalNAc), degraded to UDP and free GlcNAc, or be irrevocably committed to sialic acid biosynthesis (see Figure 3).

6. Epimerases.

$$UDP\text{-}GlcNAc \rightarrow MnAc + UDP$$

Reaction number five, Figure 3, is the first step which is unique and is not shared by other aminosugars in the biosynthetic sequence leading to the sialic acids. This reaction is catalyzed by one enzyme which apparently carries out both an epimerization of the acylamino group at C-2 and hydrolysis of N-acetylmannosamine (see Figure 4).

This reaction was first found by Cardini and Leloir (1957) using rat liver extracts, although they mistakenly identified the product of the reaction as N-acetylgalactosamine. Comb and Roseman (1958b) later identified the product of the reaction as N-acetylmannosamine. Spivak and Roseman (1966) have purified this highly labile enzyme from rat liver. It apparently requires no cofactors and the reaction was shown to be irreversible. The enzyme is widely distributed in vertebrate tissues but has not been found in bacteria.

Salo and Fletcher (1970a,b) have chemically synthesized UDP-N-acetylmannosamine and have studied the mechanism of this enzyme reaction. They found that UDP-MnNAc could be cleaved to MnAc, but the reaction could not be reversed to form UDP-GlcNAc. They also found that free UDP-MnAc is apparently not an intermediate in the reaction.

Two other enzymes have been found which lead to the synthesis of

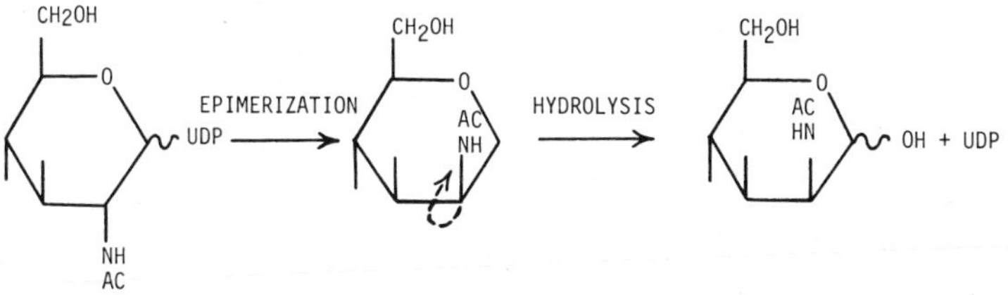

FIGURE 4. Proposed mechanism for N-acetylmannosamine synthesis from UDP-N-acetylglucosamine.

N-acetylmannosamine. The first enzyme, which has been found in bacteria, catalyzes the reaction:

$$N\text{-Acyl-D-Glucosamine-6-PO}_4 \rightleftharpoons N\text{-Acyl-D-Mannosamine-6-PO}_4$$

This enzyme has been purified from *Aerobacter aerogenes* and has been found in *E. coli* K-235, and *Clostridium perfringens* (Roseman *et al.*, 1960; Ghosh and Roseman, 1965).

The second enzyme, which was found using hog kidney extracts, catalyzes the reaction

$$\dot{N}\text{-Acyl-D-Glucosamine} \rightleftharpoons N\text{-Acyl-D-Mannosamine}$$

This enzyme has been purified by Ghosh and Roseman (1965*b*) and differs in two important aspects from the bacterial epimerase: (a) The enzyme will only utilize the free N-acyl (N-acetyl or N-glycolyl) hexosamine and not the 6-phosphate ester; (b) The enzyme shows an absolute requirement for ATP as a cofactor, although the role of ATP is not clear. There is no disappearance or isotopic exchange with Pi, PPi, ADP, or AMP. The metabolic function of this enzyme is not clear since there should be little free acylglucosamine in cells. (It should predominantly exist as the phosphate ester or as the nucleotide derivative). It may be that this enzyme scavenges any free acylmannosamine not subsequently converted into sialic acid and converts it to the more metabolically flexible acylglucosamine.

7. Acylmannosamine Kinase.

$$\text{ATP} + N\text{-Acylmannosamine} \xrightarrow{Mg^{2+}} N\text{-Acylmannosamine-6-PO}_4 + \text{ADP}$$

The major pathway for the synthesis of D-mannosamine in mammalian tissues yields the unphosphorylated derivative of acylmannosamine. The enzyme sialic acid synthetase utilizes acylmannosamine-6-phosphate. A kinase is necessary and has been found in extracts of mammalian liver and submaxillary gland (Ghosh and Roseman, 1961; Warren and Felsenfeld, 1962; Kundig *et al.*, 1966). The enzyme has been purified 2000-fold from rat liver and is specific for acylmannosamine, utilizing both N-acetyl as well as N-glycolylmannosamine. A number of tissues (lung, kidney) which are capable of synthesizing sialic-acid-containing heterosaccharides do not demonstrate this activity in crude extracts. This remains an anomaly and suggests the possibility of an alternate enzymatic pathway for synthesis of acylmannosamine-6-PO$_4$.

8. Sialic-Acid-Synthesizing Enzymes. The formation of sialic acid from its acylmannosamine precursor may be achieved by two pathways. The first of these was discovered by Heimer and Meyer (1956), who observed that extracts of *Vibrio cholerae* destroyed N-acetylneuraminic acid and produced pyruvate and an acylhexosamine.

N-Acetylneuraminic Acid ⇌ N-Acetylmannosamine + Pyruvate

Comb and Roseman (1958c, 1960) found the same enzyme using *Clostridium perfringens* cell-free medium and further identified the acylhexosamine product as N-acetyl-D-mannosamine. This elegant piece of work showed for the first time the presence of this aminosugar in nature and also led to a revision of the structure of sialic acid, as shown in Figure 5.

The enzyme "NAN-aldolase" has been purified from the medium of *C. perfringens* and has been found in *E. coli* K-235 extracts and several animal tissues. Brunetti et al. (1962) purified the enzyme 1700-fold from hog kidney cortex.

The enzyme can cleave N-acetyl, N-glycolyl, and N-acetyl-8-O-acetylneuraminic acid (bovine) at about the same rate. However, N-acetyl-7-O-acetylneuraminic acid (bovine) is cleaved at 50% the rate and N,4-O-diacetylneuraminic acid (equine) at 20% the rate of N-acetylneuraminic acid. The enzyme cannot cleave N-acetylneuraminic acid-9-phosphate. The enzyme is reversible, with an equilibrium constant (K_e = 0.064) greatly in favor of aldolytic cleavage. The enzyme can utilize both N-acyetyl- and N-glycolylmannosamine plus pyruvate to synthesize N-acylneuraminic acid but cannot utilize mannosamine or N-acylglucosa- mine. It was the inability of the enzyme to utilize N-acetylglucosamine or N-acetylgalactosamine to synthesize neuraminic acid which provided one of the clues that the acylhexosamine which resulted from

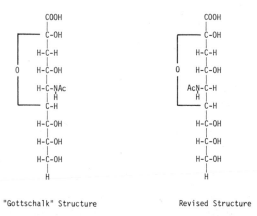

FIGURE 5. Two proposed structures for N-acetylneuraminic acid.

cleavage of NAN was not one of the known common acylhexosamines (N-acetylgluco- or galactohexosamine).

In spite of the fact that the reaction equilibrium is greatly in favor of breakdown, it was felt that *the "NAN-aldolase"* was responsible for the synthesis of N-acetylneuraminic acid by coupling to the next enzyme in the pathway (Figure 2; CMP–sialic acid synthetase). That notion was partially dispelled by the inability to detect this enzyme in mucin-producing tissues (tissues extremely rich in secreted sialic acid) (Brunetti *et al.*, 1962) and by the abundance of this enzyme in the bacteria *V. cholerae* and *C. perfingens,* which totally lack sialic acid. Thus it was felt this enzyme may primarily be a degradative enzyme, and the search continued for a second "synthetic" (anabolic) pathway.

The second enzyme was first reported by Warren and Felsenfeld (1961*a,b*; 1962) and also found by Roseman *et al.* (1961*a,b*). The reaction is

N-Acylmannosamine-6-PO$_4$ + Phosphoenolpyruvate $\longrightarrow$
N-Acylneuraminic Acid-9-PO$_4$ + Phosphate

Unlike the previous enzyme, "NAN-aldolase," this enzyme catalyzes an essentially irreversible reaction in favor of synthesis. The enzyme was first found in rat liver extracts (Warren and Felsenfeld, 1961*a*) and has been purified some 95-fold from bovine submaxillary gland (Watson *et al.*, 1966). It can utilize both N-acetyl as well as N-glycolylmannosamine-6-PO$_4$. It seems likely that this enzyme, "sialic acid–9-phosphate synthetase," catalyzes the major reaction for the synthesis of sialic acid. The product of the reaction is the 9-phosphate ester of sialic acid and must be dephosphorylated before "activation" can occur.

9. Sialic Acid-9-phosphatase.

N-Acylneuraminic Acid-9-PO$_4$ $\longrightarrow$ N-Acylneuraminic Acid + PO$_4$

A number of nonspecific phosphatases can act on NAN-9-PO$_4$ to dephosphorylate it. However, it appears that in rat liver there is a specific sialic acid-9-phosphatase (Warren and Felsenfeld, 1962). A similar enzyme has been purified some 800-fold from human erythrocyte lysates (Jourdian *et al.*, 1964) and has been shown to be highly specific for sialic acid 9-phosphate (both N-acetyl and N-glycolyl). This enzyme completes the synthesis of sialic acid in animal tissues. In bacteria, however, another pathway does exist and may be widespread. Extracts of *Neisseria meningitidis* synthesize sialic acid utilizing free N-acetylmannosamine, not the 6-phosphate ester (Blacklow and Warren, 1962). This enzyme has never been detected in animal tissue:

N-Acetyl-D-Mannosamine + PEP $\rightarrow$ N-Acetylneuraminic Acid + PO$_4$

In summary, there are three known pathways for the synthesis of sialic acid: the "NAN-aldolase," which condenses pyruvate and N-acylmannosamine (bacterial and animal); "sialic acid-9-phosphate synthetase," which condenses phosphoenolpyruvate with N-acylmannosamine-6-phosphate (animal tissue only); the enzyme from *Neisseria meningitidis*, which condenses N-acylmannosamine with phosphoenolpyruvate (bacteria only).

B. Activation

Before the sialic acids can be polymerized or incorporated in high-molecular-weight products they must be activated. This reaction is catalyzed by CMP–sialic acid synthetase.

1. Pyrophosphorylase (Activation).

N-Acylneuraminic Acid + CTP $\rightleftharpoons$ CMP-N-Acylneuraminic Acid + PPi

This reaction is unique in that the free sugar and not the sugar-1-phosphate is activated, and the nucleotide sugar is the monophosphate and not the diphosphate derivative.

The enzyme was first detected in bacteria (Warren and Blacklow, 1962) and animal tissues (Roseman, 1962b; Kean and Roseman, 1966; Spiro and Spiro, 1968). This important enzyme has been purified from hog submaxillary glands (Kean and Roseman, 1966) and *Neisseria meningitidis* (Warren and Blacklow, 1962). The equilibrium reached by this reaction, unlike that catalyzed by nucleotide diphosphate pyrophosphorylases, is greatly in favor of synthesis. The sialic acid specificity of the enzyme varies depending on the source (see Figure 1). The pig submaxillary gland CMP–sialic acid synthetase will utilize NAN and NGN (40%) as well, but not 4-O-acetyl NAN or 7,8 di-O-acetyl NAN. The bacterial enzymes will utilize NAN and NGN as well as the 4-O-acetyl derivative, but not the 7,8 di-O-acetyl derivative. Schauer *et al.* (1972) have reported that the enzyme from bovine submaxillary gland will utilize NAN, NGN, and the 7-O-acetyl and 8-O-acetyl derivatives of NAN and NGN. Thus, as might be expected, the specificity of the CMP–sialic acid synthetase roughly parallels the spectrum of neuraminic acid derivatives found in the tissue.

The amount of CMP-sialic acid in the cell is critical and may well affect the amount and composition of carbohydrate-containing heteropolymers synthesized. However, little is known about the regulation of this critical enzyme with the exception of information reported in the original papers and, more recently in an intriguing paper of Kean (1970).

C. Regulatory Problems

Kean (1970) found that CMP-sialic acid synthetase was located in the nuclei isolated from rat liver, kidney, brain, and spleen and hog retina. The subcellular location of this enzyme in the nucleus was well documented and is surprising in the light that CMP-sialic acid is utilized in a different geographic location in the cell (see Section III, C). This subcellular separation of the enzymes for synthesizing and utilizing CMP-sialic acid may represent an important control mechanism for sialic acid metabolism.

Another, more classical, control mechanism exists for the feedback regulation of key enzymes for the synthesis of $Gm-6-PO_4$ and for N-acetylmannosamine. Each of these is the first enzyme in the metabolic commitment to hexosamine and sialic acid synthesis. Kornfeld et al. (1964) showed that UDP-GlcNAc is an efficient feedback inhibitor for L-glutamine-D-fructose-6-phosphate aminotransferase and that CMP-NAN also inhibits UDP-GlcNAc-2-epimerase, which is responsible for the synthesis of N-acetylmannosamine. This is an example of the by now familar endproduct inhibition of the first enzyme of a metabolic pathway (see Figure 6). They were also able to demonstrate that *in vivo* administration of puromycin to rats, which inhibits *de novo* protein synthesis and also depresses sialic acid and hexosamine utilization, does *not* lead to an accumulation of UDP-GlcNAc. Furthermore, the turnover of the UDP-hexosamine pool was shown to be slowed down. These data suggest that impairment of the utilization of UDP-hexosamine leads to decreased synthesis of UDP-hexosamines or their precursors (i.e., classical feedback inhibition).

Winterburn and Phelps (1970) have suggested that the L-glutamine-D-fructose-6-phosphate aminotransferase does not display classical feedback kinetic behavior in the presence of UDP-GlcNAc. Furthermore, they suggest that there is a 10-fold excess of enzyme over that necessary for a steady state for hexosamine metabolism.

Similar experiments in bacteria, utilizing penicillin inhibition of the utilization of nucleotide sugars, led to an accumulation of the nucleotide sugar pools (Park and Strominger, 1957). These data suggest that bacterial systems lack a feedback mechanism. Kornfeld (1967) was not

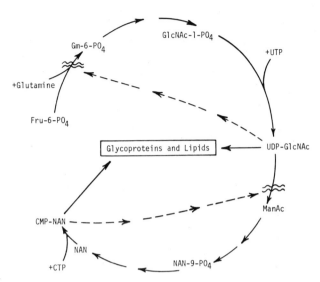

FIGURE 6. Anabolic pathway for the synthesis of activated N-acetylglucosamine and sialic acid and proposed regulatory steps (Kornfeld et al., 1964).

able to demonstrate UDP-GlcNAc inhibition of the L-glutamine-D-fructose-6-phosphate aminotransferase isolated from *Bacillus subtilis* or *Escherichia coli*, but was able to show inhibition of partially purified enzyme from rat liver and HeLa S3 tissue culture cells.

Bates et al. (1966) found that feeding rats on a high orotic acid diet raised UDP-hexosamine pools in the liver fivefold. These data suggest that the feedback control can be overcome in some unknown fashion, perhaps by orotic acid derivatives which are positive effectors. The high levels of UDP-hexosamine had no effect on glycoprotein synthesis. However, when the same authors administered the drug, Diazomycin A, which inhibits the L-glutamine-D-fructose-6-phosphate aminotransferase, the levels of UDP-hexosamine were greatly reduced. This reduction was accompanied by a 50–70% reduction in the rate of glycoprotein synthesis. This opens the possibility that nucleotide sugar pool sizes may well affect rates as well as composition of glycoproteins.

D. Other Derivatizations

Previous discussions have been primarily about the synthesis of NAN (N-acetylneuraminic acid) or its precursors, and in some cases the N-glycolyl (NGN) derivative has been mentioned. Yet, as was shown in

Figure 1, the sialic acids are made up of a family of neuraminic acid derivatives. However, until recently little was known of the metabolism of these other neuraminic acid family members.

In 1969, Schoop *et al.* found that pig submaxillary slices could synthesize NGN from acetate, N-acetylmannosamine, or N-acetylneuraminic acid, all labeled in the acetate moiety. Schauer (1970a; 1970b) demonstrated an enzyme system in high-speed supernatants of hog submaxillary gland extracts capable of oxidizing N-acetylneuraminic acid or N-acetylmannosamine to N-glycolylneuraminic acid. The enzyme is called N-acetylneuraminate:O_2-oxidoreductase and requires O_2 and ascorbate or NADPH for activity. It appeared that free N-acetylneuraminic acid is the substrate, but it is possible that CMP-NAN or the 9-phosphate ester is the substrate.

O-acetylation has also been demonstrated by Schauer (1970c,d) using slices of horse and bovine submaxillary gland. Acetyl-CoA appears to be the donor, and again free NAN or its 9-phosphate ester or CMP-NAN appear to be the acceptors. In a later paper (Schauer and Wember, 1971), both hydroxylation and O-acetylation of membrane-bound polymer NAN was demonstrated. Until these enzyme systems are solubilized and purified, it is not clear at what level derivatization (hydroxylation and O-acetylation) occurs. Schauer (1972) has presented a model which involves specific enzymes which catalyze transfer of CMP-NAN to polymers. All of the enzymes are arranged in a multienzyme membrane-bound complex. This interesting model awaits experimental verification.

III. BIOSYNTHESIS OF POLYMERS, GLYCOPROTEINS, MUCINS, AND GLYCOLIPIDS CONTAINING SIALIC ACID

Glycosyltransferases are a group of enzymes that catalyze the transfer of a sugar from its activated form (nucleotide sugar) to an appropriate acceptor molecule (usually a glycose-containing molecule). The general reaction is shown below:

Nucleoside mono(di)-PO_4-Sugar + Acceptor $\xrightarrow{metal^{2+}}$

Sugar-Acceptor + Nucleoside mono(di)-PO_4

Some general properties of glycosyltransferases are listed below:

1. They are specific for the sugar they transfer (galactose, N-acetylgalactosamine, sialic acid, etc.).
2. They are specific for the acceptor molecule (β 1→4 galactosyl residues, β 1→6 galactosyl residues, etc.).

3. They have specific metal requirements (Mn^{2+}, Mg^{2+}, no metal, etc.).
4. Sugars are added to the nonreducing terminus of glycose residues.
5. The enzymes form specific sugar linkages (α or β; 1→4, 1→3, or 1→6, etc.).
6. Many of the glycosyltransferases are membrane-bound and are located in the golgi complex, endoplasmic reticulum, and plasma membrane.

It is not within the scope of this chapter to review glycosyltransferases in general. Roseman (1968), Schachter and Rodén (1973), and R. G. Spiro (1970) have written excellent general review articles. This chapter will concentrate on the specific family of glycosyltransferases, shown in Figure 7, which transfer sialic acid to specific acceptors (the sialyltransferases). Each reaction involving the transfer of sialic acid to a *different* acceptor is catalyzed by a *different* enzyme. Also, each reaction transferring sialic acid to the *same* acceptor, but forming a different linkage (α or β; 2→2, 2→3, 2→4, 2→6), is catalyzed by a *separate* and *distinct* enzyme.

All of the glycosyltransferases that transfer sialic acid can be grouped into a family of sialyltransferases. It should be emphasized that this represents a broad, convenient classification based on acceptor specificity and the chemical linkages synthesized. However, the sialyltransferases of two separate and distinct organs (i.e., liver and mammary gland), even though they catalyze the same reactions and form identical products, may represent two totally distinct proteins (isoenzymes). At present twelve activities representative of sialyltransferases have been described. These enzymes have been described primarily on the basis of their acceptor specificities and the careful chemical characterization of the product of the reaction. These twelve activities probably represent eight separate and distinct enzymes. Based on the one linkage–one enzyme hypothesis, additional enzymes may be postulated. Undoubtedly, other undiscovered enzymes will be found in the future as

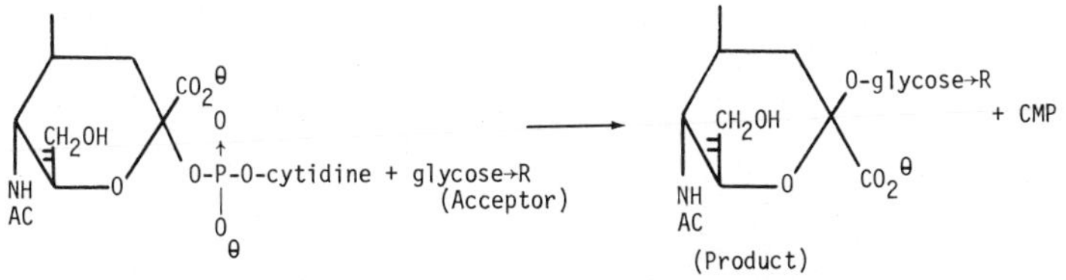

FIGURE 7. Schematic sialytransferase.

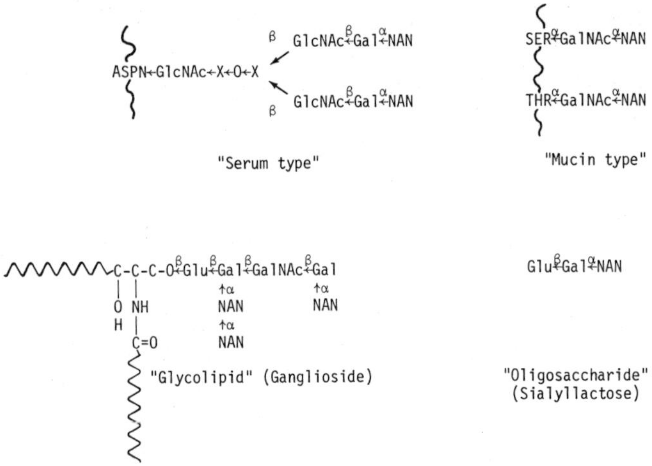

```
          ⎧                                                ⎧
          ⎪         β  GlcNAc-β→Gal1-α→NAN                 ⎪  SER-α→GalNAc-α→NAN
ASPN←GlcNAc←X←O←X ⎨                                        ⎨
          ⎪         β  GlcNAc-β→Gal1-α→NAN                 ⎪  THR-α→GalNAc-α→NAN
          ⎩                                                ⎩

              "Serum type"                                    "Mucin type"


∿∿∿∿∿C-C-O-β→Glu-β→Gal1-β→GalNAc-β→Gal          Glu-β→Gal1-α→NAN
     ‖ ‖       ↑α            ↑α
     O NH      NAN           NAN
     H ‖       ↑α
       C=O     NAN
       ⎰
       ⎱  "Glycolipid" (Ganglioside)              "Oligosaccharide"
       ⎰                                            (Sialyllactose)
       ⎱
```

FIGURE 8. Schematic representations of the major classes of acceptor molecules involved in glycosyltransferase reactions. Wavy line represents polypeptide; sharp-cornered wavy line represents fatty acid side chain.

carbohydrate analytical methods are improved and new sialic-acid-containing materials are analyzed.

Sialic acids are normally found on the nonreducing terminus of many carbohydrate-containing heteropolymers. The discovery of the biosynthetic enzymes has normally followed on the heels of the chemical isolations and characterizations of sialic-acid-containing glycoproteins, mucins, glycolipids, and oligosaccharides. Figure 8 shows schematic representations of four classes of sialic-acid-containing materials that will be discussed.

Figure 8 is a schematic simplification of glycoprotein and glycolipid structures. Marshall (1972) and Gottschalk (1972) have published comprehensive reviews of the structural complexities of this class of molecules.

In all cases sialic acid is linked to a β-galactosyl, α-N-acetylgalactosaminyl, or α-sialyl residue. Thus, logical acceptor molecules must contain a nonreducing terminus with one of these structures. However, not all of the biosynthetic work has been carried out using added acceptor molecules of known structure. For example, O'Brien *et al.* (1966) demonstrated the transfer of N-^{14}C-acetylneuraminic acid from CMP-N-^{14}C-acetylneuraminic acid to endogenous acceptors of rat liver microsomes. Upon solubilization, the ^{14}C products were characterized using immunoelectrophoresis followed by radioautography. Several rat serum proteins were demonstrated. Such studies, utilizing endogenous acceptors, are fraught with difficulty. One cannot be certain whether enzyme or acceptor is limiting, which throws the quantitation open to question. However, many workers have suggested that such measurements reflect the levels of acceptor molecules present, assuming the enzyme to be in excess (Bosmann, 1972; Wagner and Cynkin, 1971). This assumes that there is no topological separation of the enzyme and acceptor molecules.

The bulk of the work to be described has been carried out be adding exogenous acceptors of known structure to the enzyme source. This technique has been extremely valuable in defining the structural requirements of the acceptor molecule and has been of assistance in defining the chemical nature of the products synthesized. The implicit assumption of this methodology is that the glycosyltransferase is not specific with regard to the structure of the acceptor beyond the nonreducing terminal sugar and in some cases the penultimate residue. For example, the sialyltransferase that transfers sialic acid to terminal β-galactosyl residues demonstrates little preference among various glycoproteins provided they terminate with β-galactopyranosyl-R. As will be seen, this assumption is not strictly true, since the glycosyltransferases that utilize glycolipids as acceptor molecules do not seem to act on glycoprotein. Nevertheless, it has proved to be a useful assumption.

Finally, the investigation of glycosyltransferases is relatively recent and has progressed little beyond the initial discovery stage. For example, no sialyltransferase enzymes have been purified to homogeneity. As a consequence, any discussion of these enzymes raises more questions than it answers, but may stimulate further efforts to investigate these interesting enzyme systems.

A. Colominic Acid Synthesis

The first sialyltransferase described was obtained from the bacterial source, *Escherichia coli* K-235, an organism that produces a homopolymer of N-acetylneuraminic acid called colominic acid (Aminoff *et al.*, 1963; Kundig *et al.*, 1971).

$$\text{CMP-NAN}^* + (\text{NAN})_n \longrightarrow \text{NAN}^* \rightarrow (\text{NAN})_n + \text{CMP}$$

This particulate enzyme will transfer N-acetylneuraminic acid from CMP-N-acetylneuraminic acid to either exogenously added colominic acid or to endogenous colominic acid found in the particulate preparation. The transfer is apparently to the nonreducing end of the polymer. The enzyme will utilize CMP-N-glycolylneuraminic acid at only 5% the rate for CMP-N-acetylneuraminic acid. The enzyme has not been solubilized, and work has not progressed beyond the initial observations.

B. CMP-Sialic Acid:Lactose (β-Galactosyl) Sialyltransferase

$$\text{CMP-NAN} + \text{Lactose} \rightarrow \text{NAN} \rightarrow \text{Lactose} + \text{CMP}$$

Jourdian et al. (1963) and Carlson et al. (1973a,b) reported the properties of a particulate sialyltransferase prepared from rat mammary gland. The acceptor specificity is shown in Table I.

The acceptor specificity deduced from these series of experiments is that this enzyme utilizes acceptor molecules containing a nonreducing β-galactosyl residue and not an α-galactosyl residue. It shows no preference for the acceptor penultimate sugar, utilizing lactose and N-acetyllactosamine equally well. (They differ only in a Glu or GlcNAc aglycone.) Utilizing the positional isomers of N-acetyllactosamine (β-1→4, 1→6, and 1→3), there is little preference. However, this enzyme will not use acceptors of high molecular weight (greater than an oligosaccharide) such as appropriately treated glycoproteins. Neither will it utilize glycolipid acceptor molecules. However, the enzyme was not solubilized by the investigators and was not assayed in the presence

TABLE I. Substrate Specificity of Rat Mammary Gland Sialyltransferase

Acceptor	NAN incorporated, percent of lactose
Lactose (Galβ,1→4glc)	100
N-acetyllactosamine (Galβ,1→4GlcNac)	100
(Galβ,1→6GlcNAc)	91
(Galβ,1→3GlcNAc)	88
Lacto-N-tetraose (Galβ,1→3GlcNAcβ,1→3Galβ,1→4Glc)	80
Methyl β-galactoside	54
Methyl α-galactoside	0
Sialidase treated AGP[a] (Gal$\xrightarrow{\beta}$OLIGO→protein)	<5
Acid-treated ganglioside[b] (Gal$\xrightarrow{\beta}$GalNAc-R)	<5

[a] AGP = Glycoprotein (α_1-acid glycoprotein) pretreated with sialidase which exposes the penultimate β-galactosyl residue. See Figure 8.
[b] Acid-treated ganglioside conditions remove only sialic acid residues exposing terminal β-galactosyl residue. See Figure 8.

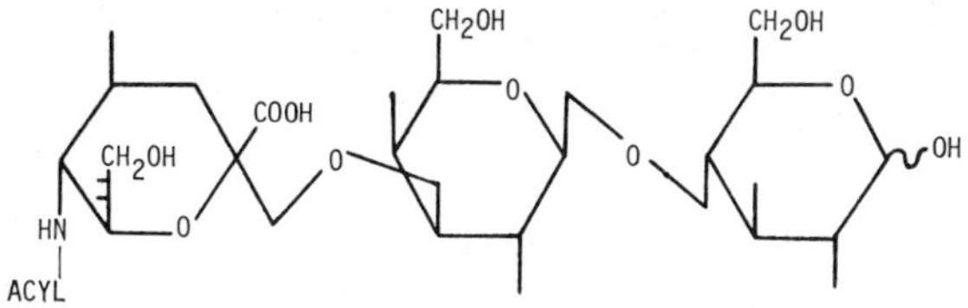

FIGURE 9. The structure of α-sialyl 2→3 lactose [β-galactosyl(1→4) glucose].

of detergents to disrupt the membrane structure of the particulate preparation.

The product of the reaction is the 2→3 isomer of sialyllactose first discovered in milk and colostrum by Kuhn and Brossmer (1958). (See Figure 9.)

Tissue surveys revealed the widespread presence of an enzyme which would transfer sialic acid from CMP-sialic acid to lactose. The enzyme was found in rat spleen, testes, kidney, lung, brain, and liver. More importantly, this work led investigators to look at milk and colostrum as sources of the glycosyltransferases. These secretions turned out to contain large amounts of *soluble* glycosyltransferases—a great boon.

The rat mammary gland sialyltransferase is probably of more historical than biochemical interest, for it not only led to the discovery of the milk and colostrum enzymes, but also set a precedent for attention to careful chemical detail which characterized this field. It was largely through the careful analysis of products, careful chemical preparation of acceptor molecules, and use of known oligosaccharide acceptor structures that some biochemical understanding has emerged.

C. CMP-Sialic Acid:Glycoprotein (β-Galactosyl) Sialyltransferases

$$\text{CMP-NAN} + \text{Gal} \xrightarrow{\beta} \text{R} \rightarrow \text{NAN} \rightarrow \text{Gal} \xrightarrow{\beta} \text{R} + \text{CMP}$$

Bartholomew and Jourdian (1966) and Bartholomew *et al.* (1973) first reported on this sialyltransferase(s), found in goat, bovine, and human colostrum. This enzyme was different in three principal ways from the previously reported mammary gland enzyme. First, in colostrum the enzyme was soluble and thus could be purified. Second, this sialyltransferase would transfer sialic acid to high-molecular-weight

TABLE II. Acceptor Specificity of Goat Colostrum Sialyltransferase

Acceptor[a]	NAN incorporated, percent of lactose
Lactose (Galβ,1→4Glc)	100
N-Acetyllactosamine (Galβ,1→4GlcNAc)	770
(Galβ,1→3GlcNAc)	146
(Galβ,1→6GlcNAc)	27
Sialidase treated "serum" glycoprotein[b]	1080
Sialidase treated "mucin" glycoprotein[c]	50
β-Galactosides	7–17
α-Galactosides	0

[a] Adapted from Schachter and Rodén (1973).
[b] Sialidase treated orosomucoid.
[c] Sialidase treated ovine submaxillary mucin.

acceptors as well as low-molecular-weight oligosaccharides containing a nonreducing terminal β-galactosyl residue. The high-molecular-weight substrates which were tested were primarily sialidase-treated "serum-type" glycoproteins such as orosomucoid (α_1-acid serum glycoprotein) which exposes a β-galactosyl residue or sialidase-treated "mucin-type" glycoproteins which expose terminal α-N-acetylgalactosaminyl residues (see Figure 8). Third, the product of the reaction is a different isomer of sialyl→galactose→R.

The enzyme(s) were purified 60-fold to 220-fold and were shown to have no metal requirement, nor did they require detergent for activity (membrane-bound glycosyltransferases require detergents). Table II shows the acceptor specificity study for this enzyme(s).

This enzyme preparation apparently demonstrates a preference for β-galactosyl termini of both high- and low-molecular-weight acceptors and shows no activity with α-galactosides. Further, it shows specificity toward the penultimate sugar, utilizing Gal β,1→4 GlcNAc almost 8-fold better than Gal β,1→4 Glc. There are also marked differences in the rate of reaction between the position isomers of galactosyl $\overset{\beta}{\rightarrow}$N-acetylglucosamine. The β,1→4 isomer is the preferred-position isomer with the β,1→3 isomer, which demonstrates an intermediate acceptor preference. This important finding demonstrates that this sialyltransferase discerns the fine structure of the acceptor molecule. The enzyme detects the anomeric linkage (α or β), the position of the linkage (1→3; 1→4; or 1→6), and the nature of the sugar to which the galactose is linked (glucose or N-acetylglucosamine). Finally, the activity demonstrated with mucin-type glycoproteins, pretreated with sialidase, is the result of low levels of β-galactosyl residues contained in the ovine submaxillary mucin preparation. When the product of this reaction is characterized, no sialyl→N-acetylgalactosamine is found. Thus, the colostrum sialyltransferase will *not* transfer to α-N-acetylgalactosaminyl residues.

The products of the reactions utilizing lactose as acceptor have been characterized as primarily the 2→6 isomer of sialyllactose, with lesser amounts of the 2→3 isomer. The finding of two products suggests that in spite of the fact that the enzyme preparation has been partially purified, it still contains a mixture of sialyltransferases. It is also possible that a single enzyme catalyzes the formation of two linkages (2→3 and 2→6), but evidence utilizing liver sialyltransferases would tend to rule out this explanation (Hudgin and Schachter, 1972).

Early progress in elucidating the details of glycoprotein biosynthesis was greatly aided by the use of a partially purified soluble transferase. Other investigators, however, took a more biological approach and pursued the site of synthesis (organ and subcellular). Miller *et al.* (1951; 1954), using perfused liver, showed that this organ is responsible for the synthesis of most plasma glycoproteins. Several groups used liver as a source of sialyltransferases. O'Brien *et al.* (1966) showed that rat liver sialyltransferase activity was located in a microsomal subcellular fraction. However, it was a fusion of the chemical and biological approaches to glycoprotein biosynthesis that led to the most significant advance. Schachter *et al.* (1970) identified the Golgi apparatus as the subcellular site of the sialyl, galactosyl, and N-acetylglucosaminyl:glycoprotein transferase in rat liver. This was accomplished by careful subcellular fractionation, utilizing an adaptation of the techniques that Morré and Mollenhauer (1964) had used to isolate the Golgi apparatus of plant tissues and subsequently a variety of other tissues including rat liver (Morré *et al.*, 1970).

Hudgin and Schachter (1971; 1972) have gone on to chemically

TABLE III. Substrate Specificity of Pork Liver Sialyltransferase

Acceptor	NAN incorporated, percent of lactose
Lactose (Galβ,1→4 Glc)	100
N-Acetyllactosamine (Gal)β,1→4GlcNAc)	760
(Galβ,1→3GlcNAc)	36
(Galβ,1→6GlcNAc)	37
Sialidase-treated "serum" glycoprotein	398

define the acceptor requirements and the nature of the products formed by the liver sialyltransferase(s). Table III shows the acceptor activity for the pork liver sialyltransferase. Again, as with the colostrum enzyme, the liver enzyme detects the anomeric linkage, position of linkage, and the sugar to which the β-galactosyl residue is linked. The liver enzyme of rat, pig, cow, and human synthesizes both isomers of sialyllactose ($2{\rightarrow}3$, $2{\rightarrow}6$), and the ratio of these two products varies with the age of the animal from which the liver is taken and with the pH of the incubation. This suggests that more than one sialyltransferase exists, each synthesizing one linkage. The ratio of activities of these two specific sialyltransferases varies with the embryonic or postnatal age of the animal.

D. CMP-Sialic Acid:Mucin (α-N-Acetylgalactosaminyl) Sialyltransferase

$$\text{CMP-NAN} + \text{GalNAc}\overset{\alpha}{\rightarrow}\text{R} \longrightarrow \text{NAN}{\rightarrow}\text{GalNAc}\overset{\alpha}{\rightarrow}\text{R} + \text{CMP}$$

Carlson *et al.* (1964; 1973) found another sialyltransferase in extracts of ovine submaxillary glands. This enzyme transfers sialic acid from CMP-sialic acid to sialidase-treated ovine submaxillary mucin and is presumably responsible for the final addition of sialic acid to this epithelial mucin (see Figure 8). This enzyme was found to exist partially in a soluble form after vigorous disruption of the tissue and as a consequence could be partially purified. The enzyme was purified 44-fold in 16% yield and was separated from an endogenous acceptor found in the extracts. Acceptor specificity studies demonstrated that this sialyltransferase was very different from any of the previously described enzymes. Only sialidase–treated mucins with exposed terminal α-N-acetylgalactosamine residues were active as acceptors. *No* low-molecular-weight oligosaccharide or sugar glycoside including methyl α-N-acetylgalactosaminide was active as an acceptor nor were sialidase-treated serum-type glycoproteins with terminal β-galactosyl residues. One exception to this was sialidase-treated fetuin, which was a good acceptor molecule for this enzyme. It was later found by Spiro (1970) that fetuin contains α-N-acetylgalactosamine residues linked to the hydroxyl groups of the amino acids serine and threonine (mucin type) as well as the serum-type oligosaccharide structures. The product of the reactions, using both sialidase-treated mucins and fetuin, was characterized by alkaline borohydride treatment and periodate oxidation as sialyl $\alpha,2{\rightarrow}6$ N-acetylgalactosamine $\alpha,{\rightarrow}$serine or threonine. This enzyme can utilize both CMP-NAN and CMP-NGN and transfers sialic acid to the

C_6-hydroxyl group of α-linked N-acetylgalactosamine of *high*-molecular-weight acceptor molecules.

In discussing sialyltransferases and ignoring the other glycosyltransferases involved in the biosynthesis of glycoconjugates, one is of necessity missing the "big picture." In the closing section of this chapter, an attempt will be made to broaden the perspective. It is necessary however, at this juncture to introduce some material not related to sialyltransferase.

While working on a mucin: galactosyltransferase, Schachter *et al.* (1971) found that this enzyme would transfer galactose to sialidase-treated ovine submaxillary mucin, but *not* to the native sialylated structure. This interesting finding suggests the possibility that the addition of sialic acid to oligosaccharide side chains may in some cases be a signal for chain termination. This addition of sialic acid would prevent further additions by other glycosyltransferases. The finding with pig submaxillary mucin glycosyltransferases is shown in Figure 10. After the addition of N-acetylgalactosamine to the polypeptide chain, two different glycosyltransferases can add a glycose residue, which can lead to different fates for the ultimate structure of the side chain. If sialic acid is added, *no* further sugar additions may occur and the final product is the disaccharide, sialyl α,2→6 N-acetylgalactosamine. If galactose is added, other sugar additions may occur, which may lead to the most complex oligosaccharide structure found in pig submaxillary mucin, a pentasaccharide (Carlson, 1968; Baig and Aminoff, 1972). This is

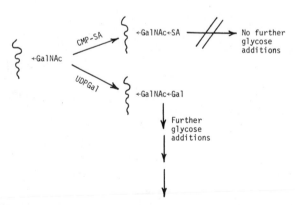

FIGURE 10. The alternative pathways for mucin biosynthesis. Sialic acid addition prevents further sugar additions. Wavy line represents polypeptide backbone of submaxillary mucin.

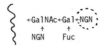

FIGURE 11. The pentasaccharide of porcine submaxillary mucin described by Baig and Aminoff (1972). Wavy line represents polypeptide backbone of mucin.

supported by the fact, that with a few exceptions, sialic acid is found at the nonreducing termini of heterosaccharide side chains. A fuller discussion of the integration of glycosyltransferases into metabolic pathways may be found in discussions by Roseman (1968) and McGuire (1970).

Based on the structure of hog submaxillary mucin oligosaccharide side chains proposed by Baig and Aminoff (1972), another sialyltransferase, responsible for the addition of sialic acid linked to β-galactosyl residues, remains to be isolated from hog submaxillary tissues (see Figure 11).

E. CMP-Sialic Acid:Ganglioside (Glycolipid) Sialytransferases

Largely through the endeavors of Kaufman and Basu (1966) and Kaufman et al. (1968) the pathways for the synthesis of the brain gangliosides (see Figure 8) have been elucidated. This work has led to the finding of three sialytransferases, each adding a specific sialic acid residue to complete the biosynthesis of the most complex ganglioside, trisialoganglioside. The enzymes were found by using embryonic chick brain homogenates. All of the glycolipid glycosyltransferases are particulate (membrane-bound) and none of the enzymes has been solubilized or purified. Keenan et al. (1974) suggested that the glycolipid glycosyltransferases are found in the Golgi subcellular fraction, as are the glycoprotein transferases.

The failure to solubilize and purify the glycolipid transferases has led to the use of indirect techniques to establish enzyme specificities. These techniques have largely been the use of differential stabilities of different enzymes toward heat or inhibitors and substrate competition studies. Although these methods are indirect, they have led to quite clear-cut results and to the formulations of the complete metabolic pathway for the biosynthesis of brain gangliosides (Kaufman et al., 1966; Roseman, 1970).

Anabolic Reactions Involving Sialic Acids

The basic findings are that three glycolipid specific sialyltransferases exist:

(1) Ceramide←Glc←Gal + CMP-NAN → Ceramide←Glc←Gal←NAN

(2) Ceramide←Glc←Gal←GalNAc←Gal + CMP-NAN →
$$\uparrow$$
$$NAN$$
→ Ceramide←Glc←Gal←GalNAc←Gal
$$\uparrow\uparrow$$
$$NANNAN$$

(3) Ceramide←Glc←Gal←GalNAc←Gal + CMP-NAN →
$$\uparrow\uparrow$$
$$NANNAN$$
→ Ceramide←Glc←Gal←GalNAc←Gal
$$\uparrow\uparrow$$
$$NANNAN$$
$$\uparrow$$
$$NAN$$

The enzymes catalyzing reactions (1) and (2) both add sialic acid to β-linked galactosyl residues; however, they are distinct and separate enzymes. Again, as with the glycoprotein and mucin sialyltransferases, the enzyme discerns the fine structure of the acceptor molecule. Further, these sialyltransferases will not transfer sialic acid to β-linked galactosyl residues of glycoproteins, mucins, or oligosaccharides. The third enzyme catalyzes the addition of sialic acid to another sialic acid residue. Little is known about this enzyme since its activity is very low. It is not clear at this time whether this enzyme catalyzes the addition as shown in reaction (3) or utilizes the product of reaction (1) as its substrate.

Finally, Grimes and Robbins (1972) have suggested that one sialytransferase can utilize desialyzed fetuin (a serum glycoprotein), desialyzed bovine submaxillary mucin (a mucin), and monosialoganglioside (a glycolipid) as acceptors. This is in contrast to the results just cited. This suggestion is based on data obtained using particulate enzyme preparations from cultured fibroblasts. It is possible that the fibroblast enzymes are less "specific" than the chick brain enzymes or that the kinetic experiments which were performed may be misleading. All of the acceptor molecules used contain more than one type of acceptor end group, which certainly complicates the interpretations. This important question remains unresolved.

IV. THOUGHTS ON PHYSIOLOGICAL FUNCTION OF SIALIC ACIDS

Eylar (1965) proposed that the function of protein-bound carbohydrate was as a signal for secretion of a particular protein species. This suggestion was based on the observation that most extracellular proteins contained carbohydrate, while intracellular proteins did not. The addition of carbohydrate would either be a signal or would be an intrinsic part of the secretion mechanism. This theory has been updated by Winterburn and Phelps (1972). The major problem with this proposal is that many proteins that are secreted do not contain carbohydrate (e.g., albumin). Also, many intracellular proteins do contain carbohydrate. Nevertheless, this proposal, in one form or another, is still the basis for current hypotheses on the role of bound carbohydrates. Schachter and Rodén (1973) have modified this theory and suggest that the addition of carbohydrate is a signal for movement of molecules across membranes (internal as well as external). Those molecules which lack carbohydrate, but nevertheless have participated in such a mechanism, would have had it removed after the membrane transport step had occurred.

Roseman (1968) proposed a general mechanism for carbohydrate assembly which suggested that membrane-bound arrays of different glycosyltransferases (multiglycosyltransferase systems) were responsible for the stepwise addition of individual glycose residues. The specificity for each addition was an inherent property of the individual transferase for the acceptor and donor molecules. The overall specificity for the glycoprotein or glycolipid was the sum of the individual specificities of the glycosyltransferase enzymes. The presence or absence of the enzymes is presumably regulated at the level of primary gene expression. An analysis of this type of control mechanism as applied to mucin biosynthesis has been given by McGuire (1970). This analysis shows that glycosyltransferases which are not expressed in the structure of the oligosaccharide side chain are present in ovine submaxillary glands. The reason that these enzymes are not functional is the presence of extremely high levels of a sialyltransferase. After the addition of sialic acid to a mucin structure, no further sugar additions may occur (see Figure 10). Thus, in this particular instance, it is the quantitative levels of glycosyltransferases that determine the ultimate structure, not the qualitative presence or absence of an enzyme. Also, this provides supportive evidence for a chain termination role for sialic acid. The addition of sialic acid in serum and mucin glycoproteins appears to prevent any further glycose additions.

Roth *et al.* (1971) found that some glycosyltransferases were located on the external cell surface. Based on this finding, they proposed that rather than serving a biosynthetic role, the enzymes may serve as a "lock" in a "lock-and-key" type mechanism for intracellular recognition. Roseman (1970) further expanded this hypothesis into a general mechanism of intercellular adhesion via carbohydrate–glycosyltransferase interactions. Jamieson et al. (1971) and Bosmann (1971) found glycosyltransferases in platelet cell membranes and suggested a role in adhesion for these enzymes. Roth and White (1972) and Bosmann (1972) have found cell surface glycosyltransferases on fibroblast cell membranes in tissue culture and have found differences in the levels and properties of the enzymes when comparing normal versus transformed cells. Both groups have suggested that the cell surface glycosyltransferases may play a role in growth regulation.

One clear-cut case of carbohydrate mediated binding to cell membranes has been obtained by Morell et al. (1968; 1971), Pricer and Ashwell (1971), and Van Lenten and Ashwell (1972). This elegant study has shown that circulating serum glycoproteins are removed from the serum by the liver, when the terminal sialic acid residue is removed, exposing a β-galactosyl residue. Isolated liver plasma membranes also will bind the desialylated glycoprotein. Presumably *in vivo* a mechanism exists for the controlled release of sialic acid. When sufficient sialic acid has been removed, the glycoprotein is bound by the liver plasma membrane, ingested, and degraded. Thus sialic acid appears to be involved in the regulation of turnover of serum glycoproteins. Interestingly, the membrane-binding site must contain sialic acid to be able to bind the desialylated glycoprotein. This represents the clearest involvement of sialic acid in the regulation of any metabolic or biological event.

All of the various pieces of data and theories can be encompassed by a broader notion of the role of bound carbohydrates in general and sialic acid, specifically. The assembled glyco moieties could possess "information" or represent a code carrying "positional" information. DNA has a code for assembling amino acids into proteins. Proteins in their linear array of amino acids can fold into the correct conformation for carrying out catalytic and structural roles; so too, may bound carbohydrates carry information. This "positional" information would be a topological concept involving the correct placement of enzymes and lipids into membranes, secretion of molecules, viral binding to cells both for infection and for release and assembly, cells homing to their correct targets, turnover of proteins and cells, hormone binding, etc.

A great effort is underway, predicated on the notion that, in some way, carbohydrates of cell surfaces are important in cell–cell interactions. A great deal of work at the phenomenological level has been carried out without an underlying hypothesis for this sense of importance. The various constellations, composed of glycose residues, provide a myriad of informational possibilities. It is hoped that this exposition on sialic acid anabolism and concluding tentative hypothesis will be of assistance in drawing together the innumerable fragments of sialic acid and carbohydrate-related research efforts.

V. REFERENCES

Allen, A., and Kent, P. W., 1968, Studies on the enzymatic N-acylation of amino sugars in the sheep colonic mucosa, *Biochem. J.* **107**:589–598.
Aminoff, D., Dodyk, F., and Roseman, S., 1963, Enzymatic synthesis of colominic acid, *J. Biol. Chem.* **238**:pc1177–pc1178.
Baig, M. M., and Aminoff, D., 1972, Glycoproteins and blood group activity; oligosaccharides of serologically inactive hog submaxillary glycoproteins, *J. Biol. Chem.* **247**:6111–6118.
Bartholomew, B. A., and Jourdian, G. W., 1966, V. Colostrum sialyltransferases, *in: Methods in Enzymology,* Vol. VIII (E. F. Neufeld and V. Ginsburg, eds.), pp. 368–372, Academic Press, New York.
Bartholomew, B. A., Jourdian, G. W., and Roseman, S., 1973, The sialic acids XV. Transfer of sialic acid to glycoproteins by a sialyltransferase from colostrum, *J. Biol. Chem.* **248**:5751–5762.
Bates, C. J., Adams, W. R., and Handschumacher, R. E., 1966, Control of the formation of uridine diphospho-N-acetylhexosamine and glycoprotein synthesis in rat liver, *J. Biol. Chem.* **241**:1705–1712.
Becker, C. E., and Day, D. G., 1953, Utilization of glucosome and the synthesis of glucosamine in the rat, *J. Biol. Chem.* **201**:795–801.
Blacklow, R. S., and Warren, L., 1962, Biosynthesis of sialic acids by *Neisseria meningitidis*, *J. Biol. Chem.* **237**:3520–3526.
Blix, G., 1936, Über die Kohlenhydrat-Gruppen des Submaxillarismucins, *Hoppe-Seyl. Z.* **240**:43–54.
Blumenthal, H. J., Horowitz, S. T., Hemerline, A., and Roseman, S., 1955, Biosynthesis of glucosamine and glucosamine polymers by molds, *Bacteriol. Proc.* 137–137.
Bosmann, H. B., 1971, Platelet adhesiveness and aggregation: The collagen: Glycosyl, polypeptide:N-acetylgalactosaminyl and glycoprotein:galactosyl transferases of human platelets, *Biochem. Biophys. Res. Commun.* **43**:1118–1124.
Bosmann, H. B., 1972, Cell surface glocosyl transferases and acceptors in normal and RNA- and DNA-virus transformed fibroblasts, *Biochem. Biophys. Res. Commun.* **48**:523–529.
Brown, D. H., 1953, Action of phosphoglucomutase on D-glucosamine-6-phosphate, *J. Biol. Chem.* **204**:877–889.
Brown, D. H., 1955, The D-glucosamine-6-phosphate N-acetylase of yeast, *Biochim. Biophys. Acta* **16**:429–431.
Brunetti, P., Jourdian, G. W., and Roseman, S., 1962, The sialic acids III. Distribution and properties of animal N-acetylneuraminic aldolase, *J. Biol. Chem.* **237**:2447–2453.
Burnet, F. M., 1948, Mucins and mucoids in relation to influenza virus action. III. Inhibition of virus haemagglutination by glandular mucins, *Aust. J. Exp. Biol. Med. Sci.* **26**:371–329.

Cabib, E., Leloir, L. F., and Cardini, C. E., 1953, Uridine diphosphate acetylglucosamine, *J. Biol. Chem.* **203**:1055–1070.
Cardini, C. E., and Leloir, L. F., 1957, Enzymatic formation of acetylgalactosamine, *J. Biol. Chem.* **225**:317–324.
Carlson, D. M., 1966, Phosphoacetylglucosamine mutase from pig submaxillary gland, *in: Methods in Enzymology,* Vol. VIII (E. F. Neufeld and V. Ginsburg, eds.), pp. 179–182, Academic Press, New York.
Carlson, D. M., 1968, Structures and immunochemical properties of oligosaccharides isolated from pig submaxillary mucin, *J. Biol. Chem.* **243**:616–626.
Carlson, D. M., McGuire, E. J., Jourdian, G. W., and Roseman, S., 1964, Submaxillary gland sialyltransferases, *Fed. Proc.* **23**:380.
Carlson, D. M., Jourdian, G. W., and Roseman, S., 1973a, The sialic acids XIV. Synthesis of sialyl-lactose by a sialyltransferase from rat mammary gland, *J. Biol. Chem.* **248**:5742–5750.
Carlson, D. M., McGuire, E. J., Jourdian, G. W., and Roseman, S., 1973b, The sialic acids XVI. Isolation of a mucin sialyltransferase from sheep submaxillary gland, *J. Biol. Chem.* **248**:5763–5773.
Castellani, A. A., and Zambotti, V., 1956, Enzymatic formation of hexosamine in epiphyseal cartilage homogenate, *Nature* **178**:313–314.
Chou, T. C., and Soodak, M., 1952, The acetylation of D-glucosamine by pigeon liver extracts, *J. Biol. Chem.* **196**:105–109.
Clark, J. S., and Pasternak, C. A., 1962, The regulation of aminosugar metabolism in *Bacillus subtillus, Biochem. J.* **84**:185–191.
Comb, D. G., and Roseman, S., 1958a, Glucosamine metabolism IV. Glucosamine-6-phosphate deaminase, *J. Biol. Chem.* **232**:807–827.
Comb, D. G., and Roseman, S., 1958b, Enzymic synthesis of *N*-acetyl-D-mannosamine, *Biochim. Biophys. Acta* **29**:653–654.
Comb, D. G., and Roseman, S., 1958c, Composition and enzymatic synthesis of *N*-acetylneuraminic acid (sialic acid), *J. Am. Chem. Soc.* **80**:497–499.
Comb, D. G., and Roseman, S., 1960, The sialic acids I. The structure and enzymatic synthesis of *N*-acetylneuraminic acid, *J. Biol. Chem.* **235**:2529–2537.
Davidson, E. A., Blumenthal, H. J., and Roseman, S., 1957, Glucosamine metabolism II. Studies on glucosamine-6-phosphate *N*-acetylase, *J. Biol. Chem.* **226**:125–133.
Eylar, E. H., 1965, On the biological role of glycoproteins, *J. Theor. Biol.* **10**:89–113.
Ghosh, S., and Roseman, S., 1961, Enzymatic phosphorylation of *N*-acetyl-D-mannosamine, *Proc. Natl. Acad. Sci. U.S.* **47**:955–958.
Ghosh, S., and Roseman, S., 1965a, The sialic acids *N*-acyl-D-glucosamine 6-phosphate 2-epimerase, *J. Biol. Chem.* **240**:1525–1530.
Ghosh, S., and Roseman, S., 1965b, The sialic acids IV. *N*-Acyl-D-glucosamine 6-phosphate 2-epimerase, *J. Biol. Chem.* **240**:1531–1536.
Ghosh, S., Blumenthal, H. J., Davidson, E., and Roseman, S., 1969, Glucosamine metabolism V. Enzymatic synthesis of glucosamine-6-phosphate. *J. Biol. Chem.* **235**:1265–1273.
Glaser, L., and Brown, D. H., 1955, The enzymatic synthesis *in vitro* of hyaluronic acid chains, *Proc. Natl. Acad. Sci. U.S.* **41**:253–260.
Gottschalk, A., 1972, Historical introduction, *in: Glycoproteins, Their Composition, Structure and Functions,* Part A (A. Gottschalk, ed.), pp. 1–23, Elsevier, Publishing Co. New York.
Gottschalk, A., and Lind, P. G., 1949, Product of the interaction between influenza virus enzyme and ovomucin, *Nature (Lond.),* **164**:232–233.

Grimes, W. J., and Robbins, P. W., 1972, Virus control of the synthesis of glucosidic linkages—glycoprotein and glycolipid sialic acid transferases from normal and SV40 transformed Balb/c cells, *in: Biochemistry of the Glycosidic Linkage* (R. Piras and H. G. Pontis, eds.), pp. 113–134, Academic Press, New York.

Hansen, R. G., Albrecht, G. J., Bass, S. T., and Seifert, L. L., 1966a, UDP-glucose pyrophosphorylase (crystalline) from liver, *in: Methods in Enzymology,* Vol. VIII (E. F. Neufeld and V. Ginsburg, eds.), pp. 248–253, Academic Press, New York.

Hansen, R. G., Verachtert, H., Rodriguez, P., and Bass, S. T., 1966b, CDP-hexose pyrophosphorylase from liver, *in: Methods in Enzymology,* Vol. VIII (E. F. Neufeld and V. Ginsburg, eds.), pp. 269–275, Academic Press, New York.

Heimer, R., and Meyer, K., 1956, Studies on sialic acid of submaxillary mucoid, *Proc. Natl. Acad. U.S.* **42:**728–734.

Hirst, G. K., 1941, Agglutination of red cells by allantoic fluid of chick embryos infected with influenza virus, *Science* **94:**22–23.

Hudgin, R. L., and Schachter, H., 1971, Porcine sugar nucleotide:glycoprotein glycosyltransferases I. Blood serum and liver sialyltransferase, *Can. J. Biochem.* **49:**829–837.

Hudgin, R. L., and Schachter, H., 1972, Evidence for two CMP-N-acetylneuraminic acid:lactose sialyltransferases in rat, procine, bovine, and human liver, *Can. J. Biochem.* **50:**1024–1028.

Hurlbert, R. B., and Potter, V. R., 1954, The conversion of orotic acid-6-C^{14} to uridine nucleotides, *J. Biol. Chem.* **209:**1–21.

Jamieson, G. A., Urban, C. L., and Barber, A. J., 1971, Enzymatic basis for platelet:collagen adhesion as the primary step in haemostasis, *Nat. New Biol.* **234:**5–7.

Jourdian, G. W., Carlson, D. M., and Roseman, S., 1963, The enzymatic synthesis of sialyl-lactose, *Biochem. Biophys. Res. Commun.* **10:**352–358.

Jourdian, G. W., Swanson, A. L., Watson, D., and Roseman, S., 1964, Isolation of sialic acid 9-phosphatase from human erythrocytes, *J. Biol. Chem.* **239:**pc2714–pc2715.

Kaufman, B., and Basu, S., 1966, Embryonic chick brain sialyl-transferases, *in: Methods in Enzymology,* Vol. VIII (E. F. Neufeld and V. Ginsburg, eds.), pp. 365–368, Academic Press, New York.

Kaufman, B., Basu, S., and Roseman, S., 1966, Studies on the biosynthesis of gangliosides, *in: Proceedings of the Third International Symposium on the Cerebral Sphingolipidoses* (S. M. Aronson and B. W. Volk, eds.), pp. 193–213, Pergamon Press, New York.

Kaufman, B., Basu, S., and Roseman, S., 1968, Enzymatic synthesis of disialogangliosides from monosialogangliosides by sialyltransferases from embryonic chick brain, *J. Biol. Chem.* **243:**5804–5807.

Kean, E. L., 1970, Nuclear cytidine 5'-monophosphate synthetase, *J. Biol. Chem.* **245:**2301–2308.

Kean, E. L., and Roseman, S., 1966, The sialic acids X. Purification and properties of cytidine 5'-monophosphosialic acid synthetase, *J. Biol. Chem.* **241:**5643.

Keenan, T. W., Morré, D. J., and Basu, S., 1974, Ganglioside biosynthesis: Concentration of glycosphingolipid glycosyltransferases in golgi apparatus from rat liver, *J. Biol. Chem.* **249:**310–315.

Kizer, D. H., and McCoy, T. A., 1959, The synthesis of hexosamines in tumor homogenates, *Cancer Res.* **19:**307–310.

Klenk, E., 1941, Neuraminsäure, das spaltprodukt eines neuen gehirnlipoids, *Z. Physiol. Chem.* **268:**50–58.

Kornberg, A., 1950, Reversible enzymatic synthesis of diphosphopyridine nucleotide and inorganic pyrophosphate, *J. Biol. Chem.* **182:**779–793.

Kornfeld, R., 1967, Studies on L-glutamine: D-fructose 6-phosphate amidotransferase I. Feedback inhibition by uridine diphosphate-N-acetylglucosamine, *J. Biol. Chem.* **242:**3135–3141.

Kornfeld, S., Kornfeld, R., Neufeld, E., and O'Brien, P. J., 1964, The feedback control of sugar nucleotide biosynthesis in liver, *Proc. Natl. Acad. Sci. U.S.* **52:**371–379.

Kuhn, R., and Brossmer, R., 1958, Die Konstitution der Lactaminsäurelactose; α-Ketosidase-Wirkung von Viren der Influenza-Gruppe, *Angew. Chem.* **70:**25–26.

Kundig, W., Ghosh, S., and Roseman, S., 1966, The sialic acids VII. N-acyl-D-mannosamine kinase from rat liver, *J. Biol. Chem.* **241:**5619–5626.

Kundig, F. D., Aminoff, D., and Roseman, S., 1971, The sialic acids XII. Synthesis of colominic acid by a sialyltransferase from *Escherichia coli* K-235, *J. Biol. Chem.* **246:**2543–2550.

Lehrs, H., 1930, Ueber, Gruppenspezifische Eigenschaften des menschlichen Speichels, *Z. Immunitätsforschung* **66:**175–192.

Leloir, L. F., and Cardini, C. E., 1953, The biosynthesis of glucosamine, *Biochim. Biophys. Acta.* **12:**15–22.

Leloir, L. F., and Cardini, C. E., 1956, Enzymes acting on glucosamine phosphates, *Biochim. Biophys. Acta* **20:**33–42.

Lowther, D. A., and Rodgers, H. J., 1953, The relation of glutamine to the synthesis of hyaluronate or hyaluronate-like substances by *Streptococci*, *Biochem. J.* **53:**XXXIX.

Lowther, D. A., and Rodgers, H. J., 1955, Biosynthesis of hyaluronate, *Nature* **175:**435.

Lowther, D. A., and Rodgers, H. J., 1956, The role of glutamine in the biosynthesis of hyaluronate by *Streptococcal* suspensions, *Biochem. J.* **62:**304–314.

Maley, F., and Lardy, H. A., 1956, Formation of UDPGla and related compounds by the soluble fraction of liver, *Science* **124:**1207–1208.

Maley, F., Maley, G. F., and Lardy, H. A., 1956, The synthesis of α-D-glucosamine-1-phosphate and N-acetyl-α-D-glucosamine-1-phosphate. Enzymatic formation of uridine diphosphoglucosamine, *J. Am. Chem. Soc.* **78:**5303–5307.

Marshall, R. D., 1972, Glycoproteins, in: *Annual Review of Biochemistry,* Vol. 41 (E. E. Snell, ed.), pp. 673–702, Annual Reviews, Palo Alto.

McClelland, L., and Hare, R., 1941, Adsorption of influenza virus by red cells and a new *in vitro* method of measuring antibodies for influenza virus, *Can. J. Pub. Health* **32:**530–538.

McGuire, E. J., 1970, Biosynthesis of submaxillary mucins, in: *Blood and Tissue Antigens* (David Aminoff, ed.), pp. 461–478, Academic Press, New York.

Miller, L. L., and Bale, W. F., 1954, Synthesis of all plasma protein fractions except gamma globulins by the liver. The use of zone electrophoresis and lysine-ϵ-C[14] to define the plasma proteins synthesized by the isolated perfused liver, *J. Exp. Med.* **99:**125–132.

Miller, L. L., Bly, C. G., Watson, M. L., and Bale, W. F., 1951, The dominant role of the liver in plasma protein synthesis. A direct study of the isolated perfused rat liver with the aid of lysine-ϵ-C[14], *J. Exp. Med.* **94:**431–453.

Morell, A. G., Irvine, R. A., Sternlieb, I. H., and Ashwell, G., 1968, Physical and chemical studies on ceruloplasmin V. Metabolic studies on sialic acid-free ceruloplasmin *in vivo*, *J. Biol. Chem.* **243:**155–159.

Morell, A. G., Gregoridadis, G., Scheinberg, I. H., Hickman, J., and Ashwell, G., 1971, The role of sialic acid in determining the survival of glycoproteins in the circulation, *J. Biol. Chem.* **246:**1461–1467.

Morré, D. J., and Mollenhauer, H. H., 1964, Isolation of the golgi apparatus from plant cells, *J. Cell Biol.* **23:**295–305.

Morré, D. J., Hamilton, R. L., Mollenhauer, H. H., Mahley, R. W., Cunningham, W. P., Cheetham, R. D., and Lequire, V. S., 1970, Isolation of a golgi apparatus-rich fraction from rat liver. I. Method and morphology, *J. Cell Biol.* **44**:484–491.

O'Brien, P. J., Canady, M. R., Hall, C. W., and Neufield, E. F., 1966, Tranfer of *N*-acetylneuraminic acid to in complete glycoproteins associated with microsomes, *Biochim. Biophys. Acta* **117**:331–341.

Park, J. T., and Strominger, J. L., 1957, Mode of action of penicillin, *Science*, **125**:99–101.

Pogell, B. M., and Gryder, R. M., 1957, Enzymatic synthesis of glucosamine-6-phosphate in rat liver, *J. Biol. Chem.* **228**:701–712.

Pogell, B. M., and Gryder, R. M., 1969, Further studies on glucosamine-6-phosphate synthesis by rat liver, *J. Biol. Chem.* **235**:558–562.

Pricer, W. E., Jr., and Ashwell, G., 1971, The binding of desialylated glycoproteins by plasma membranes of rat liver, *J. Biol. Chem.* **246**:4825–4833.

Putkonen, T., 1930, Über die blutgruppen-spezifizität des fruchtwassers, *Acta Path. Microbiol. Scand.* **5**:64–65.

Reissig, J. L., 1956, Phosphoacetylglucosamine mutase of *Neurospora*, *J. Biol. Chem.* **219**:753–767.

Rieder, S. V., and Buchanan, J. M., 1958, Studies on the biological formation of glucosamine *in vivo*. I. Origin of the carbon chain, *J. Biol. Chem.* **232**:951–957.

Roseman, S., 1962a, Metabolism of sialic acids and D-mannosamine, *Fed. Proc.* **21**:1075–1083.

Roseman, S., 1962b, Enzymatic synthesis of cytidine 5'-monophosphosialic acids, *Proc. Natl. Acad. Sci. U.S.* **48**:437–441.

Roseman, S., 1968, Biochemistry of glycoproteins and related substances, in: *Proceedings of the Fourth International Conference on Cystic Fibrosis of the Pancreas (Mucoviscidosis)*, Part II, (E. Rossi and E. Stoll, eds.), pp. 244–269, Karger, Basel.

Roseman, S., 1970, The synthesis of complex carbohydrates by multiglycosyltransferase systems and their potential function in intercellular adhesion, *Chem. Phys. Lipids.* **5**:270–297.

Roseman, S., Moses, F. E., Ludowieg, J., and Dorfman, A., 1953, The biosynthesis of hyaluronic acid by group A *Streptococcus* I. Utilization of 1-C^{14}-glucose, *J. Biol. Chem.* **203**:213–225.

Roseman, S., Davidson, E., Blumenthal, H. J., and Dockrill, M., 1958, Conversion of fructose-6-phosphate to glucosamine-6-phosphate by microbial and mammalian enzymes, *Bacteriol. Proc.* 107.

Roseman, S., Hayes, F., and Ghosh, S., 1960, Enzymatic synthesis of *N*-acetylmannosamine 6-phosphate, *Fed. Proc.* **19**:85.

Roseman, S., Distler, J. J., Moffatt, J. G., and Khorana, H. G., 1961a, Nucleoside polyphosphates. XI. An improved general method for the synthesis of nucleotide coenzymes. Synthesis of uridine-5', cytidine-5', and guanosine-5' diphosphate derivatives, *J. Am. Chem. Soc.* **83**:659–663.

Roseman, S., Jourdian, G. W., Watson, D., and Rood, R., 1961b, Enzymatic synthesis of sialic acids 9-phosphates, *Proc. Natl. Acad. Sci. U.S.* **47**:958–961.

Roth, S., and White, D., 1972, Intercellular contact and cell-surface galactosyl transferase activity, *Proc. Natl. Acad. Sci. U.S.* **69**:485–489.

Roth, S., McGuire, E. J., and Roseman, S., 1971, Evidence for cell-surface glycosyltransferases: Their potential role in cellular recognition, *J. Cell Biol.* **51**:536–547.

Salo, W. L., and Fletcher, H. G., Jr., 1970a, Synthesis of 2-acetamido-2-deoxy-α-D-mannopyranosyl phosphate and uridine 5'-(2-acetamido-2-deoxy-α-D-mannopyranosyl dipotassium pyrophosphate), *Biochemistry* **9**:878–881.

Salo, W. L., and Fletcher, H. G., Jr., 1970b, Studies on the mechanism of action of uridine diphosphate N-acetylglucosamine 2-epimerase, *Biochemistry* **9**:882–885.

Schachter, H., and Rodén, L., 1973, The biosynthesis of animal glycoproteins, *in: Metabolic Conjugation and Metabolic Hydrolysis*, Vol. III (W. H. Fishman, ed.), pp. 1–149, Academic Press, New York.

Schachter, H., Jabbal, I., Hudgin, R. L., Pinteric, L., McGuire, E. J., and Roseman, S., 1970, Intracellular localization of liver sugar nucleotide glycoprotein glycosyltransferases in a golgi-rich fraction, *J. Biol. Chem.* **245**:1090–1100.

Schachter, H., McGuire, E. J., and Roseman, S., 1971, Sialic acids XIII. A uridine diphosphate D-galactose:mucin galactosyltransferase from porcine submaxillary gland, *J. Biol. Chem.* **246**:5321–5328.

Schauer, R., 1970a, The biosynthesis of N-glycoloylneuraminic acid by an ascorbate- or NADPH-dependent, N-acetyl hydroxylating "N-acetylneuraminate:O_2-oxidoreductase" in homogenates of porcine submaxillary gland, *Hoppe-Seyl. Z.* **351**:783–791.

Schauer, R., 1970b, Studies on the subcellular site of the biosynthesis of N-glycoloylneuraminic acid in porcine submaxillary gland, *Hoppe-Seyl. Z.* **351**:1353–1358.

Schauer, R., 1970c, Biosynthesis of N-acetylneuraminic acids I: Incorporation of [^{14}C] acetate into slices of the submaxillary salivary glands of ox and horse, *Hoppe-Seyl. Z.* **351**:595–602.

Schauer, R., 1970d, Biosynthesis of N-acetyl-O-acetylneuraminic acid II: Studies on the substrate and intracellular localization of the bovine acetyl-coenzyme A:N-acetylneuraminate-7-and 8-O-acetyltransferase, *Hoppe-Seyl. Z.* **351**:749–758.

Schauer, R., 1972, Biosynthesis of glycoprotein precursors and the mechanism of their assembly, *Biochem. J.* **128**:112p–114p.

Schauer, R., and Wember, M., 1971, Hydroxylation and O-acetylation of N-acetylneuraminic acid bound to glycoproteins of isolated subcellular membranes from porcine and bovine submaxillary glands, *Hoppe-Seyl. Z.* **352**:1282–1290.

Schauer, R., Wember, M., and Ferreira do Amaral, C., 1972, Synthesis of CMP-glycosides of radioactive N-acetyl-, N-glycoloyl-, N-acetyl-7-O-acetyl- and N-acetyl-8-O-acetylneuraminic acids by CMP-sialate synthetase from bovine submaxillary glands, *Hoppe-Seyl. Z.* **353**:883–886.

Schoop, H. J., Schauer, R., and Faillard, H., 1969, Die oxydative entstehung von N-glykolyl-neuraminsäure aus N-acetyl-neuraminsäure, *Hoppe-Seyl. Z.* **350**:155–162.

Schrecker, A. W., and Kornberg, A., 1950, Reversible enzymatic synthesis of flavin-adenine dinucleotide, *J. Biol. Chem.* **182**:795–803.

Smith, E. E. B., Munch-Peterson, A., and Mills, G. T., 1953, Pyrophosphorolysis of uridine diphosphoglucose and "UDPX" by a rat liver nuclear fraction, *Nature* **172**:1038–1039.

Spiro, R. G., 1959, Studies on the biosynthesis of glucosamine in the intact rat, *J. Biol. Chem.* **234**:742–748.

Spiro, R. G., 1970, Glycoproteins, *in: Annual Review of Biochemistry*, Vol. 39 (E. E. Snell, ed.), pp. 599–638, Annual Reviews, Palo Alto.

Spiro, M. J., and Spiro, R. G., 1968, Glycoprotein biosynthesis: studies on thyroglobulin thyroid sialyltransferase, *J. Biol. Chem.* **243**:6520–6528.

Spivak, C. T., and Roseman, S., 1966, UDP-N-acetyl-D-glucosamine 2'-epimerase, *in: Methods in Enzymology*, Vol. IX (W. A. Wood, ed.), pp. 612–615, Academic Press, New York.

Tabor, H., Mehler, A. H., and Stadtman, E. R., 1953, The enzymatic acetylation of amines, *J. Biol. Chem.* **204**:127–138.

Topper, Y. J., and Lipton, M. M., 1953, The biosynthesis of a *Streptococcal* capsular polysaccharide, *J. Biol. Chem.* **203:**135–142.

Van Lenten, L., and Ashwell, G., 1972, The binding of desialylated glycoproteins by plasma membranes of rat liver—development of a quantitative inhibition assay, *J. Biol. Chem.* **247:**4633–4640.

Wagner, R. R., and Cynkin, M. A., 1971, Glycoprotein biosynthesis: Incorporation of glycosyl groups into endogenous acceptors in a golgi-apparatus-rich fraction of liver, *J. Biol. Chem.* **246:**143–151.

Warren, L., 1972, The biosynthesis and metabolism of amino sugars and amino-sugar-containing compounds, in: *Glycoproteins, Their Composition, Structure, and Function,* Part B (A. Gottschalk, ed.), pp. 1097–1126, Elsevier Publishing Co., New York.

Warren, L., and Blacklow, R. S., 1962, The biosynthesis of cytidine 5′-monophospho-N-acetylneuraminic acid by an enzyme from *Neisseria meningitidis, J. Biol. Chem.* **237:**3527–3534.

Warren, L., and Felsenfeld, H., 1961a, The biosynthesis of N-acetylneuraminic acid, *Biochem. Biophys. Res. Commun.* **4:**232–235.

Warren, L., and Felsenfeld, H., 1961b, N-acetylmannosamine-6-phosphate and N-acetylneuraminic acid-9-phosphate as intermediates in sialic acid biosynthesis, *Biochem. Biophys. Res. Commun.* **5:**185–190.

Warren, L., and Felsenfeld, H., 1962, The biosynthesis of sialic acids, *J. Biol. Chem.* **237:**1421–1431.

Watkins, W. M., 1966, Blood-group substances, *Science* **152:**172–181.

Watson, D. R., Jourdian, G. W., and Roseman, S., 1966, The sialic acids VIII. Sialic acid-9-phosphate synthetase, *J. Biol. Chem.* **241:**5627–5636.

Winterburn, P. J., and Phelps, C. F., 1970, Relevance of feedback inhibition applied to the biosynthesis of hexosamines, *Nature (Lond.)* **228:**1311–1313.

Winterburn, P. J., and Phelps, C. F., 1971a, Purification and some kinetic properties of rat liver glucosamine synthetase, *Biochem. J.* **121:**701–709.

Winterburn, P. J., and Phelps, C. F., 1971b, Studies on the control of hexosamine biosynthesis by glucosamine synthetase, *Biochem. J.* **121:**711–720.

Winterburn, P. J., and Phelps, C. F., 1971c, Binding of substrates and modifiers to glucosamine synthetase, *Biochem. J.* **121:**721–730.

Winterburn, P. J., and Phelps, C. F., 1972, The significance of glycosylated proteins, *Nature (Lond.),* **236:**147–151.

Chapter 5

Catabolism of Sialyl Compounds in Nature

Kunihiko Suzuki

I. INTRODUCTION

While sialic acid exists in free form, or as a part of relatively small heterosaccharide molecules, the major portion of naturally occurring sialic acid is present as a constituent of either glycoproteins or gangliosides.* Distribution of these sialic-acid-containing compounds in nature and their biosynthesis, are dealt with in Chapters 2 and 4. As for any other chemical constituents of biological systems, sialic acids exist in a metabolically dynamic state which, naturally, is the net result of constant biosynthesis and degradation.

This chapter examines current knowledge concerning the enzymatic machinery of degradation of sialoglycoproteins and sialoglycolipids (gangliosides). The subject will be reviewed in three successively more complex and integrated phases: (1) enzymatic pathways of degradation, (2) the intracellular mechanism of degradation, and (3) the potential physiological or functional implications of degradation of sialic-acid-containing compounds.

---

* The nomenclature of gangliosides used in this chapter is that of Svennerholm (1963), see Chapter 2.

KUNIHIKO SUZUKI • The Saul R. Korey Department of Neurology, Department of Neuroscience, and the Rose F. Kennedy Center for Research in Mental Retardation and Human Development, Albert Einstein College of Medicine, Bronx, New York 10461.

II. PATHWAYS OF DEGRADATION

A. Degradation of Gangliosides

1. **Degradative Pathways.** Gangliosides are glycosphingolipids containing sialic acid. The pathway of physiological degradation of gangliosides and related hexosylceramides has been under intensive study in recent years, and has largely been clarified through characterization of enzymes involved in individual degradative steps. The sialic acid may be N-acetylneuraminic acid or N-glycolylneuraminic acid. The structures of gangliosides are diverse, the simplest containing one residue each of ceramide, hexose, and sialic acid—sialylgalactosylceramide (Kuhn and Wiegandt, 1964; Siddiqui and McCluer, 1968; Ledeen et al., 1973), and the most complex to date, containing ceramide, four neutral sugars, and five sialic acid residues (Ishizuka and Wiegandt, 1972) (cf. Chapter 2). Degradation of gangliosides appears to proceed exclusively through sequential removal of terminal carbohydrate moieties from the nonreducing end (Figure 1). The first step in degradation of all gangliosides containing more than one sialic acid residue is removal of sialic acid by action of sialidase. This process continues until the last sialic acid moiety is left attached to the internal galactose residue as in G_{M1}- or G_{M2}-ganglioside (see Chapter 2). When the galactose residue is not internal, as in G_{M3}- or G_{D3}-gangliosides, cleavage of sialic acid by sialidase continues until sialic acid is completely removed. There is no evidence at the present time that different sialidases are required for sialic acid removal from polysialogangliosides. However, when sialidase can cleave either of two terminal sialic acid residues, such as those in G_{T1}-ganglioside, that on the terminal galactose appears to be preferentially cleaved (Svennerholm, 1963; Öhman et al., 1970). Most preparations of sialidase, including those from mammalian sources, are unable to cleave the sialic acid moiety of G_{M1}- and G_{M2}-gangliosides. This resistance to cleavage by sialidase is probably due to steric hindrance by the N-acetylgalactosamine, which is also on the internal galactose, but see Chapter 10. This supposition is supported by the fact that, when galactose and N-acetylgalactosamine are removed from the internal galactose, the remaining sialic acid residue becomes susceptible to sialidase. Kolodny et al., (1971), however, described an unusual sialidase from mammalian tissues that is capable of cleaving sialic acid from G_{M2}-ganglioside to form its asialo derivative, N-acetylgalactosaminylgalactosyl-glucosylceramide. The level of total activity was much lower than that of the more common sialidase, and it is not clear whether or

Catabolism of Sialyl Compounds in Nature

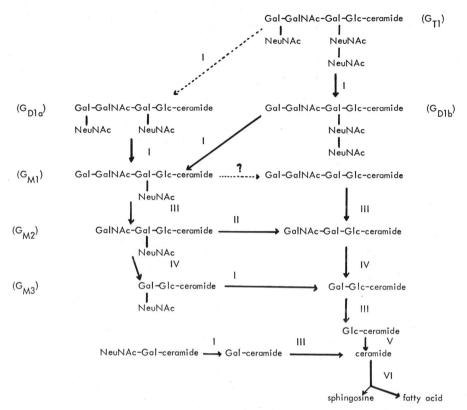

FIGURE 1. Degradative pathway of major gangliosides. Trivial names of gangliosides according to the Svennerholm nomenclature (see Chapter 4) are given in parentheses. The hydrolytic enzymes involved are I, sialidase; II, sialidase (Kolodny et al., 1971); III, β-galactosidase; IV, N-acetyl-β-galactosaminidase; V, β-glucosidase; VI, ceramidase. Refer to text for details.

not this sialidase is active toward other gangliosides or sialic-acid-containing glycoproteins (cf. Chapter 10).

Normal degradation of G_{M1}-ganglioside appears to proceed by sequential removal of the terminal galactose, and then of N-acetylgalactosamine, prior to cleavage of the last sialic acid. This can be inferred from the fact that, in genetic disorders in which β-galactosidase or N-acetyl-β-galactosaminidase are lacking, abnormal accumulations of the gangliosides, G_{M1} and G_{M2}, respectively, occur with only minor increases of their asialo derivatives (refer to Chapter 6). Once the ganglioside molecule is degraded to G_{M3}-ganglioside (sialylgalactosyl-glucosylceramide or hematoside), the last sialic acid is removed by sialidase presumably because the steric block due to substitution of the internal galactose by N-acetylgalactosamine no longer exists. At this step, the compound loses its last sialic acid to become a neutral glycosphingolipid, lactosylceramide. Lactosylceramide is further degraded to glucosylceramide by β-galactosidase (Sandhoff et al., 1964; Gatt and Rapport, 1966a,b; Radin et al., 1969; Suzuki and Suzuki, 1974a; Miyatake and Suzuki, 1974a). Glucosylceramide in turn is degraded to ceramide by β-glucosidase (Brady et al., 1965; Gatt, 1966a). An amidase, known as ceramidase, then cleaves ceramide to its two constituents, sphingosine and fatty acid (Gatt, 1966b; Yavin and Gatt, 1969).

There are a few gangliosides that have unusual structures and do not fit the general degradation scheme given above. For example, gangliosides containing N-acetyl-D-glucosamine, instead of the usual N-acetylgalactosamine, have been found in human plasma (Yu and Ledeen, 1972), platelets (Marcus et al., 1972), bovine spleen and kidney (Wiegandt and Bücking, 1970), and peripheral nerve (Li et al., 1973a). can reasonably be assumed that the degradative pathway for these glucosamine-containing gangliosides is essentially the same as for the galactosamine-containing analogs. Sialylgalactosylceramide is susceptible to sialidase and yields galactosylceramide (Ledeen et al., 1973). Galactosylceramide is then degraded to galactose and ceramide by a specific β-galactosidase (Bowen and Radin, 1968a,b, 1969).

2. Degradative Enzymes. While the degradative pathway of gangliosides has been reasonably well elucidated as outlined above, there are still a number of points of uncertainty regarding the enzymes that catalyze their degradation. Since the subject of sialidase is dealt with in detail in Chapter 10, other enzymes involved in ganglioside degradation will be discussed here.

β-Galactosidase. As is clear from the above outline and Figure 1, there are three steps in ganglioside degradation in which β-galactosidase is involved: removal of galactose from G_{M1}-ganglioside, from lactosylceramide, and from galactosylceramide. β-Galactosidase, partially purified from rat brain by Gatt and Rapport (1966a), showed hydrolytic activities toward G_{M1}-ganglioside, its asialo derivative, and lactosylceramide (Gatt and Rapport, 1966b; Gatt, 1967). More detailed studies on G_{M1}-ganglioside β-galactosidase have been reported for human liver and brain (Sloan et al., 1971; Ho, et al., 1973; Suzuki and Suzuki, 1974a; Norden and O'Brien, 1974), rat brain and other organs (Miyatake and Suzuki, 1974a), and rabbit brain (Callahan and Gerrie, 1974). At present, available evidence is consistent with presumption of the occurrence of a single β-galactoside galactosylhydrolase catalyzing degradation of both G_{M1}-ganglioside and its asialo derivative. The β-galactosidase that hydrolyzes lactosylceramide was studied by Gatt (1966b) and Radin et al. (1969) in mammalian brains. The enzyme has received closer attention more recently, and partial purification and characterization has been described for the enzyme from human liver (Suzuki and Suzuki, 1974a), rat brain and other organs (Miyatake and Suzuki, 1975) and rabbit brain (Callahan and Gerrie, 1974). Galactosylceramide β-galactosidase of the brain was characterized by Bowen and Radin (1968a,b, 1969). Recent investigations have provided evidence that the same enzyme also catalyzes hydrolysis of galactosylsphingosine (Miyatake and Suzuki, 1972a,b, 1974b) and monogalactosyldiglyceride (Wenger et al., 1973).

What is not clear at the present time is the possible interrelationship among the three β-galactosidases. The strongest evidence for three distinct enzymes for hydrolysis of G_{M1}-ganglioside (and asialo G_{M1}-ganglioside), lactosylceramide, and galactosylceramide comes from the fact that three distinct genetic disorders are known, each showing specific deficiency of a particular β-galactosidase: globoid cell leukodystrophy (galactosylceramide β-galactosidase) (Suzuki and Suzuki, 1970), lactosylceramidosis (lactosylceramide β-galactosidase) (Dawson and Stein, 1970), and G_{M1}-gangliosidosis (G_{M1}-ganglioside β-galactosidase) (Okada and O'Brien, 1968). However, abnormal electrofocusing patterns of presumably unaffected β-galactosidases were found in livers of patients with globoid cell leukodystrophy (Suzuki and Suzuki, 1974b) and G_{M1}-gangliosidosis (Suzuki and Suzuki, 1974c), suggesting a close interrelationship among the β-galactosidases in these diseases. Furthermore, hepatic lactosylceramide β-galactosidase was found to be deficient in G_{M1}-gangliosidosis and nearly normal in globoid cell leukodystrophy by one laboratory (Suzuki and Suzuki, 1974b,c), while another laboratory found exactly the opposite (Wenger et al., 1974). This apparent contradiction was resolved most recently by a finding of two genetically distinct lactosylceramide β-galactosidases, one related to, if not identical with, galactosylceramide β-galactosidase, and the other, genetically related to G_{M1}-ganglioside β-galactosidase (Tanaka and Suzuki, 1975). (See also Note Added in Proof, p. 175).

N-Acetyl-β-hexosaminidase. Degradation of G_{M2}-ganglioside and its asialo derivative requires the action of N-acetyl-β-hexosaminidase. N-acetyl-β-hexosaminidase from most mammalian sources can be separated into two major fractions, A and B (Robinson and Stirling, 1968). In contrast with G_{M1}-ganglioside β-galactosidase, the enzymes show different substrate specificities. The asialo derivative of G_{M2}-ganglioside is degraded readily by either hexosaminidase A or B, while G_{M2}-ganglioside is hydrolyzed only by the A component, and at a much slower rate (Sandhoff and Wässle, 1971). The question of the substrate specificity and the exceedingly low rate of G_{M2}-ganglioside hydrolysis, however, is not yet resolved. Bach and Suzuki (1974) provided evidence that, for hexosaminidase A, only the most acidic subfraction is capable of degrading G_{M2}-ganglioside. In contrast, Tallman et al. (1974) reported that hexosaminidase B can hydrolyze G_{M2}-ganglioside under certain conditions. The maximum rate of G_{M2}-ganglioside hydrolysis so far achieved in *in vitro* assay systems is less than 1% of that for G_{M1}-ganglioside, G_{M3}-ganglioside, or lactosylceramide hydrolysis. Therefore, degradation of G_{M2}-ganglioside to G_{M3}-ganglioside may well represent the rate-limiting step for ganglioside degradation in general. An interesting finding related to this problem is the report by Li et al. (1973b) on human liver enzymes that a heat-stable factor, probably a glycoprotein, is required for active hydrolysis of G_{M2}-ganglioside.

N-Acetyl-β-hexosaminidase A is itself a sialylated glycoprotein, and many reports have indicated that hexosaminidase A can be converted by sialidase to another form which behaves similarly to hexosaminidase B in electrophoresis, and that hexosaminidase B might be a precursor of hexosaminidase A (Robinson and Stirling, 1968; Goldstone et al., 1971; Sandhoff et al., 1971; Murphy and Craig, 1972; Tallman and Brady, 1972; Snyder et al., 1972). However, characterization of the desialylated product did not include substrate specificity studies, and there are instances where sialidase treatment failed to affect the conversion (Srivastava et al., 1974a, Suzuki, unpublished observation). While such failure might be due to variations in sialidase preparation (Murphy and Craig, 1974), a recent series of studies by Beutler and coworkers appears to argue strongly that hexosaminidase B is not the desialylated form of hexosaminidase A because the two hexosaminidases have different amino acid compositions and different carboxy-terminal amino acids (Srivastava et al., 1974a,b; Srivastava and Beutler, 1974). These authors have concluded that hexosaminidase A and B share a common subunit, and that the other subunits in the two hexosaminidases are different.

β-Glucosidase. Hydrolysis of glucosylceramide by β-glucosidase was investigated by Brady et al. (1965) in human spleen, and by Gatt (1966a) in bovine brain. More recently the enzyme from human placenta was purified 4000-fold to homogeneity (Pentchev et al., 1973). It was most active toward the natural substrate, glucosylceramide. Ho (1973) reported that glucosylceramide β-glucosidase from human spleen is associated with the "acid" β-glucosidase fraction. For full activity, two components, a membrane-bound factor C, and a heat-stable, soluble, acidic glycoprotein factor P were required. Either component was

inactive by itself but the activity could be reconstituted by combining the two factors. Such reconstitution of the activity appeared to be dependent on acidic phospholipids (Ho and Light, 1973).

B. Degradation of Glycoproteins

Glycoproteins constitute a second major class of naturally occurring sialic-acid-containing compounds. Unlike gangliosides, which generally have chemical structures systematically built up from a relatively small number of simple basic compounds, the structures of glycoproteins are exceedingly diverse and complex. Therefore, it is difficult to depict glycoproteins as a series of mutually interrelated compounds. The protein portions of glycoproteins are as diverse and complex as carbohydrate-free proteins in general. The chemical nature of the protein–carbohydrate linkage is similarly diverse, having an O-glycosidic linkage to glutamic acid, aspartic acid, serine, threonine, hydroxyproline, and hydroxylysine, an amide linkage to asparagine or others. Then, there are numerous variations in the number, length, and the constituent monosaccharides. Thus, discussion of glycoprotein degradation will have to be somewhat abstract and general in nature.

1. Degradative Pathways. The carbohydrate content of glycoproteins ranges from only a few percent to nearly half of the total weight of the molecule. A large portion of glycoproteins contains sialic acid, and it is common to subdivide glycoproteins into sialic-acid-containing and sialic-acid-free. The noteworthy characteristic of sialic acid in glycoproteins is that it is always at the nonreducing terminal of the carbohydrate chains, unless it is penultimate to another sialic acid which is at the terminal. Since the principle of degradation of glycoprotein carbohydrate chains is similar to that for gangliosides—sequential removal of constituent monosaccharides from the nonreducing end—the hydrolysis of sialic acid is the first step for degradation of such sialic-acid-containing carbohydrate chains. Other monosaccharides commonly found in glycoproteins are N-acetylglucosamine, N-acetylgalactosamine, galactose, mannose, and fucose. Fucose is also characteristically a terminal sugar. Glucose is rarely found as a constituent of glycoproteins. Xylose is found, however, but only at the carbohydrate–protein linkage regions.

Figure 2 depicts schematically some of the typical carbohydrate chains of glycoproteins and the enzymes involved in each step of degradation. As the figure shows, there are two major ways to initiate glycoprotein degradation. It can be started either by a sequential hydrolysis of the saccharide portion or by degradation of the protein backbone by the action of a variety of proteolytic enzymes. In fact,

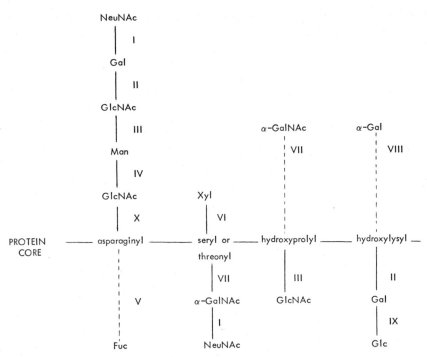

FIGURE 5. Schematic representation of glycoprotein degradation. The enzymes involved are I, sialidase; II, β-galactosidase; III, N-acetyl-β-glucosaminidase; IV, α-mannosidase; V, α-fucosidase; VI, β-xylosidase; VII, N-acetyl-α-galactosaminidase; VIII, α-galactosidase; IX, β-glucosidase; X, aspartyl-glycosylamine amidase. Cathepsins A, B, C, and D, collagenase, and acid carboxypeptidases all participate in the degradation of the protein backbone.

release of free amino acids, in the form of ninhydrin-positive material, and of free sialic acid appear to occur almost simultaneously, followed by the later appearance of N-acetylglucosamine, when fetuin or orosomucoid is incubated with an hepatic lysosomal fraction at an acidic pH (Aronson and de Duve, 1968). Therefore, under appropriate conditions and with a full complement of the required enzymes, degradation of glycoproteins can proceed simultaneously through the two available routes. However, such simultaneous degradation of the carbohydrate chains and the protein backbone is not universal, as shown by the behavior of ovine submaxillary mucin towards hydrolytic enzymes. The compound has approximately 800 disaccharide units, N-acetylneuraminyl-N-acetylgalactosamine, glycosidically linked to serine and threonine in the protein backbone (Graham and Gottschalk, 1960; Carubelli *et al.*,

1965). Most of the sialic acid in the molecule is readily hydrolyzed by sialidase (Graham and Gottschalk, 1960), and once sialic acid is removed, most of the N-acetylgalactosamine becomes susceptible to N-acetyl-α-galactosaminidase (Bhavanandan et al., 1964). Despite the facile hydrolysis of monosaccharides, the protein backbone of ovine submaxillary mucin appears quite resistant to a mixture of lysosomal hydrolytic enzymes. Release of ninhydrin-positive materials does not occur until a substantial portion of the carbohydrates are removed, and even then, the rate of hydrolysis is very slow (Aronson and de Duve, 1968). It appears that the high concentration of sialic acid with its strongly negative charge renders the intact submaxillary mucin resistant to proteolytic enzymes.

2. Degradative Enzymes. *Sialidase.* As stated above, sialic acid when present always occupies the terminal position of carbohydrate chains of glycoproteins, thereby making action of sialidase prerequisite for sialoglycoprotein degradation. Sialidases from a variety of sources are active toward sialic acid in glycoproteins, including those from *Vibrio cholerae, Clostridium perfringens,* mammalian liver or kidney (Taha and Carubelli, 1967; Mahadevan and Tappel, 1967a,b; Horvat and Touster, 1968). Since these sialidases are also active in cleaving most of the sialic acid from gangliosides, they are of a relatively broad specificity. For further discussion of sialidase, refer to Chapter 10.

α-Fucosidase. α-Fucose is another sugar which apparently always occurs at the nonreducing terminal of a carbohydrate chain. Conchie and Hay (1963) described an α-fucosidase in mammalian liver and spleen that is capable of cleaving the terminal fucosyl residues of glycoproteins. Inability to degrade fucose-containing carbohydrate chains does not appear to be incompatible with life, because a state of genetic deficiency of this enzyme, "fucosidosis," is known. Patients affected with this disorder exhibit neurological and systemic signs and symptoms with abnormal accumulation of fucose-containing glycoproteins and glycosphingolipids (van Hoof and Hers, 1968; Durand et al., 1969; Loeb et al., 1969; van Hoof, 1973).

α-Mannosidase. Mannose is another common monosaccharide constituent in glycoproteins. α-D-Mannosidase capable of cleaving mannose residues from glycoproteins has been demonstrated in liver, spleen, kidney, and circulating leukocytes (Sellinger et al., 1960; Conchie and Hay, 1963; Bowers and de Duve, 1967; Baggiolini et al., 1969). Similar to the α-fucosidase deficiency, a genetic disorder characterized by deficiency of α-mannosidase is known. Affected patients show slowly progressive mental and neurological deterioration, visceromegaly, and other systemic signs. As a result of the enzymatic deficiency, abnormal accumulations of mannose-containing glycoproteins and their partially degraded fragments occur (Öckerman, 1967, 1969, 1973; Norden et al., 1973; Autio et al., 1973; Masson and Lundblad, 1974; Norden et al., 1974).

Other Glycosidases. Other glycosidases involved in degradation of glycoproteins include β-D-galactosidase, α-D-galactosidase, β-D-glucosidase, α-D-glucosidase, N-acetyl-β-D-hexosaminidase, N-acetyl-α-D-glucosaminidase, and N-acetyl-α-D-galactosaminidase. Action of these enzymes are required at appropriate steps of degradation of the glycoprotein carbohydrate chains. Most of these enzymes have not yet been purified to homogeneity from sources in which they are known to participate in glycoprotein degradation. Consequently considerable uncertainty remains regarding the extent of the specificity of these enzymes. A good example is β-galactosidase. The question of specificity toward natural glycosphingolipids has been discussed earlier, but activity toward natural glycoproteins has not been considered. Does only one β-galactosidase catalyze hydrolysis of all β-D-galactose residues in glycoproteins, or is there more than one β-galactosidase, each exhibiting specificity toward a different β-D-galactose residue. What is the relationship between β-galactosidase that cleaves β-galactose from glycoproteins and that active toward glycolipids? Answers to these questions are still largely obscure, but it is known that some β-galactosidases have an overlapping specificity for glycolipid and glycoprotein substrates. Such information comes from studies of G_{M1}-gangliosidosis. The disease is caused by a genetic defect of β-galactosidase demonstrable with artificial substrates such as p-nitrophenyl or 4-methylumbelliferyl β-galactoside, or natural substrates such as G_{M1}-ganglioside, asialo G_{M1}-ganglioside or, with an appropriate assay system, lactosylceramide (Okada and O'Brien, 1968; Suzuki and Suzuki, 1974c; Tanaka and Suzuki, 1975). In addition to abnormal accumulation of G_{M1}-ganglioside, the patients accumulate galactose-rich keratan sulfate-like glycoproteins in systemic organs (Suzuki, 1968; Suzuki et al., 1969; Callahan and Wolfe, 1970). Such accumulation appears to be due also to inability to cleave galactose residues from the glycoproteins (McBrinn et al., 1969). This provides circumstantial evidence that the β-galactosidase which is genetically deficient in this disorder normally cleaves galactose from both glycosphingolipids and glycoproteins.

Cathepsin. The cathepsins are a group of proteolytic enzymes that are found widely distributed among animal tissues. Because of their low pH optimum and their localization within lysosomes together with the other glycosidases discussed above, they are considered to be the most important group of proteolytic enzymes primarily responsible for physiological degradation of glycoproteins. While four different cathepsins are recognized (Cathepsin A, B, C, and D), they are for the most part categorized by operational definitions based on substrate specificity and behavior to sulfhydryl reagents.

III. CELLULAR MECHANISM OF DEGRADATION

A. Lysosomes

It is a well-established concept that essentially all naturally occurring sialic-acid-containing macromolecules undergo continual synthesis and degradation. As discussed above, degradation of these complex compounds requires a multitude of hydrolytic enzymes. If the enzymes are present in close proximity to each other within the cell, degradation of the macromolecules may proceed much more smoothly and efficiently than if the enzymes were widely and randomly dispersed. The concept of the lysosome as a biochemical entity, originally introduced by de Duve *et al.* (1955), soon received morphological delineation (Novikoff *et al.,* 1956). There are several pertinent review articles on the current state of lysosome research (e.g., de Duve and Wattiaux, 1966; Straus, 1967; Novikoff, 1973). These subcellular organelles contain 40 or more hydrolytic enzymes with certain common properties. Morphologically, the organelle is unit membrane-bound, and the shape, size, and ultrastructure vary considerably according to state of activity: (a) hydrolases inactive (primary lysosomes), (b) hydrolases active (secondary lysosomes), or (c) undigested materials accumulating (residual bodies). Lysosomes appear to contain a full complement of the hydrolytic enzymes necessary for complete degradation of proteins, glycolipids, glycoproteins, nucleic acids, mucopolysaccharides, and polysaccharides. All of the enzymes discussed earlier for degradation of gangliosides and glycoproteins are found within lysosomes. These enzymes are characterized by unusual stability and optimal activities at acidic pH, usually between 4.0 and 5.0. They appear to be synthesized in the rough endoplasmic reticulum by the conventional mechanism for protein synthesis, and are transported through the channels of endoplasmic reticulum, eventually to the Golgi apparatus where they are incorporated into the primary lysosomes, ready to function as the cellular digestive machinery.

B. Uptake and Disposition of Substrates

Assuming that lysosomes are the main site for the physiological degradation of gangliosides and glycoproteins, and knowing that they are discrete intracytoplasmic structures with limiting membranes, one may assume that the mere presence of lysosomes would not be sufficient for degradation of complex molecules to occur. An appropriate cellular mechanism would be required to bring the substrates to be degraded into contact with the hydrolytic enzymes packaged within the lysosomes. In some specific and rare instances, degradation of tissue components appears to be initiated by release of lysosomal hydrolytic enzymes into extracellular regions. Such is the case in digestion and absorption of bones. In most instances, however, degradation of complex molecules takes place within the cell, particularly within the lysosomes. de Duve and Wattiaux (1966) divided the process of substrate uptake into two major categories, depending on the origin of the materials to be digested. When such materials originate outside the cell, the process is termed heterophagy. In this process, the exogenous material to be digested is first taken up by the cell by endocytosis and once intracellular is surrounded by a membrane which was originally a part of the cell plasma membrane. The term "endocytosis" includes both phagocytosis and pinocytosis. This intracytoplasmic organelle containing exogenous materials is called a heterophagic phagosome. The next step of heterophagy is fusion of the phagosome with a lysosome which may be primary or secondary. Heterophagy is most commonly observed in all types of phagocytic cells, including macrophages, polymorphonuclear leukocytes, right down to the protozoa.

Perhaps the physiologically most significant mode of substrate uptake is the process of autophagy, in which cells degrade their own constituents. Intensive cytochemical and electron microscopic investigations revealed that autophagy is observed in practically all types of eukaryotic cells. Its occurrence appears to be relatively rare in normal cells but increases in frequency under conditions of metabolic and other stresses. In the autophagic process of digestion, a portion of the cells' own cytoplasm becomes segregated from the rest of the cell and surrounded by a membrane, thus forming autophagic phagosomes. Digestion of the content of autophagic vacuoles is initiated when they fuse with primary or secondary lysosomes. A special form of autophagy, first observed by Smith and Farquhar (1966) and Farquhar (1969) in secretory cells in the pituitary, has been termed by de Duve (1969) as "crinophagy." In this process, secretory granules arising from portions of the Golgi apparatus or the so-called GERL system (Pelletier and Novikoff, 1972) fuse with a secondary lysosome, in contrast with typical autophagy in which organelles are sequestered and retain their membrane until they are digested.

If lysosomes are able to degrade the materials to small enough constituents, the products of degradation could diffuse or be transported out of the lysosomes and reenter the cytoplasmic metabolic pools. However, when the endproducts of lysosomal digestion cannot be eliminated through the lysosomal membrane, they remain within the lysosomes. As these materials accumulate, the lysosomes assume more and more of an electron-dense appearance. They are then called dense bodies or residual bodies (Palade, 1956). According to de Duve and Wattiaux (1966), "undigested residues (membrane fragments, myelin figures, amorphous aggregates, and ferritin-like particles) are the morphological hallmark of residual bodies." In lower organisms, such as protozoa, these residual bodies are eliminated directly to the outside of the cell by exocytosis. In contrast, mammalian cells do not appear to be capable of eliminating residual bodies through such a mechanism, and they remain within the cell. Occasional reports of exocytosis of residual bodies by mammalian cells, such as that by Bruni and Porter (1965) for hepatocytes, are not without controversy. Recycling of once-active lysosomes, however, does occur in mammalian cells. For example, Gordon et al. (1965) clearly demonstrated that residual bodies in fibroblasts still containing undigested materials could readily fuse with newly formed endocytotic vacuoles. A similar phenomenon was also demonstrated by Porter et al. (1971). When sulfatide is added to the culture medium containing skin fibroblasts from patients with metachromatic leukodystrophy, sulfatide accumulates within the residual bodies because of the genetic lack of arylsulfatase A. When purified arylsulfatase A is added to the culture medium, it enters the cell by pinocytosis. The residual bodies loaded with undigested sulfatide then fuse with the pinocytotic vacuoles containing arylsulfatase A. This fusion results in elimination of sulfatide by the exogenous arylsulfatase A.

As more and more undigested materials accumulate within lysosomes, filling their entire volume, the lysosomes apparently become inactive. Accumulation of such residual bodies, particularly in the form of lipofuscin, appears to be one of the important aspects of the cellular aging process. Another example of the end-stage of residual bodies can be seen in many of the so-called lysosomal diseases, in which specific lysosomal enzymes are genetically deficient. In these conditions, greatly accelerated accumulation of residual bodies, containing particular chemical compounds that cannot be degraded because of the enzymatic deficiencies, occur (refer to Chapter 6). Whether such residual bodies at the terminal stage lose their hydrolytic enzymes remains unclear. However, lipofuscin granules, which are considered to be typical residual bodies, clearly retain acid phosphatase activity, demonstrable by conventional histochemical means. Furthermore, a highly purified fraction of membranous cytoplasmic bodies from the cerebral cortex of a patient with Tay–Sachs disease (N-acetyl-β-D-hexosaminidase A deficiency) was shown to be enriched with nine acid hydrolases normally present in lysosomes (Tallman *et al.*, 1971). The membranous cytoplasmic bodies of Tay–Sachs disease are a classical example of residual bodies occurring in genetic lysosomal storage disorders. These findings indicate that at least in some instances, residual bodies retain their acid hydrolases even at the terminal stage where they no longer appear to be participating in the intracellular digestive process.

IV. FUNCTIONAL IMPLICATIONS

So far we have discussed the cellular enzymatic machinery and the enzymatic pathways for degradation of sialyl compounds. These lead to a more elusive but important question regarding factors which set the cellular and enzymatic machinery in motion to degrade particular cellular constituents. Perhaps the question is less important for the heterophagic function because what is to be digested in that case is foreign material to the cell. Still, as discussed below, a distinct recognition process appears to take place for foreign materials to be taken up and degraded. The question is most pertinent for the autophagic process in which the cell digests some of its own constituents. The general answer to this question is far from clear and an exhaustive review will not be attempted. However, there are certain aspects of sialyl compound degradation which appear to be closely related to this question.

It is often suggested and suspected that one of the biological functions of the carbohydrate chains of glycolipids and glycoproteins might be as the recognition site in a variety of biological phenomena, such as cell to cell contact, secretion of proteins from the cell, etc. (Eylar, 1965). Carbohydrate chains of glycoproteins and hormones (van der Hamer *et al.*, 1970; Gregoriadis *et al.*, 1970; Morell *et al.*, 1971) may endow uptake specificity. When ceruloplasmin, desialylated by sialidase, was injected into rats, it was rapidly removed from circulation and taken up by hepatocytes in which it was degraded by lysosomes. The endproduct no longer contained galactose, copper, or immunological activity. Essentially similar phenomena were observed with sialidase-treated orosomucoid, fetuin, hepatoglobin, α_2-macroglobulin, thyroglobulin, lactoferrin, human chorionic gonatotropin, and follicle-stimulating hormone. When intact glycoproteins were similarly injected, their survival time was much longer.

The recognition of desialylated glycoproteins by the cellular degradative machinery is not due to the absence of sialic acid. It is the galactose moieties, formerly covered but now exposed by the removal of sialic acid, that apparently are directly involved in the recognition process. Morell et al. (1968) and van den Hamer et al. (1970) showed that when galactose moieties are removed from desialylated glycoproteins, e.g., ceruloplasmin, by galactosidase, or modified by galactose oxidase, the remainder had a survival time similar to that of intact ceruloplasmin. Interestingly, the binding site of the hepatocyte plasma membrane for such asialoglycoproteins must contain sialic acid and calcium ion for proper recognition to occur (Pricer and Ashwell, 1971). These studies concern the mechanism for initiation of degradation in the heterophagic process. Whether or not similar process of recognition takes place when a cell digests some of its own constituents in the autophagic process has not been clarified.

Presence of sialic acid appears to be essential for certain of the biological functions of glycoproteins. Thus, experimental removal of sialic acid destroyed the biological activities of erythropoietin (Lowy et al., 1959; Winkert and Gordon, 1960), human chorionic gonadotropin (Whitten, 1948) and follicle-stimulating hormone (Gottschalk, 1960). The surfaces of erythrocytes and lymphocytes carry relatively high concentrations of sialic acid. When sialic acid is removed from the red cell surface membrane, the cells became more susceptible to phagocytosis by macrophages (Lee, 1968). Similarly, neuraminidase treatment of lymphocytes prevented them from distributing normally within the body. Such lymphocytes instead tended to be trapped in the liver (Woodruff and Gesner, 1969). Removal of sialic acid from the surface of L-1210 mouse leukemia cells resulted in inhibition of potassium transport across the membrane (Glick and Githens, 1965), and neuraminidase treatment of HeLa cells caused a reduction in the uptake of α-aminoisobutyric acid (Brown and Michael, 1969), both instances suggesting participation of cell surface sialic acid in the plasma membrane transport process. Removal of sialic acid from the surface of the plasma membrane often results in other modifications of cellular properties, such as loss of adhesion capacity (Weiss, 1961), deformation (Weiss, 1965), inhibition of or spontaneous occurrence of aggregation (Kemp, 1970; Hovig, 1965). Generally, sialic acids at the end of carbohydrate chains of glycoproteins and gangliosides appear to reduce antigenicity of the compounds (Currie and Bagshawe, 1969; Apffel and Peters, 1970; Uhlenbruck and Gielen, 1970). On the other hand, erythrocyte glycoprotein-bound sialic acid determines the MN blood groups, and removal or modification of sialic acids results in loss of antigenecity (Ebert and Metz, 1972). The somewhat controversial findings of conversion of hexosaminidase A to a hexosaminidase-B-like form by neuraminidase were discussed earlier.

Changes in stability and other physicochemical and enzymatic properties of enzymes by neuraminidase treatment have been reported also for purified α-galactosidase (Faillard and Schauer, 1972) and acid and alkaline phosphatases (Dziembor et al., 1970; Moss, 1969; Saraswathi and Bachhawat, 1968). For further details see Chapter 10.

The functional role of sialic acid in gangliosides is even more obscure. Wooley and Gommi (1965) suggested that ganglioside G_{D3}(disialyllactosylceramide) might be the specific serotonin receptor of smooth muscles, and Wesemann and Zilliken (1968) demonstrated that removal of sialic acid from rat smooth muscle by neuraminidase destroyed the serotonin receptor activity. Because of the high concentration of gangliosides in the synaptic membranes (Lapetina et al., 1967; Wiegandt, 1967; Morgan et al., 1972), and because of their unusual molecular structures, with both long hydrophilic and lipophilic chains, they have been the subject of speculation that they might play some important roles in transmission of nervous impulses. Very little direct evidence exists to substantiate this theory however. Some toxins, such as tetanus toxin, cholera toxin, or botulinum toxin, are known to bind specifically to gangliosides (van Heyningen, 1959; van Heyningen and Miller, 1961; Simpson and Rapport, 1971; Cuatrecasas, 1973; Pierce, 1973; van Heyningen, 1974). Such binding capacity is lost when sialic acid is removed from ganglioside molecules.

V. CONCLUDING REMARKS

We have briefly reviewed here the present status of our knowledge concerning the catabolism of naturally occurring sialic-acid-containing compounds. Advances in the past decade have clarified a great deal of the problems concerning the structural chemistry, distribution, and biosynthetic and degradative pathways of these compounds. These chemical and enzymological investigations have also been greatly aided by the parallel development of the concept of lysosomes and their functional role as the digestive machinery of the cell. In contrast, little is yet known of the functional importance of the catabolic process of sialyl compounds. This is directly related to the functional role of sialic acid in intact molecules because a substantial part of the functional implications of the degradation of sialyl compounds lies in the loss of functions carried out by sialic acid. That such negative aspects are not the only functional meaning of catabolism of sialyl compounds is amply demonstrated by the role played by exposed galactose as the signal for further degradation, or by the enhanced antigenecity of desialylated compounds. Knowledge in this area is at present inevitably tentative.

Note Added in Proof

Since the manuscript of this chapter was completed, the only patient reported as a case of lactosylceramidosis has been conclusively disproven (Wenger, D. A., Sattler, M., Clark, C., Tanaka, H., Suzuki, K., and Dawson, G., 1975, Lactosylceramidosis: Normal activity for two lactosylceramide β-galactosidases, *Science* **188**:1310–1312). Therefore, there is no evidence at present for a β-galactosidase specific for lactosylceramide hydrolysis.

VI. REFERENCES

Apffel, C. A., and Peters, J. H., 1970, Regulation of antigenic expression, *J. Theor. Biol.* **26**:47–59.

Aronson, N. N., and de Duve, C., 1968, Digestive activity of lysosomes. II. The digestion of macromolecular carbohydrates by extracts of rat liver lysosomes, *J. Biol. Chem.* **243**:4564–4573.

Autio, S., Norden, N. E., Öckerman, P. A., Riekkinen, P., Rapola, J., and Louhimo, T., 1973, Mannosidosis: Clinical, fine structural and biochemical findings in three cases, *Acta Paediat. Scand.* **62**:555–565.

Bach, G., and Suzuki, K., 1975, Heterogeneity of human hepatic N-acetyl-β-D-hexosaminidase A activity toward natural glycosphingolipid substrates, *J. Biol. Chem.* **250**:1328–1332.

Baggiolini, M., Hirsch, J. G., and de Duve, C., 1969, Resolution of granules from rabbit heterophil leukocytes into distinct populations by zonal sedimentation, *J. Cell Biol.* **40**:529–541.

Bhavanandan, V. P., Buddecke, E., Carubelli, R., and Gottschalk, A., 1964, The complete enzymic degradation of glycopeptides containing O-seryl and O-threonyl linked carbohydrate, *Biochem. Biophys. Res. Commun.* **16**:353–357.

Bowen, D. M., and Radin, N. S., 1968a, Purification of cerebroside galactosidase from rat brain, *Biochim. Biophys. Acta* **152**:587–598.

Bowen, D. M., and Radin, N. S., 1968b, Properties of cerebroside galactosidase, *Biochim. Biophys. Acta* **152**:599–610.

Bowen, D. M., and Radin, N. S., 1969, Cerebroside galactosidase: A method for determination and a comparison with other lysosomal enzymes in developing rat brain, *J. Neurochem.* **16**:501–511.

Bowers, W. E., and de Duve, C., 1967, Lysosomes in lymphoid tissue II. Intracellular distribution of acid hydrolases, *J. Cell Biol.* **32**:339–348.

Brady, R. O., Kanfer, J., and Shapiro, D., 1965, The metabolism of glucocerebroside. I. Purification and properties of glucocerebroside-cleaving enzyme from spleen tissue, *J. Biol. Chem.* **240**:39–43.

Brown, D. M., and Michael, A. F., 1969, Effect of neuraminidase on the accumulation of alpha-aminoisobutyric acid in HeLa cells, *Proc. Soc. Exp. Biol. Med.* **131**:568–570.

Bruni, C., and Porter, K. R., 1965, The fine structure of the parenchymal cell of the normal rat liver. I. General Observations, *Am. J. Path.* **46**:691–755.

Callahan, J., and Gerrie, J., 1974, Substrate specificity of a rabbit β-galactosidase, *Trans. Am. Soc. Neurochem.* **5**:129.

Callahan, J. W., and Wolfe, L. S., 1970, Isolation and characterization of keratan sulfates from the liver of a patient with G_{M1}-gangliosidosis type I, *Biochim. Biophys. Acta* **215**:527–543.

Carubelli, R., Bhavanadan, V. P., and Gottschalk, A., 1965, Studies on glycoproteins XI. The O-glycosidic linkage of N-acetylgalactosamine to seryl and threonyl residues in ovine submaxillary gland glycoprotein, *Biochim. Biophys. Acta* **101**:67–82.

Conchie, J., and Hay, A. J., 1963, Mammalian glycosidases 4. The intracellular localization of β-galactosidase, α-mannosidase, β-N-acetylglucosaminidase and α-L-fucosidase in mammalian tissues, *Biochem. J.* **87**:354–361.

Cuatrecasas, P., 1973, Gangliosides and membrane receptors for cholera toxin, *Biochemistry* **12**:3558–3566.

Currie, G. A., and Bagshawe, K. D., 1969, Tumor specific immunogenecity of methylcholanthrene-induced sarcoma cells after incubation in neuraminidase, *Brit. J. Cancer* **23**:141–149.

Dawson, G., and Stein, A. O., 1970, Lactosylceramidosis: Catabolic enzyme defect of glycosphingolipid metabolism, *Science* **170**:556–558.

de Duve, C., 1969, The lysosome in retrospect, *in: Lysosomes in Biology and Pathology* (J. T. Dingle and H. B. Fell, eds.), Vol. 1, pp. 3–40, North-Holland Publishing Co., Amsterdam.

de Duve, C., and Wattiaux, R., 1966, Functions of lysosomes. *Ann. Rev. Physiol.* **28**:435–492.

de Duve, C., Pressman, B. C., Gianetto, R., Wattiaux, R., and Appelmans, F., 1955, Tissue fractionation studies, 6. Intracellular distribution patterns of enzymes in rat liver tissue. *Biochem. J.* **60**:604–617.

Durand, P., Borrone, C., and Della Cella, G., 1969, Fucosidosis. *J. Pediat.* **75**:665–674.

Dziembor, E., Gryskiewicz, J., and Ostrowski, W., 1970, The role of neuraminic acid in the stability and enzymic activity of acid phosphatase of the human prostate gland, *Experientia* **26**:947–948.

Ebert, W., and Metz, J., 1972, Modifications of N-acetylneuraminic acid and their influence on the antigen activity of erythrocyte glycoproteins, *Europ. J. Biochem.* **27**:470–472.

Eylar, E. H., 1965, On the biological role of glycoproteins, *J. Theor. Biol.* **10**:89–113.

Faillard, H., and Schauer, R., 1972, Glycoproteins as lubricants, protective agents, carriers, structural proteins and as participants in other functions, *in: Glycoproteins, Their Composition, Structure and Function* (A. Gottschlak, ed.) pp. 1246–1267, Elsevier Publishing Co., Amsterdam.

Farquhar, M. G., 1969, Lysosome function in regulating secretion: Disposal of secretory granules in cells of the anterior pituitary gland, *in: Lysosomes in Biology and Pathology* (J. T. Dingle and H. B. Fell, eds.), Vol. II, pp. 462–483, North-Holland Publishing Co., Amsterdam.

Gatt, S., 1966a, Enzymatic hydrolysis of sphingolipids. Hydrolysis of ceramide glucoside by an enzyme from ox brain, *Biochem. J.* **101**:687–691.

Gatt, S., 1966b, Enzymatic hydrolysis of sphingolipids. I. Hydrolysis and Synthesis of ceramides by an enzyme from rat brain, *J. Biol. Chem.* **241**:3724–3730.

Gatt, S., 1967, Enzymatic hydrolysis of sphingolipids. V. Hydrolysis of monosialoganglioside and hexosylceramides by rat brain β-galactosidase, *Biochim. Biophys. Acta* **137**:192–195.

Gatt, S., and Rapport, M. M., 1966a, Isolation of β-galactosidase and β-glucosidase from brain, *Biochim. Biophys. Acta* **113**:567–576.

Gatt, S., and Rapport, M. M., 1966b, Enzymic hydrolysis of sphingolipids. Hydrolysis of ceramide lactoside by an enzyme from rat brain, *Biochem. J.* **101**:680–686.

Glick, J. L., and Githens, S., 1965, Role of sialic acid in potassium transport of L-1210 leukaemia cells, *Nature* **208**:88.

Goldstone, A., Konecny, P., and Koenig, H., 1971, Lysosomal hydrolases: Conversion of acidic to basic forms by neuraminidase, *FEBS Lett.* **13**:68–72.

Gordon, G. B., Miller, L. R., and Bensch, K. G., 1965, Studies on the intracellular digestive process in mammalian tissue culture cells, *J. Cell Biol.* **25**:Suppl. 41–55.

Gottschalk, A., 1960, Sialic acids: Their molecular structure and biological function in mucoproteins, *Bull. Soc. Chim. Biol.* **42**:1387–1393.

Graham, E. R. B., and Gottschalk, A., 1960, Studies on mucoproteins. I. The structure of the prosthetic group of ovine submaxillary gland mucoprotein, *Biochim. Biophys. Acta* **38**:513–524.

Gregoriadis, G., Morell, A. G., Sternlieb, I., and Scheinberg, I. H., 1970, Catabolism of desialylated ceruloplasmin in the liver, *J. Biol. Chem.* **245**:5833–5837.

Ho, M. W., 1973, Identity of "acid" β-glucosidase and glucocerebrosidase in human spleen, *Biochem. J.* **136**:721–729.

Ho, M. W., and Light, N. D., 1973, Glucocerebrosidase: Reconstitution from macromolecular components depends on acidic phospholipids, *Biochem. J.* **136**:821–823.

Ho, M. W., Cheetham, P., and Robinson, D., 1973, Hydrolysis of G_{M1}-ganglioside by human liver β-galactosidase isoenzymes, *Biochem. J.* **136**:351–359.

Horvat, A., and Touster, O., 1968, On the lysosomal occurrence and the properties of the neuraminidase of rat liver and of Ehrich ascites tumor cells, *J. Biol. Chem.* **243**:4380–4390.

Hovig, T., 1965, The effect of various enzymes on the ultrastructure, aggregation and clot retraction ability of rabbit blood platelets, *Thromb. Diath. Haemorrhag.* **13**:84–113.

Ishizuka, I., and Wiegandt, H., 1972. An isomer of trisialoganglioside and the structure of tetra- and pentasialogangliosides from fish brain, *Biochim. Biophys. Acta* **260**:279–289.

Kemp, R. B., 1970, The effect of neuraminidase on the aggregation of cells dissociated from embryonic chick muscle tissue, *J. Cell Sci.* **6**:751–766.

Kolodny, E. H., Kanfer, J. N., Quirk, J. M., and Brady, R. O., 1971, Properties of a particle-bound enzyme from rat intestine that cleaves sialic acid from Tay–Sachs ganglioside, *J. Biol. Chem.* **246**:1426–1431.

Kuhn, R., and Wiegandt, H., 1964, Weitere Ganglioside aus Menschenhirn, *Z. Naturforsch.* **19B**:256–257.

Lapetina, E. G., Soto, E. F., and de Robertis, E., 1967, Gangliosides and acetylcholinesterase in isolated membranes of the rat brain cortex, *Biochim. Biophys. Acta* **135**:33–43.

Ledeen, R. W., Yu, R. K., and Eng, L. F., 1973, Gangliosides of human myelin: Sialosylgalactosylceramide (G_7) as a major component, *J. Neurochem.* **21**:829–839.

Lee, A., 1968, Effect of neuramindase on the phagocytosis of heterologous red cells by mouse peritoneal macrophages, *Proc. Soc. Exp. Biol. Med.* **128**:891–894.

Li, Y.-T., Mansson, J.-E., Vanier, M.-T., and Svennerholm, L., 1973a, Structure of the major glucosamine-containing ganglioside of human tissues, *J. Biol. Chem.* **248**:2634–2636.

Li, Y.-T., Mazzotta, M. Y., Wan, C.-C., Orth, R., and Li, S.-C., 1973b, Hydrolysis of Tay–Sachs ganglioside by β-hexosaminidase A of human liver and urine, *J. Biol. Chem.* **248**:7512–7515.

Loeb, H., Tondeur, M., Jonniaux, G., Mockel-Pohl, S., and Vamos-Hurwitz, E., 1969, Biochemical and ultrastructural studies in a case of mucopolysaccharidosis "F" (fucosidosis), *Helv. Paediat. Acta* **24**:519–537.

Lowy, P. H., Keighley, G., Borsook, H., 1959, Inactivation of erythropoietin by neuraminidase and by mild substitution reactions, *Nature* **185**:102–103.

Mahadevan, S., and Tappel, A. L., 1967a, Arylamidase of rat liver and kidney, *J. Biol. Chem.* **242**:2369–2374.

Mahadevan, S., and Tappel, A. L., 1967b, β-Aspartylglucosamine amido hydrolase of rat liver and kidney, *J. Biol. Chem.* **242**:4568–4576.

Marcus, A. J., Ullman, H. L., and Saifer, L. B., 1972, Studies on human platelet gangliosides, *J. Clin. Invest.* **51**:2602–2612.

Masson, P. K., and Lundblad, A., 1974, Mannosidosis: Detection of the disease and of heterozygotes using serum and leukocytes, *Biochem. Biophys. Res. Commun.* **56**:296–303.

MacBrinn, M. C., Okada, S., Ho, M. W., Hu, C. C., and O'Brien, J. S., 1969, Generalized gangliosidosis: Impaired cleavage of galactose from a mucopolysaccharide and a glycoprotein, *Science* **163**:946–947.

Miyatake, T., and Suzuki, K., 1972a, Galactosylsphingosine galactosyl hydrolase: Partial purification and properties of the enzyme in rat brain, *J. Biol. Chem.* **247**:5398–5403.

Miyatake, T., and Suzuki, K., 1972b, Globoid cell leukodystrophy: Additional deficiency of psychosine galactosidase, *Biochem. Biophys. Res. Commun.* **48**:538–543.

Miyatake, T., and Suzuki, K., 1974a, Glycosphingolipid β-galactosidases in rat brain: Properties and standard assay procedures of the enzymes in whole homogenate, *Biochim. Biophys. Acta* **337**:333–342.

Miyatake, T., and Suzuki, K., 1974b, Galactosylsphingosine galactosyl hydrolase in rat brain: Probable identity with galactosylceramide galactosyl hydrolase, *J. Neurochem.* **22**:231–237.

Miyatake, T., and Suzuki, K., 1975, Partial purification and characterization of β-galactosidase from rat brain hydrolyzing glycosphingolipids, *J. Biol. Chem.* **250**:585–592.

Morgan, I. G., Reith, M., Marinari, U., Breckenridge, W. C., and Gombos, G., 1972, The isolation and characterization of synaptosomal plasma membranes, in: *Glycolipids, Glycoproteins, and Mucopolysaccharides of the Nervous System* (V. Zambotti, G. Tettamanti, and M. Arrigoni, eds.), pp. 209–228, Plenum Press, New York.

Morell, A. G., Irvine, R. A., Sternlieb, I., Scheinberg, I. H., and Ashwell, G., 1968, Physical and chemical studies on ceruloplasmin V. Metabolic studies on sialic acid-free ceruloplasmin in vivo, *J. Biol. Chem.* **243**:155–159.

Morell, A. G., Gregoriadis, G., Scheinberg, I. H., Hickman, J., and Ashwell, G., 1971, The role of sialic acid in determining the survival of glycoproteins in the circulation, *J. Biol. Chem.* **246**:1461–1467.

Moss, D. W., 1969, Biochemical studies on phosphohydrolase isoenzymes. *Ann. N.Y. Acad. Sci.*, **166**:641–652.

Murphy, J. V., and Craig, L., 1972, Neuraminidase induced changes in white blood cell hexosaminidase A., *Clin. Chim. Acta* **42**:267–272.

Murphy, J. V., and Craig, L., 1974, Effect of human cerebral neuraminidase on hexosaminidase A, *Clin. Chim. Acta* **51**:67–73.

Norden, A. G. W., and O'Brien, J. S., 1973, Ganglioside G_{M1} β-galactosidase: Studies in human liver and brain, *Arch. Biochem. Biophys.* **159**:383–392.

Norden, N. E., Öckerman, P. A., and Szabo, L., 1973, Urinary mannose in mannosidosis, *J. Pediat.* **82**:686–688.
Norden, N. E., Lundblad, A., Svensson, S., and Autio, S., 1974, Characterization of two mannose-containing oligosaccharides isolated from the urine of patients with mannosidosis, *Biochemistry* **13**:871–874.
Novikoff, A. B., 1973, Lysosomes: A personal account, in: *Lysosomes and Storage Diseases* (H. G. Hers and F. van Hoof, eds.), pp. 1–41, Academic Press, New York.
Novikoff, A. B., Beaufay, H., and de Duve, C., 1956, Electron microscopy of lysosome-rich fractions from rat liver, *J. Biophys. Biochem. Cytol.* **2**, Suppl.: 179–184.
Öckerman, P. A., 1967, A generalized storage disorder resembling Hurler's syndrome, *Lancet* **11**:239–241.
Öckerman, P. A., 1969, Mannosidosis: Isolation of oligosaccharide storage material from brain, *J. Pediat.* **75**:360–373.
Öckerman, P. A., 1973, Mannosidosis, in: *Lysosomes and Storage Diseases* (H. G. Hers and F. van Hoof, eds.), pp. 292–304, Academic Press, New York.
Öhman, R., Rosenberg, A., and Svennerholm, L., 1970, Human brain sialidase, *Biochemistry* **9**:3774–3782.
Okada, S., and O'Brien, J. S., 1968, Generalized gangliosidosis: Beta-galactosidase deficiency, *Science* **160**:1002–1004.
Palade, G. E., 1965, The endoplasmic reticulum, *J. Biophys. Biochem. Cytol.* **2**, Suppl.:85–98.
Pelletier, G., and Novikoff, A. B., 1972, Localization of phosphatase activities in the rat anterior pituitary gland, *J. Histochem. Cytochem.* **20**:1–12.
Pentchev, P. G., Brady, R. O., Hibbert, S. R., Gal, A. E., and Shapiro, D., 1973, Isolation and characterization of glucocerebrosidase from human placental tissue, *J. Biol. Chem.* **248**:5256–5261.
Pierce, N. F., 1973, Differential inhibitory effects of cholera toxoids and ganglioside on the enterotoxins of *Vibrio cholerae* and *Escherichia coli*, *J. Exp. Med.* **137**:1009–1023.
Porter, M. T., Fluharty, A. L., and Kihara, H., 1971, Correction of abnormal cerebroside sulfate metabolism in cultured metachromatic leukodystrophy fibroblasts, *Science* **172**:1263–1265.
Pricer, W. E., and Ashwell, G., 1971, The binding of desialylated glycoproteins by plasma membranes of rat liver, *J. Biol. Chem.* **246**:4825–4833.
Radin, N. S., Hof, L., Bradley, R. M., and Brady, R. O., 1969, Lactosylceramide galactosidase: Comparison with other sphingolipid hydrolases in developing rat brain, *Brain Res.* **14**:497–505.
Robinson, D., and Stirling, J. L., 1968, N-acetyl-β-glucosaminidase in human spleen, *Biochem. J.* **107**:321–327.
Sandhoff, K., and Wässle, W., 1971, Anreicherung und Charakterisierung zweier Formen der menschlichen N-Acetyl-β-D-Hexosaminidase, *Z. Physiol. Chem.* **352**:1119–1133.
Sandhoff, K., Pilz, H., and Jatzkewitz, H., 1964, Über den enzymatischen Abbau von N-acetylneuraminsaurenfreien Gangliosidresten (Ceramidoligosacchariden), *Z. Physiol. Chem.* **338**:281–293.
Sandhoff, K., Harzer, K., Wässle, W., and Jatzkewitz, H., 1971, Enzyme alterations and lipid storage in three variants of Tay–Sachs disease, *J. Neurochem.* **18**:2469–2489.
Saraswathi, S., and Bachhawat, B. K., 1968, Role of neuraminic acid in the heterogeneity of alkaline phosphatase in sheep brain, *Biochem. J.* **107**:185–190.
Sellinger, O. Z., Beaufay, H., Jacques, P., Doyen, A., and de Duve, C., 1960, Tissue fractionation studies, 15. Intracellular distribution and properties of β-N-acetylglucosaminidase and β-galactosidase in rat liver, *Biochem. J.* **74**:450–456.

Siddiqui, B., and McCluer, R. H., 1968, Lipid component of sialosylgalactosylceramide of human brain, *J. Lipid Res.* **9**:366–370.

Simpson, L. L., and Rapport, M. M., 1971, Ganglioside inactivation of botulinum toxin, *J. Neurochem.* **18**:1341–1343.

Sloan, H. R., Uhlendorf, B. W., Jacobson, C. B., and Frederickson, D. S., 1971, β-Galactosidase in tissue culture derived from human skin and bone marrow: Enzyme defect in G_{M1}-gangliosidosis, *Pediat. Res.* **3**:532–537.

Smith, R. E., and Farquhar, M. G., 1966, Lysosome function in the regulation of the secretory process in cells of the anterior pituitary gland, *J. Cell Biol.* **31**:319–347.

Snyder, P. D., Jr., Krivit, W., and Sweeley, C. C., 1972, Generalized accumulation of neutral glycosphingolipids with G_M-ganglioside accumulation in the brain, *J. Lipid Res.* **13**:128–136.

Srivastava, S. K., and Beutler, E., 1974, Studies on human β-D-N-acetylhexosaminidases III. Biochemical genetics of Tay–Sachs and Sandhoff's diseases, *J. Biol. Chem.* **249**:2054–2057.

Srivastava, S. K., Awasthi, Y. C., Yoshida, A., and Beutler, E., 1974a, Studies on human β-D-N-acetylhexosaminidases. I. Purification and properties, *J. Biol. Chem.* **249**:2043–2048.

Srivastava, S. K., Yoshida, A., Swasthi, Y. C., and Beutler, E., 1974b, Studies on human β-D-N-acetylhexosaminidases. II. Kinetic and structural properties, *J. Biol. Chem.* **249**:2049–2053.

Straus, W., 1967, Lysosomes, phagosomes and related particles, *in: Enzyme Cytology* (D. B. Roodyn, ed.), pp. 239–319, Academic Press, New York.

Suzuki, K., 1968, Cerebral G_{M1}-gangliosidosis: Chemical pathology of visceral organs, *Science* **159**:1471–1472.

Suzuki, K., and Suzuki, Y., 1970, Globoid cell leukodystrophy (Krabbe's disease): Deficiency of galactocerebroside β-galactosidase, *Proc. Natl. Acad. Sci. U.S.* **66**:302–309.

Suzuki, Y., and Suzuki, K., 1974a, Glycosphingolipid β-galactosidases I. Standard assay procedures and characterization by electrofocusing and gel filtration of the enzymes in normal human liver, *J. Biol. Chem.* **249**:2098–2104.

Suzuki, Y., and Suzuki, K., 1974b, Glycophingolipid β-galactosidases. II. Electrofocusing characterization of the enzymes in human globoid cell leukodystrophy (Krabbe's disease), *J. Biol. Chem.* **249**:2105–2108.

Suzuki, Y., and Suzuki, K., 1974c, Glycosphingolipid β-galactosidases. IV. Electrofocusing characterization in G_{M1}-gangliosidosis, *J. Biol. Chem.* **249**:2113–2117.

Suzuki, Y., Suzuki, K., and Kamoshita, S., 1969, Chemical pathology of G_{M1}-gangliosidosis (generalized gangliosidosis), *J. Neuropath. Exp. Neurol.* **28**:25–73.

Svennerholm, L., 1963, Chromatographic separation of human brain gangliosides, *J. Neurochem.* **10**:612–623.

Taha, B. H., and Carubelli, R., 1967, Mammalian neuraminidase: Intracellular distribution and changes of enzyme activity during lactation, *Arch. Biochem. Biophys.* **119**:55–61.

Tallman, J. F., and Brady, R. O., 1972, The catabolism of Tay–Sachs ganglioside in rat brain lysosomes, *J. Biol. Chem.* **247**:7570–7575.

Tallman, J. F., Jr., Brady, R. O., and Suzuki, K., 1971, Enzymic activities associated with membranous cytoplasmic bodies and isolated brain lysosomes, *J. Neurochem.* **18**:1775–1777.

Tallman, J. F., Brady, R. O., Quirk, J. M., Villalba, M., and Gal, A. E., 1974, Isolation and relationship of human hexosaminidases, *J. Biol. Chem.* **249**:3489–3499.

Tanaka, H., and Suzuki, K., 1975, Lactosylceramide β-galactosidase in human sphingolipidoses: Evidence for two genetically distinct enzymes, *J. Biol. Chem.* **250**:2324–2332.

Uhlenbruck, G., and Gielen, W., 1970, Immunobiologie der Neuraminsäure: ein Beitrag zur Antigenität der Hirntumoren, *Fortschr. Neurol.-Psych. Grenzg.* **38**:202–218.
van den Hamer, C. J. A., Morell, A. G., Scheinberg, I. H., Hickman, J., and Ashwell, G., 1970, Physical and chemical studies on ceruloplasmin IX. The role of galactosyl residues in the clearance of ceruloplasmin from the circulation, *J. Biol. Chem.* **245**:4397–4402.
van Heyningen, S., 1974, Cholera toxin: Interaction of subunits with ganglioside G_{M1}, *Science* **183**:656–657.
van Heyningen, W. E., 1959, Tentative identification of the tetanus toxin receptor in nervous tissue, *J. Gen. Microbiol.* **20**:310–320.
van Heyningen, W. E., and Miller, P. A., 1961, The fixation of tetanus toxin by ganglioside *J. Gen. Microbiol.* **24**:107–119.
van Hoof, F., 1973, Fucosidosis, *in: Lysosomes and Storage Diseases* (H. G. Hers and F. van Hoof, eds.), pp. 277–290, Academic Press, New York.
van Hoof, F., and Hers, H. G., 1968, The abnormalities of lysosomal enzymes in mucopolysaccharidoses, *Europ. J. Biochem.* **7**:34–44.
Weiss, L., 1961, Sialic acid as a structural component of some mammalian cell surfaces, *Nature* **191**:1108–1109.
Weiss, L., 1965, Studies on cell deformability I. Effect of surface charge, *J. Cell Biol.* **26**:735–739.
Wenger, D. A., Sattler, M., and Markey, S. P., 1973, Deficiency of monogalactosyl diglyceride β-galactosidase activity in Krabbe's disease, *Biochem. Biophys. Res. Commun.* **53**:680–685.
Wenger, D. A., Sattler, M., and Hiatt, W., 1974, Globoid cell leukodystrophy: Deficiency of lactosylceramide β-galactosidase, *Proc. Natl. Acad. Sci. U.S.* **71**:854–857.
Wesemann, W., and Zilliken, F., 1968, Rezeptoren der Neurotransmitter, IV. Serotoninrezeptor und Neuraminsäurestoffwechsel der glatten Muskulatur. *Z. Physiol. Chem.* **349**:823–830.
Whitten, W. K., 1948, Inactivation of gonadotrophins II. Inactivation of pituitary and chorionic gonadotrophins by influenza virus and receptor-destroying enzyme, *Austr. J. Sci. Res. B* **1**:388–390.
Wiegandt, H., 1966, Ganglioside, *Ergeb. Physiol. Biol. Chem. Exp. Pharmakol.* **57**:190–222.
Wiegandt, H., 1967, The subcellular localization of gangliosides in the brain, *J. Neurochem.* **14**:671–674.
Wiegandt, H., and Bücking, H. W., 1970, Carbohydrate components of extraneural gangliosides from bovine and human spleen, and bovine kidney. *Europ. J. Biochem.* **15**:287–292.
Winkert, J. W., and Gordon, A. S., 1960, Enzymic actions on the human urinary erythropoietic-stimulating factor, *Biochim. Biophys. Acta* **42**:170–171.
Woodruff, J. J., and Gesner, B. M., 1969, The effect of neuraminidase on the fate of transfused lymphocytes, *J. Exp. Med.* **129**:551–567.
Wooley, D. W., and Gommi, B. W., 1965, Serotonin receptors, VII. Activities of various pure gangliosides as the receptors. *Proc. Natl. Acad. Sci.* **53**:959–963.
Yavin, E., and Gatt, S., 1969, Enzymatic hydrolysis of sphingolipids. VIII. Further purification and properties of rat brain ceramidase. *Biochemistry* **8**:1692–1698.
Yu, R. K., and Ledeen, R. W., 1972, Gangliosides of human, bovine and rabbit plasma. *J. Lipid Res.* **13**:680–686.

Chapter 6

Disorders of Ganglioside Catabolism

John F. Tallman and Roscoe O. Brady

I. INTRODUCTION—THE CATABOLISM OF GANGLIOSIDES

Reactions involved in the catabolism of gangliosides and related neutral glycosphingolipids have been intensively investigated in recent years because deficiencies in certain of the enzymes catalyzing these reactions lead to characteristic neurological disorders known as the lipidoses (Brady, 1966, 1972, 1973). It is presently believed that the catabolism of a ganglioside proceeds stepwise from the nonreducing terminus of the carbohydrate portion of the molecule to the ceramide moiety* (Figure 1), (Tallman and Brady, 1973). The enzymes involved in this process are located in the lysosomes and exhibit properties characteristic of such enzymes. Of the six different enzymes known to participate in ganglioside breakdown, clinical disorders associated with deficiencies in five of these separate reactions have been found. Only a neuraminidase deficiency has not been described (see below). Deficiencies in glucocerebrosidase [Gaucher's disease (Brady et al; 1965)], and lactosylceramide-β-galactosidase (Dawson and Stein, 1970) are known and discussed elsewhere (Brady, 1973). In this chapter we shall focus on the deficiencies leading to ganglioside accumulation. (See Chapter 2, Table I, for a listing of ganglioside structures and abbreviations.)

---

* The nomenclature used is that used in Tallman and Brady (1972).

JOHN F. TALLMAN AND ROSCOE O. BRADY • Developmental and Metabolic Neurology Branch, National Institute of Neurological Diseases and Stroke, National Institutes of Health, Bethesda, Maryland 20014.

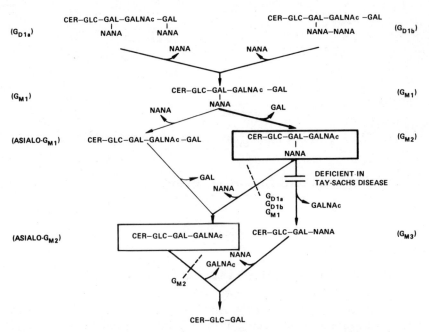

FIGURE 1. The catabolism of gangliosides and its impairment in Tay–Sachs disease.

II. TAY–SACHS DISEASE (TYPE I G_{M2}-GANGLIOSIDOSIS)

A. Clinical Aspects

Tay–Sachs disease was first described in a child of Ashkenazi Jewish descent by Warren Tay (1881) and Bernard Sachs (1887), and since these initial descriptions, numerous cases have been reported. Motor weakness is one of the earliest signs and may be manifested between the third and sixth postnatal month. This weakness becomes progressively more obvious, and hyperacusis may also be present. After about one year, progressive mental retardation becomes apparent, the status of the patient declines, with the onset of blindness, deafness, and convulsions. Affected children commonly die between age three and four.

B. Pathology

Patients with Tay–Sachs disease are characterized by the presence of a cherry-red spot in the macular region of the eye. Macrocephaly occurs after about one year. Visceral changes such as hepatosplenomegaly or bony changes do not occur in patients with type I G_{M2}-gangliosidosis.

Microscopic changes characteristic of acute lipidosis are found in neuronal cells in various organs. The cytoplasm of these cells is distended and contains numerous granules. On electron microscopic examination, these granules consist of spiral lamellar structures with a cross section of about 50 Å (Samuels *et al.*, 1963) (Figure 2). These bodies have been isolated by sucrose gradient centrifugation, and by chemical analysis they have been shown to contain large quantities of G_{M2} (Samuels *et al.*, 1963) (Figure 3). Enzymatic analysis indicates that these isolated membranous cytoplasmic bodies contain high levels of lysosomal enzymes and points out their probable lysosomal origin (Tallman *et al.*, 1970). Axonal degeneration, followed by demyelination, may also occur (Terry and Weiss, 1963).

C. Chemistry of the Storage Material

A change in the chemical composition of the brain was first detected by Klenk in 1942, who showed an increased ganglioside content in Tay–Sachs patients. This report was confirmed, and the particular ganglioside present in elevated quantities was identified as G_{M2} (Svennerholm, 1962). Its structure is (Ledeen and Salsman, 1965)

N-Acetylgalactosaminyl $(1\xrightarrow{\beta}4)$ (N-Acetylneuraminosyl) $(2\xrightarrow{\alpha}3)$

Galactosyl $(1\xrightarrow{\beta}4)$ Glucosyl $1\xrightarrow{\beta}1$ Ceramide

The total ganglioside content is four to five times normal, and G_{M2} represents 80% of the ganglioside present. The asialo derivative of G_{M2} (G_{A2}) is present in the brains of these patients (about 1 mole G_{A2}/8 moles G_{M2}) and is elevated approximately 20–50 times normal (Sandhoff *et al.*, 1971).

D. Nature of the Metabolic Defect

Based on previous information concerning the nature of enzymatic defects in related lipid storage diseases (Brady, 1966), it was logical to look for a defect in the catabolism of the accumulating ganglioside and its corresponding asialo derivative. Since the structure of G_{M2} is branched at its nonreducing terminus, catabolism of G_{M2} could conceivably proceed either through the removal of N-acetylneuraminic acid to yield G_{A2} or N-acetylgalactosamine to yield G_{M3} (Figure 1). Initial

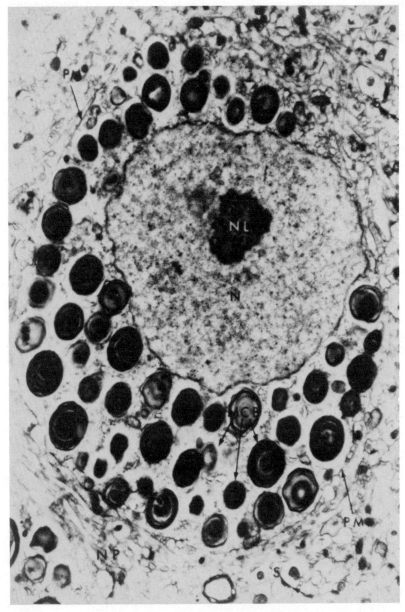

FIGURE 2. An electron micrograph of a typical neuronal cell from the brain of a patient with type I G_{M2}-gangliosidosis. Numerous membranous cytoplasmic bodies (MCB) may be seen in the cytoplasm. From Terry and Weiss, 1963.

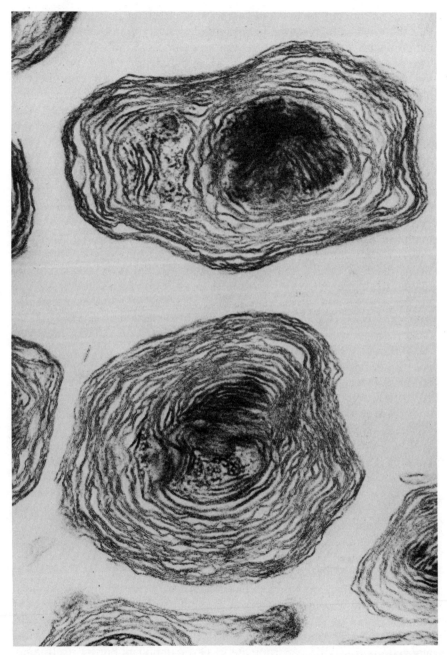

FIGURE 3. Typical isolated membranous cytoplasmic bodies (MCB). Magnification × 34,000. From Suzuki *et al.*, 1969.

studies on human hexosaminidases using artificial substrates showed that there was a higher total hexosaminidase activity in brain tissue from patients with Tay–Sachs disease than in controls (Sandhoff *et al.*, 1968). The possibility that a G_{M2}-neuraminidase deficiency was responsible for the ganglioside accumulation had to be considered. To examine this alternative, G_{M2} was biosynthetically labeled in the *N*-acetylneuraminosyl portion of the molecule from precursor *N*-acetylmannosamine (Kolodny *et al.*, 1971; Tallman and Brady, 1972), and the activity of this neuraminidase was determined in skeletal muscle (Kolodny *et al.*, 1969) and in brain tissue (Tallman *et al.*, 1972) from normal and Tay–Sachs patients. The specific activity of this neuraminidase was similar in tissue specimens from normal humans and patients with Tay–Sachs disease (Kolodny *et al.*, 1969; Tallman *et al.*, 1972). The nature of the enzymatic defect was then partially resolved by the demonstration of two hexosaminidase isozymes (A and B) in human tissues (Robinson and Stirling, 1968). Hexosaminidase A is deficient in tissue from patients with Tay–Sachs disease (Okada and O'Brien, 1969). At that time, the activity of the hexosaminidases with regard to their ability to break down G_{M2} had not been thoroughly investigated. The release of *N*-acetylgalactosamine from doubly labeled G_{M2} was diminished in skeletal muscle preparations from patients with Tay–Sachs disease (Kolodny *et al.*, 1969); however, the possibility of a two-step reaction in which *N*-acetylneuraminic acid was first removed could not be ruled out. Using G_{M2}, specifically labeled in the *N*-acetylgalactosaminyl portion of the molecule (Quirk *et al.*, 1972), it was possible to show a G_{M2}-hexosaminidase in brain (Tallman and Brady, 1972) which was deficient in patients with Tay–Sachs disease (Tallman *et al.*, 1972) (Table I).

In order to obtain a full understanding of the pathological biochemistry of Tay–Sachs disease, several considerations must be taken into account. Hexosaminidase B, which is present in the tissues of patients with G_{M2}-gangliosidosis types I and III, can catalyze the hydrolysis of G_{A2} produced by the G_{M2}-neuraminidase. Since the specific activity of this neuraminidase is normal in Tay–Sachs patients and the activity of the B-like enzyme is elevated (Sandhoff *et al.*, 1971), why does G_{M2} accumulate at all?

This question may be partly answered by the finding that in type II G_{M2} gangliosidosis, in which patients lack both hexosaminidase A and B isozymes, the absolute quantity of G_{A2} accumulating in brain is higher, and the ratio of G_{A2} to G_{M2} is larger (one-third) than in Tay–Sachs patients (type I) (Sandhoff *et al.*, 1971). This finding suggests that the alternate pathway does participate in G_{M2} catabolism but by itself is not sufficient to handle the load of G_{M2} caused by rapid ganglioside turnover

TABLE I. Ganglioside Catabolism by Human Brain Lysosomes

	Enzyme activities		
	Hexosaminidase		
Source of tissue	G_{M2} [^{14}C]Ganglioside[a]	Artificial substrate[b]	Sialidase G_{M2} [^{3}H]Ganglioside[a]
Control series (7)	140 ± 53	134 ± 11	219 ± 33
G_{M2}-Gangliosidosis patients			
VD (Type I)	0	513	225
JK (Type I)	0	1416	Not determined
DT (Type II)	6	3	232
Mixed experiment			
Control + TS	138 (Theory = 98)	Not determined	210 (Theory = 224)

[a] pmoles G_{M2} hydrolyzed/mg protein/hr.
[b] nmoles 4-methylumbelliferyl-N-acetyl-D-glucosaminide hydrolyzed/mg protein/h. Data from Tallman et al., 1972.

in early life. The insufficiency of this pathway may be due to a number of factors, including nonoptimum spatial arrangement of the hexosaminidase B relative to the neuraminidase, inhibition of G_{A2} hydrolysis in later stages of the disease by G_{M2} (Sandhoff, 1968), inhibition of the neuraminidase by the di- and trisialogangliosides (Tallman and Brady, 1973), or nonoptimum pH conditions for one or both of the alternate enzymes (Tallman and Brady, 1972). The existence of this alternate pathway may also account for the relatively insignificant accumulation of G_{M2} outside the central nervous system.

E. Enzymology of Type I G_{M2}-Gangliosidosis

We have very recently purified both forms of hexosaminidase to homogeneity from human placental tissue (Tallman et al., 1974b). They have similar molecular weights. In other studies, these isozymes, purified from beef spleen, had almost identical amino acid composition (Verpoorte, 1972). The hexosaminidases also are immunologically related (Carroll and Robinson, 1972; Srivastava and Beutler, 1973). Antibodies prepared to hexosaminidase B were totally adsorbed by A; however, specific anti-A antibodies could be prepared by adsorbing anti-A with hexosaminidase B. There was no cross-reacting material in patients with the classic form of Tay–Sachs disease (type I) using this specific anti-A antibody. However, in the O-variant form of this disorder, material cross-reacting with both the specific and other anti-A antibodies was present (Srivastava and Beutler, 1973).

The subunit structure of these enzymes is quite complex. In the presence of reducing agent and denaturing conditions the enzymes seem to possess four subunits to equal size with a mass of 30,000 daltons after electrophoresis in the presence of SDS (Tallman *et al.*, 1974*b*). In the presence of sodium dodecylsulfate alone, hexosaminidase A seems to be composed of two hydrophobically associated units of 60,000 molecular weight. In contrast, hexosaminidase B retains the full molecular weight of the native enzyme (130,000) and does not separate into its subunits. From this information, we deduce that a disulfide or other covalent bond participates in the association of hexosaminidase B units. This bond is not present or is quite labile in isozyme A. It is possible to convert the A form of the enzyme into B by heating highly purified hexosaminidase A at 50° in pH 6.0 buffer. Here, instead of the denaturation which usually accompanies heating when selected conditions of the serum assay are used (Robinson and Stirling, 1968), conversion of A occurs to give a B-like enzyme which also has a B-characteristic SDS pattern (Tallman *et al.*, 1974*b*).

F. Prenatal Diagnosis and Treatment

Screening for carriers of type I G_{M2}-gangliosidosis has been successfully undertaken utilizing a heat denaturation method in serum. Large-scale testing programs for high-risk populations are currently being implemented (Kaback *et al.*, 1972; Graves *et al.*, 1973), and successful intrauterine diagnosis of Tay–Sachs disease has been made directly on amniotic fluid samples (Schneck *et al.*, 1970) and cultured amniotic cells (O'Brien *et al.*, 1971*a*.) There is no current therapy for Tay–Sachs disease, and the success of enzyme-replacement trials seems slight because of the blood–brain barrier (Tallman *et al.*, 1974*a*).

III. TYPE II G_{M2}-GANGLIOSIDOSIS

A. Clinical and Pathological Aspects

These patients have neurological signs similar to those of patients with type I G_{M2}-gangliosidosis. Additionally, there are changes in visceral organs attendant on the accumulation of the neutral glycosphingolipid, globoside: *N*-acetylgalactosaminyl-galactosyl-galactosyl-glucosylceramide in both liver and spleen (Sandhoff *et al.*, 1968). The progression of this disease is similar to type I, although it may occur more rapidly. There is no predilection to Jewish ancestry in these patients.

Microscopically, there is little difference between the neuronal inclusions in types I and II G_{M2}-gangliosidosis. The predominant difference is the presence of numerous membranous cytoplasmic bodies in the liver of type II patients. On electron microscopic examination these bodies appear quite similar to those described for the brains of patients with type I disease.

B. Chemistry of the Storage Material

Patients with type II G_{M2}-gangliosidosis are characterized by the accumulation of G_{M2} and its asialo derivative in neuronal cells. It is interesting that the amount of G_{A2} accumulation in relation to G_{M2} is greater than in type I disease. This finding may be related to the nature of the metabolic defect discussed in Section II, D. In type II patients, there is also a significantly higher accumulation of G_{M2} in liver when compared to Tay–Sachs patients, where G_{M2} accumulation is minimal (Kolodny, 1972).

C. Metabolic Defect—Diagnosis and Treatment

Patients with type II G_{M2}-gangliosidosis are characterized by a total lack of hexosaminidase activity when measured with artificial substrates (Sandhoff et al., 1971). These patients also lack the ability to degrade G_{M2} (Tallman et al., 1972) and neutral glycosphingolipids containing a terminal molecule of N-acetylgalactosamine (Sandhoff et al., 1971; Kolodny, 1972). Detection of heterozygotes for type II G_{M2}-gangliosidosis is possible, and prenatal detection of this disease has been carried out (Kolodny, 1972).

Since there seems to be nonactive cross-reacting hexosaminidase protein to antibodies prepared against hexosaminidases A and B and to specific anti-A antibodies (Srivastava and Beutler, 1973), it seems likely that the basic defect in this disorder is a modification of the active site of both hexosaminidases. Here the substitution of an amino acid involved in the actual hydrolysis of N-acetylhexosamines seems to be implicated (Tallman et al., 1973b).

Treatment of patients with type II G_{M2}-gangliosidosis has been attempted by the intravenous injection of highly purified hexosaminidase A (Johnson et al., 1973). Although some lowering of the serum globoside levels was obtained, there was no increase in hexosaminidase activity in the brain. Thus, this therapeutic effort must be regarded as unsuccessful.

IV. OTHER VARIANT FORMS

A. Type III G_{M2}-Gangliosidosis

A patient with type III G_{M2}-gangliosidosis has been described who appeared clinically as a type I patient and did not show any visceral changes. The amount of both G_{M2} and G_{A2} accumulated in the brain of this patient were higher than in type I patients, although the relative proportions are close to those of type I patients (Sandhoff et al., 1971).

When artificial substrates are used to monitor activity, both hexosaminidases A and B appear to be present although the ability of the patient to break down G_{M2} is diminished (Sandhoff et al., 1971). Neither form of hexosaminidase is able to catalyze the breakdown of G_{M2} although the conformational stability of A is not affected in this genetic variant.

B. Hexosaminidase-A-Deficient Adults

Since both hexosaminidase A and B possess activity toward G_{M2}, one might predict on theoretical grounds that normal adults may exist who are totally lacking in hexosaminidase A yet retain the ability to degrade G_{M2} in their hexosaminidase B. Recently, a number of these patients have been described (Navon et al., 1973; Vidgoff et al., 1973). In these cases, prenatal diagnosis is impossible using current procedures with artificial substrates.

V. GENERALIZED GANGLIOSIDOSIS (G_{M1}-GANGLIOSIDOSIS)

A. Clinical Aspects

Patients with this disorder have been divided into two general classes depending on the time of onset and rapidity of progression of the signs and symptoms of the disorder (O'Brien et al., 1971b). Type I is the more fulminating, and the affected infants show motor and mental impairment shortly after birth. About one-half of the patients have a cherry-red spot in the macular region of the eye. Convulsions, rigidity, deafness, and blindness follow, and the patients die by 2 years of age. In type II, motor and mental development is generally normal in the first year; around the first birthday the patients become ataxic, lethargic, progressively spastic, and frequent convulsions set in. Death occurs generally around the fifth year.

B. Pathology

Patients with G_{M1}-gangliosidosis have rarefaction of the bones and deformed vertebrae. The neuronal cells in the brain are distended with membranous cytoplasmic bodies which resemble in some degree the inclusions found in nerve cells of Tay-Sachs patients. The spleen and liver are enlarged, and the bone marrow contains foam cells distended with stored material. The glomeruli of the kidneys are also swollen with accumulating substance.

C. Chemistry of the Stored Material

The major accumulating substance is ganglioside G_{M1} [galactosyl $(1\xrightarrow{\beta}3)$ N-acetylgalactosaminyl $(1\xrightarrow{\beta}4)$ (N-acetylneuraminyl) $(2\xrightarrow{\alpha}3)$ galactosyl $(1\xrightarrow{\beta}4)$ glucosyl $(1\xrightarrow{\beta}1)$ ceramide]. The major fatty acids of the ceramide moiety have been reported to be stearic (70%), palmitic (13%), and nonadecanoic acid (19:0) (6%). The increase in G_{M1} in brain is about tenfold over that in normal gray matter and from 20 to 59 times that in normal liver.

It is of considerable interest that mucopolysaccharides also accumulate in the systemic organs in patients with generalized gangliosidosis (Suzuki et al., 1969). The structure of this accumulating material is said to be similar to keratan sulfate and contains equimolar quantities of galactose and glucosamine. Accumulation of an acidic mucopolysaccharide with N-acetylneuraminic acid as well as galactose and glucosamine has also been reported in tissues of these patients.

D. Metabolic Defect

Based on previous work on Gaucher's disease and Niemann–Pick disease, it was anticipated that the metabolic defect in G_{M1}-gangliosidosis would be a deficiency of the galactosidase which catalyzes the cleavage of the terminal molecule of galactose of G_{M1} (Brady, 1966):

Cer-Glc-Gal-(NeuNAc)-GalNAc-Gal

$+ \; H_2O \xrightarrow{G_{M1}\text{-Ganglioside-}\beta\text{-Galactosidase}}$ Cer-Glc-Gal-(NeuNAc)-GalNAc + Gal

The correctness of this supposition was demonstrated by Okada and O'Brien (1968), who showed that G_{M1}-β-galactosidase activity was very low in the tissues of patients with this disorder. Furthermore, the same investigators found that total β-galactosidase activity in the tissues of these patients was greatly reduced, as determined with p-nitrophenyl-β-D-galactopyranoside or 4-methylumbelliferyl-β-D-galactopyranoside as substrates. Thus, the diagnosis of homozygotes (the patients) and the detection of unaffected heterozygous carriers can be carried out with these artificial β-galactopyranoside substrates using various tissue sources such as washed white blood cells or cultured skin fibroblasts (Wolfe *et al.*, 1970). This information has provided the basis for the successful antenatal detection of fetuses afflicted with G_{M1}-gangliosidosis (Kaback *et al.*, 1972, Lowden *et al.*, 1973).

The accumulation of galactose-containing mucopolysaccharides in patients with G_{M1}-gangliosidosis is consistent with the drastic decrease in total β-galactosidase activity in the tissues of these patients. A β-galactosidase which catalyzes the hydrolysis of G_{M1} has been partially purified from rat liver lysosomes (Cumar *et al.*, 1972). We do not yet known whether the same enzyme catalyzes the hydrolysis of mucopolysaccharides. In view of the reduction in total β-galactosidase activity in tissues of patients with G_{M1}-gangliosidosis, it might be expected that other glycolipids with a terminal β-galactose such as galactocerebroside [$Cer^{(1\overset{\beta}{\leftarrow}1)}Gal$] and ceramidelactoside [$Cer^{(1\overset{\beta}{\leftarrow}1)}Glc^{(4\overset{\beta}{\leftarrow}1)}Gal$] might also accumulate. This is not the case, and in fact the activity of the galactosidases which catalyze the cleavage of these compounds is greatly increased in the brain of patients with G_{M1}-gangliosidosis compared with the levels in control human specimens (Table II).

TABLE II. *p*-Nitrophenylgalactosidase and Glycolipid Hydrolase Activities in Aqueous Extracts of Acetone Powders of Human Cerebral Gray Matter[a]

			Substrate nmoles hydrolyzed per mg of protein per hr	
Tissue source	Age, years	p-Nitrophenyl-β-D-galacto-pyranoside	Galacto-cerebroside	Ceramide-lactoside
Control	3	109	0.12	0.46
Control	2	117	0.20	0.51
G_{M1}-Gangliosidosis	2/3	7.8	1.03	1.9
G_{M1}-Gangliosidosis	2	11	0.65	1.8

[a] Summary of data from Brady *et al.*, 1970.

VI. POTENTIALLY RELATED DISORDERS

A. Hematoside (G_{M3})-Gangliosidosis

A modest amount of G_{M3} accumulates in the brain in many of the patients with G_{M2} gangliosidosis (Suzuki, 1967). However, no specific human disorder has been described in which the accumulation of G_{M3} is preponderant. An increase in brain G_{M3} (and G_{M2}) has been reported in a patient in which the largest elevation of accumulating lipid was ceramidelactoside (Pilz *et al.*, 1966). Very little is known about the neuraminidase which catalyzes the hydrolysis of G_{M3}, although its activity has been amply demonstrated in homogenates of mammalian brain tissue (Tettamanti and Zambotti, 1968) and cultured mouse cell lines (Cumar *et al.*, 1970). In view of the fact that highly purified preparations of rat heart muscle neuraminidase catalyze the cleavage of sialic acid from G_{M2}, G_{D1a}, G_{D1b}, G_{T1}, and fetuin, it seems likely that G_{M3} may also be catabolized by this enzyme (Tallman and Brady, 1973). Unfortunately an examination of this activity was not carried out with this enzyme. In view of the potential multiplicity of substrates hydrolyzed by this neuraminidase, it might be expected that a modification of the primary structure of this enzyme causing diminution of its catalytic activity would be a lethal mutation.

B. Animal Model Gangliosidoses

Reports occasionally have appeared in the literature of G_{M2}-gangliosidosis in certain strains of dogs (Frankhauser, 1965; Karbe and Schiefer, 1967; Bernheimer and Karbe, 1970). It is a great misfortune that the carriers of this condition were not identified and maintained, since an animal model for investigating many aspects of G_{M2}-gangliosidosis is highly desirable. Since it is now perfectly feasible to identify human heterozygotes for this disorder, it seems equally probable that canine carriers could also be detected. Hopefully this oversight will be corrected if and when a case of G_{M2}-gangliosidosis is identified in another animal.

Recently a disorder highly suggestive of G_{M1}-gangliosidosis has been reported in cats (Handa and Yamakawa, 1971; Farrell *et al.*, 1973). Here again it is easy to identify human carriers of the trait and the test should be applicable to felines. In fact, this logic, and that indicated in the preceding paragraph, can be extended to the concept that screening for carrier animals for both of these conditions might well be undertaken with automated techniques for hexosaminidase and β-galactosidase assays which are now available.

C. *In Vitro* Model Studies

The recent demonstration of the reversal of sulfatide accumulation in cultured skin fibroblasts from patients with metachromatic leukodystrophy upon the addition of arylsulfatase A to the culture medium (Porter *et al.*, 1971; Weisman *et al.*, 1972) has added impetus to the search for similar systems for investigations of ganglioside storage diseases. Several years ago, Dr. Donald Silberberg in the Department of Neurology of the University of Pennsylvania School of Medicine undertook a series of collaborative experiments with us on the induction of ganglioside accumulation by the addition of inhibitors of ganglioside catabolism in the fluid of cultured brain explants. We selected a derivative of isoquinoline which had been shown to be an inhibitor of neuraminidase activity (Tallman and Brady, 1973). We added a small quantity of this compound to the organotypic cultures of neonatal rat cerebellar tissue. A significant accumulation of G_{D1a} was observed in these explants compared with the virtually undetectable quantity of this ganglioside in a similar group of control explants. More surprising was the observation that rounded cytoplasmic inclusion bodies appeared in these inhibited cells which bore a marked resemblance to those seen in the neurons in patients with ganglioside storage diseases. However, the excitement provoked by these experiments was short-lived. Additional controls were examined which contained equimolar quantities of isoquinoline itself, which is not an inhibitor of neuraminidase activity. The latter explants also contained rather similar intracytoplasmic inclusions, albeit without the attendant accumulation of G_{D1a}.

Another approach to this problem was undertaken by Stern and coworkers (1972). These investigators added a mixture of gangliosides to the culture fluid of a system similar to that described in the preceding paragraph. This supplementation also caused the appearance of membranous cytoplasmic bodies but in both glial and neuronal cells in their cultures. Furthermore, these investigators observed that the inclusions gradually disappeared from the cells when the gangliosides were removed from the culture medium. The latter finding is especially significant since it implies that the accumulation of gangliosides in the neuronal cells of patients with ganglioside storage diseases might be reversed if a procedure could be devised to provide the missing enzyme to the cells in the central nervous system.

A final aspect of this topic which should be mentioned is the apparent failure of hexosaminidase which is present in high concentration in fetal calf serum (~2000 units/ml) to cross the plasma membrane of cultured skin and amniotic cells. Fetal calf serum is usually added to the culture medium used for growing such cells in about 20% by volume. However, the reliability of hexosaminidase assays in harvested washed cells from these sources for the diagnosis of G_{M2}-gangliosidoses has been abundantly documented. If the enzyme were taken up by such cells, it is assumed that such diagnostic tests would have been seriously compromised. Studies with cultured cells from patients with ganglioside storage diseases provide a unique opportunity to explore the structural makeup of enzymes and components on the surface of cells involved in the uptake of exogenous enzymes.

VII. REFERENCES

Bernheimer, H., and Karbe, E., 1970, Morphologische und neurochemische Untersuchungen von 2 Formes der amaurotischen Idiotie des Hunds: Nachweis einer G_{M2}-Gangliosidose, *Acta Neuropath.* **16**:243.
Brady, R. O., 1966, the Sphingolipidoses, *N. Engl. J. Med.* **275**:312.
Brady, R. O., 1972, Lipidoses, *Biochimie* **54**:723.
Brady, R. O., 1973, Hereditary diseases—Causes, cures, and problems, *Angew Chem. Intern. Ed.* **12**:1.
Brady, R. O., Kanfer, J. N., and Shapiro, D., 1965, The metabolism of glucocerebrosides II. Evidence of an enzymatic deficiency in Gaucher's disease, *Biochem. Biophys. Res. Commun.* **18**:221.
Brady, R. O., O'Brien, J. S., Bradley, R. M., and Gal, A. E., 1970, Sphingolipid hydrolases in brain tissue of patients with generalized gangliosides, *Biochem. Biophys. Acta* **210**:194.
Carroll, M., and Robinson, D., 1973, Immunological properties of N-acetyl-β-D-glucosaminidase of normal human liver and of G_{M2}-gangliosidosis liver, *Biochem. J.* **131**:91.
Cumar, F. A., Brady, R. O., Kolodny, E. H., McFarland, V. W., and Mora, P. T., 1970, Enzymatic block in the synthesis of gangliosides in DNA virus-transformed tumorgenic mouse cell lines, *Proc. Natl. Acad. Sci. U.S.* **67**:757.
Cumar, F. A., Tallman, J. F., and Brady, R. O., 1972, The biosynthesis of a disialylganglioside by galactosyltransferase from rat brain tissue, *J. Biol. Chem.* **247**:2322.
Dawson, G., and Stein, A. O., 1970, Lactosylceramidosis: Catabolic enzyme defect of glycosphingolipid metabolism, *Science* **170**:556.
Farrell, D. F., Baker, H. J., Herndon, R. M., Lindsey, J. R., and McKhann, G. M., 1973, Feline G_{M1}-gangliosidosis: Biochemical and ultrastructural comparisons with the disease in man, *J. Neuropath. Exp. Neurol.* **32**:1.
Frankhauser, Van R., 1965, Degenerative, lipidiotische Erkrankung des Zentral Nerven Systems bei Zwei Hunden, *Schwerz. Arch. Tierheilk* **107**:73.
Graves, R., Mamunes, P., and Bakerman, S., 1973, Screening for Tay–Sachs disease carriers, *Fed. Proc.* **32**:866 Abs.
Handa, S., and Yamakama, T., 1971, Biochemical studies in cat and human gangliosidosis, *J. Neurochem.* **18**:1275.
Johnson, W. G., Desnick, R. J., Long, S. M., Sharp. H. L., Krivit, W., Brady, B., and Brady, R. O., 1973, Intravenous injection of purified hexosaminidase into a patient with Tay–Sachs disease. in: *Enzyme therapy in genetic diseases: Birth Defects*. Original article series IX, 120–125.

Kaback, M. M., and Zeiger, R., 1972, Heterozygote detection in Tay–Sachs disease: A prototype community screening program for the prevention of recessive genetic disorders, *in: Sphingolipids, Sphingolipidoses and Allied Diseases* (B. Volk and S. Aronson, eds.), pp. 613–632, Plenum Press, New York.

Kaback, M. M., Sloan, H. R., and Percy, A. K., 1972, G_{M1}-gangliosidosis type 1: *In utero* detection and fetal manifestations, *Pediatric Res.* **6**:357.

Karbe, E., and Schiefer, B. Familial amaurotic idiocy in male German Shorthair pointers, *Path. Vet.* **4**:223.

Klenk, E., 1942, Uber die ganglioside des gehirns bei der infantlen amaurotischen Idiotie von Typus Tay–Sachs, *Ber Dtsch. Chem. Ges.* **75**:1632.

Kolodny, E. H., 1972, Sandhoff's disease: Studies on the enzyme defect in homozygotes and detection of heterozygotes, *in:* Sphingolipids, Sphingolipidoses and Allied diseases (B. Volk and S. Aronson, eds.), pp. 221–341, Plenum Press, New York.

Kolodny, E. H., Brady, R. O., and Volk, B. W., 1969, Demonstration of an alternation of ganglioside metabolism in Tay–Sachs disease, *Biochem. Biophys. Res. Commun.* **37**:526.

Kolodny, E. H., Brady, R. O., Quirk, J. M., and Kanfer, J. N., 1970, Preparation of radioactive Tay–Sachs ganglioside labeled in the sialic acid moiety, *J. Lipid Res.* **11**:144.

Kolodny, E. H., Kanfer, J. N., Quirk, J. M., and Brady, R. O., 1971, Properties of a particle-bound enzyme from rat intestine that cleaves sialic acid from Tay–Sachs ganglioside, *J. Biol. Chem.* **246**:1426.

Ledeen, R., and Salsman, K., 1965, Structure of the Tay–Sachs ganglioside, *Biochemistry* **4**:2225.

Lowden, J. A., Cutz, E., Conen, P. E., Rudd, N., and Doran, T. A., 1973, Prenatal diagnosis of G_{M1}-gangliosidosis, *New Engl. J. Med.* **288**:225.

Navon, R., Padeh, B., and Adam, A., 1973, Apparent deficiency of hexosaminidase A in healthy members of a family with Tay–Sachs disease, *Am. J. Hum. Genet.* **25**:287.

O'Brien, J. S., Okada, S., Fillerup, D. L., Veath, M. L., Adornato, B., Brenner, P., and Leroy, J., 1971a, Tay–Sachs disease—Prenatal diagnosis, *Science* **172**:61.

O'Brien, J. S., Okada, S., Ho, M. W., Fillerup, D. L., Veath, M. L., and Adams, K., 1971b, Ganglioside storage diseases, *Fed. Proc.* **30**:956.

Okada, S., and O'Brien, J. S., 1968, Generalized gangliosidosis: Beta-galactosidase deficiency, *Science* **160**:1002.

Okada, S., and O'Brien, J. S., 1969, Tay–Sachs disease: Generalized absence of a β-D-N-acetylhexosaminidase component, *Science* **165**:698.

Pilz, H., Sandhoff, K., and Jatzkewitz, H., 1966, Ein gangliostoffwechselstorung mit Anhaufung von ceramidlactosid im Genirm, *J. Neurochem.* **13**:1273.

Porter, M. T., Fluharty, A. L., and Kihara, N., 1971, Correction of abnormal cerebroside sulfate metabolism in cultured metachromatic leukodystrophy fibroblasts, *Science* **172**:1263.

Quirk, J. M., Tallman, J. F., and Brady, R. O., 1972, Preparation of trihexosyl-and tetrahexosyl-gangliosides specifically labelled in the N-acetylgalactosamine moiety, *J. Labelled Compounds* **8**:484.

Robinson, D., and Stirling, J. L., 1968, N-acetyl-β-glucosaminidases in human spleen, *Biochem. J.* **107**:321.

Sachs, B., 1887, On arrested cerebral development with special references to its cortical pathology, *J. Nerv. Ment. Dis.* **14**:541.

Samuels, S., Korey, S., Gonatas, J., Terry, R., and Weiss, M., 1963, Studies on Tay–Sachs disease IV. Membranous cytoplasmic bodies, *J. Neuropath. Exp. Neurol.* **22**:81.

Sandhoff, K., 1968, Auftrennung der Sanger N-Acetyl-β-Hexosaminidase in multiple formen durch elektrofokussierung *Hoppe-Seyl. Z.* **349**:1095.
Sandhoff, K., Andrea, U., and Jatzkewitz, H., 1968, Deficient hexosaminidase activity in an exceptional use of Tay–Sachs disease with additional storage of globoside in visceral organs, *Life Sci.* **7**:283.
Sandhoff, K., Harzer, K., Wassle, W., and Jatzkewitz, H., 1971, Enzyme alterations and lipid storage in three variants of Tay–Sachs disease, *J. Neurochem.* **18**:2469.
Srivastava, S., and Beutler, E., 1973, Hexosaminidase A and hexosaminidase B Tay–Sachs and Sandhoff's disease, *Nature* **241**:463.
Stern, J., 1972, The induction of ganglioside storage in nervous system cultures, *Lab. Invest.* **25**:509.
Stern, J., Novikoff, A. B., and Terry, R. D., 1972, The induction of sulfatide ganglioside and cerebroside storage in organized nervous system cultures, in: *"Sphingolipids, Sphingolipidoses and Allied Disorders* (B. W. Volk and S. Aronson, eds.), pp. 651–660, Plenum Press, New York.
Suzuki, K., 1967, Ganglioside patterns of normal and pathological brains, in: *Inborn Disorders of Sphingolipid Metabolism* (S. M. Aronson and B. W. Volk, eds.), pp. 215–230, Pergamon Press, New York.
Suzuki, K., Suzuki, K., and Kamoshita, S., 1969, Chemical pathology of G_{M1}-gangliosidosis, *J. Neuropath Exp. Neurol.* **28**:25.
Svennerholm, L., 1962, The chemical structure of normal human brain and Tay–Sachs gangliosides, *Biochem. Biophys. Res. Commun.* **9**:436.
Tallman, J. F., and Brady, R. O., 1972, The catabolism of Tay–Sachs ganglioside in rat brain lysosomes, *J. Biol. Chem.* **247**:7570.
Tallman, J. F., and Brady, R. O., 1973, The purification and properties of a mammalian neuraminidase (sialidase), *Biochim. Biophys. Acta* **293**:434.
Tallman, J. F., Brady, R. O., and Suzuki, K., 1971, Enzymic activities associated with membranous cytoplasmic bodies and isolated brain lysosomes, *J. Neurochem.* **18**:1775.
Tallman, J. F., Johnson, W. G., and Brady, R. O., 1972, The metabolism of Tay–Sachs ganglioside: Catabolic studies with lysosomal enzymes from normal and Tay–Sachs brain tissue, *J. Clin. Invest.* **51**:2339.
Tallman, J. F., Pentchev, P. G., and Brady, R. O., 1973a, An enzymological approach to the lipidoses, *Enzyme:* In Press.
Tallman, J. F., Brady, R. O., Quirk, J. M., Villalba, M., and Gal, A. E., 1973b, Isolation and relationship of human hexosaminidases, *J. Biol. Chem.* In Press.
Tay, W., 1881, Symmetrical changes in the region of the yellow spot in each eye of an infant, *J. Ophth. Soc. U. K.* **1**:55.
Tettamanti, G., and Zambotti, V., 1968, Purification of neuraminidase from pig brain and its action on different gangliosides, *Enzymology* **18**:61.
Terry, R. D., and Weiss, R., 1963, Studies In Tay–Sachs disease II. Ultrastructure of the Cerebrum, *J. Neuropath Exp. Neurol.* **22**:18.
Verpoorte, J., 1972, Purification of two β-N-acetyl-D-glucosaminidases, *J. Biol. Chem.* **247**:4787.
Vidgoff, J., Buist, N., and O'Brien, J. S., 1973, Absence of β-N-acetyl-D hexosaminidase A activity in a healthy woman, *Am. J. Hum. Genet.* **25**:372.
Weismann, U. N., Rossi, E. E., and Herschkowitz, N. N., 1972, Correction of the defective sulfatide degradation in cultured fibroblasts from patients with metachromatic leukodystrophy, *Acta Pediat. Scand.* **61**:296.
Wolfe, L. S., Callahan, J., Fawcett, J. S., Anderman, F., and Scriver, C. R., 1970, G_{M1}-Gangliosidoses without chondrodystrophy or visceromegaly, *Neurology* **20**:23.

Chapter 7

The Biological Role of Sialic Acid at the Surface of the Cell

Roger W. Jeanloz and John F. Codington

I. INTRODUCTION

Hirst (1945; 1948) was the first to suggest that the substrate for the "receptor destroying enzyme" of influenza virus, later recognized as sialidase (neuraminidase, see Chapter 10), was a carbohydrate compound located in the outer membrane of the red blood cell after observing that periodate oxidation "destroys" the cellular receptor. This suggestion was supported by Fazekas de St. Groth (1949), who was able to eliminate by mild periodate oxidation the substrate activity without impairing the adsorption of the influenza virus. In the following years, the cellular receptors for the influenza virus were shown to contain galactose, galactosamine, and sialic acid (McCrea, 1954; Klenk and Stoffel, 1956; Yamakawa et al., 1956) and, in 1957, Klenk and Lempfrid isolated pure, crystalline N-acetylneuraminic acid after neuraminidase treatment of human erythrocytes. Since erythrocyte stroma contains sialic acid residues linked to gangliosides or glycoproteins, it may be assumed that sialic acid exists as a component of these two classes of compounds at the surface of the cell.

The importance of sialic acid residues at the surface of the cell in the phenomenon of cell electrophoretic mobility was only slowly accepted (see review by Cook and Stoddart, 1973). This relationship was

ROGER W. JEANLOZ and JOHN F. CODINGTON • Laboratory for Carbohydrate Research, Departments of Biological Chemistry and Medicine, Harvard Medical School, and Massachusetts General Hospital, Boston, Massachusetts 02114.

well established by the work of Cook *et al.* (1960) on the tryptic degradation of erythrocytes and by subsequent work with purified neuraminidases (Cook *et al.*, 1961; Eylar *et al.*, 1962).

Visual detection of sialic acid at the surface of the cell with light and electron microscopes, after chemical treatment, presents numerous difficulties owing to the lack of specificity of the procedures. Although periodate oxidation, followed by treatment with Schiffs' reagent, occurs preferentially with the side chain of the sialic acid residues, the galactose and *N*-acetylgalactosamine residues are also susceptible to rapid periodate oxidation at the *cis*-vicinal C_3- and C_4-hydroxyl groups, as shown by Gasic and Gasic (1963) with the TA3-Ha ascitic carcinoma cell. Attempts to identify further the components oxidized are complicated by the resistance of some sialic acid residues to neuraminidase treatment and by the difficulty of obtaining glycosidases active on cell surface components. Similarly, detection with the Hale stain is difficult to interpret because of its reactivity with carboxyl groups of both uronic and sialic acid residues (see Cook and Stoddard, 1973). For details on the chemistry of sialic acid, see Chapter 1.

II. OCCURRENCE, FORMS, AND AMOUNTS OF SIALIC ACID RESIDUES AT THE SURFACE OF THE CELL

Following the early observation of Klenk and Yamakawa and their associates, the presence of sialic acid residues at the surface of numerous cells was established, either by action of neuraminidase followed by determination of the cell electrophoretic mobility or by determination of the sialic acid released. For example, sialic acid residues were detected at the surface of Ehrlich ascites tumor cells (Wallach and Eylar, 1961; Cook *et al.*, 1962), hamster kidney fibroblasts transformed by polyoma virus (Forrester *et al.*, 1964), squamous epithelial cells (Berwick and Coman, 1962), solid and ascites sarcoma cells (Cook *et al.*, 1963; Wallach and de Perez Esandi, 1964), normal and malignant rat liver cells (Kalant *et al.*, 1964), cells from human bronchial carcinoma, cells from rat myeloma, HeLa cells, and L-strain mouse fibroblasts (Fuhrmann *et al.*, 1962).

Because of the lack of susceptibility to neuraminidase of some of the sialic acid residues at the surface of the cell, quantitative determination has been successful only when the major proportion of these residues is released. These amounts vary widely, from 3×10^6 residues of sialic acid per human erythrocyte (Madoff *et al.*, 1964), to 3×10^7 residues per human platelet (Bray and Alexander, 1969), to approximately 10^9 for the Ha subline, 4×10^8 for the St subline of the TA3 murine cancer cell (Codington *et al.*, 1974), and 9×10^{18} for the L cell (Glick, 1974).

The electrophoretic mobility of the cell increases greatly with age, but this variation could not be related to variations in the amount of sialic acid at the surface (see review by Balazs and Jacobson, 1966). During the cell cycle, a definite increase of neuraminidase-susceptible sialic acid was observed by Rosenberg and Einstein (1972) in human lymphoid cells, whereas Kraemer (1967) could not observe this change in osteosarcoma cells.

Both forms of sialic acid, *N*-acetyl- and *N*-glycolylneuraminic acid (cf. Chapter 1), may coexist at the surface of the same cell, as shown in the murine TA3-Ha cancer cell by Codington *et al.* (1970). In the erythrocyte, the proportion of *N*-acetyl- and *N*-glycolylneuraminic acid is similar to that of these compounds in other tissues, and varies with the animal species (Klenk, 1958*a*). No information is available at the present time on the occurrence and relative proportions of the *O*-acetyl derivatives of sialic acid at the surface of the mammalian cell. The absolute and relative amounts of sialic acid residues linked either to glycoproteins or to glycolipids vary widely and have been determined only on the erythrocytes of a few animal species (Uhlenbruck and Wintzer, 1970).

It is evident that the sialic acid residue may have an indirect biological role, in addition to its direct effect as a strongly anionic carbohydrate residue located at the extremity of the carbohydrate moiety of a glycolipid or a glycoprotein embedded in the plasma membrane. This indirect role may consist in the masking of the effects of neighboring carbohydrate or protein moieties either by direct chemical linkage or through a charge effect. Recent reviews of the chemistry and biology of the surface of the eukaryotic and tumor cell have been published (Rapin and Burger, 1974; Lee and Smith, 1974).

III. THE MASKING OF CELL-SURFACE ANTIGENS BY SIALIC ACID

The role played by sialic acid in masking cell-surface antigens may be ascertained only by the enhanced antigenicity created by its removal from the cell surface. Masking was first noted by Thomsen (1927) in the study of the erythrocyte. In a phenomenon which came to be known as the "Thomsen effect," Friedenreich (1930) observed the agglutination of red blood cells under the influence of a substance later shown to be the enzyme neuraminidase. In 1947, Burnet and Anderson showed that washed human erythrocytes could be agglutinated by neuraminidase only in the presence of serum. This serum factor, present in the blood of every mammalian species so far examined (Uhlenbruck et al., 1969), has been partially characterized and will be discussed later. The receptor sites exposed on the cell surface by the action of neuraminidase are called Friedenreich antigens and are characterized by terminal β-D-galactopyranosyl residues that had been in penultimate position in the chain prior to enzyme treatment. A structure for the Friedenreich antigen has been suggested by Springer and Desai (1974).

Studying a phenomenon perhaps related to the Thomsen effect, Sanford (1967) reported that incubation with neuraminidase of the mammary carcinoma ascites cell, TA3-Ha, reduced its transplantability in an allogeneic mouse strain (Sanford, 1967). Other examples of the reduced transplantability of neuraminidase-treated tumor cells, both in allogeneic (Currie, 1967; Currie and Bagshawe, 1968a,b) and syngeneic (Simmons et al., 1971a,b; Rios and Simmons, 1973; Sethi and Brandis, 1973; Bekesi et al., 1971) systems were soon reported.

The reduced capacity of neuraminidase-treated tumor cells to proliferate *in vivo* has been partly attributed to the presence of a serum factor (Sanford and Codington, 1971; Rosenberg and Schwarz, 1974). This factor was shown to be present in the serum of most strains of mice (Sanford and Codington, 1971; Rosenberg and Schwarz, 1974), of the guinea pig (Sanford and Codington, 1971; Hughes et al., 1973), of the rabbit (Rosenberg et al., 1972), and of man (Rosenberg and Rogentine, 1972; Grothaus et al., 1971; Rogentine and Plocinik, 1974). It is cytotoxic to neuraminidase-treated cells but not to untreated cells. Both in the transplantation phenomenon (Hughes et al., 1973; Sethi and Brandis, 1972) and in the Thomsen effect (Uhlenbruck et al., 1969), the active serum factor had properties similar to a γ_M-type antibody. The similarity between the two phenomena extends at least to the terminal component of the carbohydrate receptor, which in the agglutination reaction (Uhlenbruck et al., 1969) and in the transplantation effect (Hughes et al., 1973) was shown to be a β-D-galactopyranosyl residue. In each case N-acetylneuraminic acid is probably linked $\alpha\text{-}(2\rightarrow3)$ to a penultimate galactopyranosyl residue.

In an attempt to determine whether this effect of neuraminidase could be applied to the clinical management of human cancer, Simmons and his collaborators (Simmons and Rios, 1971, 1972; Rios and Simmons, 1974) and others (Holland et al., 1972) inoculated tumor-bearing rodents with neuraminidase-treated tumor cells and used this procedure in combination with immunotherapy (Simmons and Rios, 1971; Holland *et al.*, 1972), chemotherapy (Holland *et al.*, 1972), and surgery (Rios and Simmons, 1974). The results were generally encouraging.

The exposure of new galactosyl residues to the action of the cytotoxic "natural" serum antibodies is, probably, not solely responsible for transplantation rejection. Upon inoculation of neuraminidase-treated cells, the immune response by the host, as characterized by an enhanced lymphoblastic effect, increased (Watkins *et al.*, 1971; Novogrodsky and Katchalski, 1973*a*). Furthermore, palpable tumors in syngeneic mice completely regressed upon inoculation of a small number of neuraminidase-treated cells of the same tumor, a result inexplicable on the basis of the cytotoxic effect alone (Simmons *et al.*, 1971*a*). Finally, it was early established that neither thymectomized nor irradiated animals were able to reject neuraminidase-treated tumor cells (Currie and Bagshawe, 1968*b*). Whether these results are due to an enhancement of both the humoral and cell-mediated immune responses or to an enhancement of the more effectual cell-mediated route alone, as suggested by Simmons *et al.*, (1971*a*), has not been determined. Recent attempts to explain this phenomenon (Rosenberg and Schwarz, 1974; Simmons and Rios, 1973; Simmons *et al.*, 1970) have implicated both an enhanced immune response and the cytotoxic action of a natural γ_M-type antibody. The results in syngeneic systems are more difficult to explain, particularly if tumor specific transplantation antigens are not involved, a view expressed by Simmons *et al.* (1970). The possible effect on the enhancement of the immune response of a decrease of the negative surface charge, due to the loss of sialic acid (Weiss and Hauschka, 1970; Sachtleben *et al.*, 1973), cannot be totally discounted. This would permit an easier approach by negatively charged lymphocytes, as suggested by Currie (1967). It does not seem probable, however, that the removal of surface charge alone could have a pronounced effect upon the adsorption of antibodies. As described below, the removal of cell-surface sialic acid had no significant effect upon the adsorption of antisera to surface histocompatibility antigens in some systems (Currie and Bagshawe, 1968*b*; Sanford *et al.*, 1973). An observation that further complicates the picture is the rapid regeneration (6–8 hr) of cell-surface sialic acid after its removal with neuraminidase (Hughes *et al.*, 1972; Warren and Glick, 1968).

Another possible explanation (Simmons and Rios, 1972) is that loss of sialic acid would lower the negative charge at the cell surface, thus increasing the deformability of the cell (Weiss, 1965) and enhancing its susceptibility to phagocytosis (Lee, 1968). Furthermore, no activity change could be detected, after removal of sialic acid residues with neuraminidase, in the *in vitro* determination of inhibition of isolated mouse (H-2) or human (HL-A) fragments (Nathenson et al., 1970). In an apparent contradiction, however, human lymphocytes treated with neuraminidase were more readily lysed by antiserum directed against HL-A antigens than were untreated cells (Grothaus et al., 1971).

Although Currie and Bagshawe (1967) reported that removal of sialic acid from the surface of mouse trophoblasts exposed previously masked histocompatibility antigens, Simmons et al. (1971c) were unable to detect an antigen after treatment of mouse trophoblasts with neuraminidase.

The strain-specific TA3-St ascites cell adsorbs far more histocompatibility (H-2) antisera than the non-strain-specific, TA3-Ha cell, which was originally derived from the same spontaneous mammary adenocarcinoma of the strain A mouse (Sanford et al., 1973; Friberg, 1972). It has been suggested (Sanford et al., 1973; Codington et al., 1973; Codington, 1975) that this difference is due to the masking of cell-surface histocompatibility antigens by mucin-type sialic-acid-containing glycoprotein molecules of high molecular weight that are present at the surface of the TA3-Ha cell but absent from the surface of the TA3-St cell. Although these molecules contain approximately 12% of N-acetylneuraminic acid (Codington et al., 1975a), this component did not appear to play any direct role in the masking process, since its removal did not significantly alter the adsorption of H-2 antisera (Sanford and Codington, 1971; Sanford et al., 1973).

The essential role played by sialic acid in the homing pattern of lymphocytes to the lymph nodes was reported by Woodruff and Gesner (Woodruff and Gesner, 1969; Gesner and Woodruff, 1969), who labeled neuraminidase-treated and -untreated small lymphocytes with ^{51}Cr and transfused the desialylated lymphocytes back into the hosts. They observed that a proportion of neuraminidase-treated lymphocytes larger than that of untreated lymphocytes was adsorbed by the liver, and that a smaller proportion migrated normally to the lymph nodes. Sialic acid thus appears to perform an essential function in the circulation of small lymphocytes in the lymph and blood. A similar loss of homing characteristics by lymphocytes incubated with trypsin was recently observed by Woodruff (1974). It is probable that this enzyme cleaved sialic-acid-containing glycopeptides from the cell surface. Whether the function of cell-surface sialic acid on lymphocytes is due to its negative charge or to a masking effect is not known. As a masking component, sialic acid might serve to protect penultimate galactose residues that could serve as potential receptors, which would allow the cells to be bound by the liver or other organs. In this respect, surface sialic acid may serve in lymphocytes a function somewhat similar to that which it serves in serum glycoproteins. In investigations performed mainly by Ashwell and coworkers (Ashwell and Morrell, 1971; 1974), most serum glycoproteins were observed to be bound to the liver shortly after neuraminidase treatment and to be lost to the circulatory system; untreated sialoglycoproteins continued to circulate in the blood for long periods.

IV. SIALIC ACID AS A RECEPTOR AT CELL SURFACES

Glycoproteins at cell surfaces are involved in various activities that affect cellular behavior or survival (Burger, 1969; Apffel and Peters, 1970; Sanderson *et al.*, 1971; van Beek *et al.* 1973). They may participate in the response of a cell to other cells, to viruses, and other microorganisms, or to other macromolecules. Although the mechanisms by which "messages" are transmitted by surface glycoproteins and responded to by the cell are not generally well understood, the involvement of specific carbohydrate receptors has been established in many cases. Since sialic acid occupies a terminal position in carbohydrate chains of mammalian glycoproteins, it might be expected that it would play an important role as a receptor at the cell surface. Yet, as described in Section III, on masking of cell-surface antigens, it has been demonstrated that this component may function in a different manner by masking potentially antigenic galactose units at the cell surface (Sanford and Codington, 1971; Rosenberg and Schwarz, 1974) or that it may be involved in nonspecific repulsion of cells or macromolecules by virtue of its negative charge (Weiss, 1965; Lee, 1968). Nevertheless, glycoprotein-bound sialic acid may serve at the cell surface as a receptor for lectins, virus particles, mycoplasma, possibly hormones, and antibodies.

A. Receptor for Lectins

Two lectins, one isolated from the horseshoe crab, *Limulus polyphemus* (Pardoe and Uhlenbruck, 1970; Marchalonis and Edelman, 1968), and the other from wheat germ, *Triticum vulgaris* (Greenaway and LeVine, 1973; Bhavanandan and Davidson, 1975; Matsumoto and Codington, unpublished results), may attach at the cell periphery to receptor sites that include terminal glycoprotein-bound sialic acid units. The horseshoe crab lectin, purified and characterized by Marchalonis and Edelman (1968), was found by Pardoe and Uhlenbruck (1970) to be specific for terminal sialic acid residues.

The specificity of wheat germ agglutinin has been attributed to bound *N*-acetylglucosamine residues, since this component, both as a monosaccharide or as an oligosaccharide of two to five units, effectively inhibits agglutination of cells by this lectin (LeVine et al., 1972; Allen et al., 1973) far better than other simple compounds tested; the monosaccharide, N-acetylneuraminic acid, exhibited no inhibitory activity (Greenaway and LeVine, 1973). In the same report (Greenaway and LeVine, 1973), however, it was shown that a sialic-acid-containing glycopeptide effectively inhibited the agglutination of ascites hepatoma cells, and that the degree of inhibition was proportional to the concentration of bound sialic acid. Similarly, the inhibition of the agglutination by wheat germ agglutinin of B16 mouse melanoma cells was effectively performed by Bhavanandan and Davidson (1975), who employed low concentrations of a sialic-acid-containing glycoprotein with no detectable N-acetylglucosamine which had been isolated from the surfaces of these cells. In like fashion, Matsumoto, Codington, and Jeanloz (unpublished results) observed that the facile agglutination of murine mammary-carcinoma ascites cells by wheat germ agglutinin was inhibited by low concentration of epiglycanin, a glycoprotein derived from the surface of the TA3-Ha cell (Codington et al., 1972). This compound contains about 12% sialic acid, whereas N-acetylglucosamine, which is totally in a nonterminal position, represents only about 3–4% of the glycoprotein. With neither the B16 nor the TA3-Ha glycoproteins was a similar inhibitory activity observed after removal of sialic acid with neuraminidase. Reviews of the physicochemical properties and immunochemical specificities of lectins have appeared (Sharon and Lis, 1972; Nicolson, 1974).

B. Receptor for Viruses

In 1941, Hirst, and McClelland and Hare demonstrated that myxoviruses and paramyxoviruses are able to agglutinate chicken erythrocytes. It was later shown (Hirst, 1942; Henle et al., 1944) that the virus itself was the hemagglutinating agent, and agglutination occurred by attachment of the virus to receptor sites at the surfaces of two cells to form a bridge. Intact mammalian cells, in addition to erythrocytes, were also susceptible to agglutination by a variety of related viruses, but, unlike erythrocytes, were often infected by the virus (Hirst, 1943). Hirst (1943) recognized that the agglutinating viruses contained an enzyme, which was first called "receptor-destroying-enzyme" and later neuraminidase (cf. Chapter 10), that was capable of eliminating the capability of the cells to adsorb viruses. The pioneer work in this field has been discussed by Burnet (1951). The purification and elucidation of the complete structure of the carbohydrate component, N-acetylneuraminic acid, released from the host cell by the viral neuraminidase, required a number of years. Discussions of the physicochemical investigations leading to the determination of the structure of sialic acid have been published (Blix and Jeanloz, 1969; Tuppy and Gottschalk, 1972); see Chapter 1 for further details.

It was not until many years after the brilliant early work in this field that major advances in our understanding of the nature of infections by the myxo- and paramyxoviruses were developed. A brief description of some of these findings is presented here. In 1971, Tiffany and Blough reported that fetuin, linked to artificial membranes, binds myxoviruses. More recently, Huang *et al.* (1973) found that several varieties of myxovirus particles would attach to neuraminidase-treated erythrocytes that had been coated with any of a number of different glycoproteins, among them fetuin, α_1-acid glycoprotein, and α_2-macroglobulin. Thus, it was established that the host-cell glycoprotein may not require structural specificity. In all cases, however, the presence of glycoprotein-bound N-acetylneuraminic acid residues was required for adsorption. Suttajit and Winzler (1971) concluded, on the basis of mild periodate oxidation followed by borohydride reduction, that carbons-7, -8, and -9 of the N-acetylneuraminic acid residue were involved as the receptor site. Evidence that sialic acids other than N-acetylneuraminic acid may be involved in some systems was obtained by Levinson *et al.* (1969), who reported that the inhibition of agglutination by A_2 influenza virus of erythrocytes by horse serum requires a terminal 4-O-acetyl-N-acetylneuraminic acid residue. However, the active inhibitor of agglutination by influenza B virus in the same system was the N-acetylneuraminic acid residue. Additional evidence for specificity in the host-cell receptor was obtained by Drzeniek and Gauhe (1970), who found that neuraminidases isolated from different myxoviruses possess markedly different specificities.

Lukert (1972) made the interesting observation, which may not, however, be directly related to substrate specificity, that the attachment of avian, infectious bronchitis virus to monolayers of chicken embryo kidney cells was inhibited not only by the addition of free or bound sialic acid, but by sulfhydryl-containing compounds. Receptors could be destroyed by neuraminidase or p-hydroxymercuribenzoate treatment. The adsorption of Newcastle disease virus, however, was not affected by the addition of sulfhydryl compounds.

The role of viral neuraminidase in viral adsorption, hemagglutination, infectivity, and virus release has been studied extensively. Studies of sialic acid release have revealed two distinct stages in infectivity (Fresen and Ubendorfer, 1973; Tsvetkova and Lipkind, 1973). For adsorption of the virus, however, enzyme activity is not required. This has been demonstrated by the use of antineuraminidase antibodies (Brecht *et al.*, 1971) and specific neuraminidase inhibitors (Palese *et al.*, 1974*a*; Zakstelskaya *et al.*, 1972; Meindl *et al.*, 1974). Inhibition of enzyme activity with univalent antineuraminidase antibodies did not prevent the multiplication of fowl plague virus, although no virus was released after inhibition of neuraminidase activity with bivalent antibodies (Brecht *et al.*, 1971). By the use of specific inhibitors, Zakstelskaya *et al.* (1972) were able to block separately neuraminidase and hemagglutinating activity; inhibition of enzyme activity did not prevent the adsorption of virus, infectivity, or elution of virus from the cells. As a result of these findings, it was concluded that the role of neuraminidase is to regulate the hemagglutination reaction. Palese *et al.* (1974*b*), on the other hand, concluded that viral neuraminidase was essential for the replication of influenza viruses and for the removal of sialic acid from a sialic-acid-containing glycoprotein in the viral envelope, thus preventing aggregation of the progeny-virus particles. Hemagglutination activity appears to be more directly involved in infectivity, since its inhibition with specific inhibitors (Zakstelskaya *et al.*, 1972) blocked the capability of the virus to be adsorbed on cells and to infect them.

Recent reports have further elucidated the nature of the macromolecular reactions involved in the adsorption and infectivity steps. For parainfluenza viruses, the hemagglutinating and neuraminidase activities appear to reside on a single glycoprotein having a molecular weight of 65,000, although, in the case of the influenza viruses, the activities may reside in different glycoproteins (Schied and Choppin, 1973, 1974). After the adsorption step, Choppin and Schied (1974) found that the infectivity is initiated by a protease at the surface of the host cell. This protease reduces the size of the active, viral glycoprotein to a molecular weight of approximately 53,000. This newly formed protein initiates cell fusion, a step that appears to be necessary for cell penetration by the virus and cell infectivity.

C. Receptor for Mycoplasma

The nature of mycoplasma receptor sites on erythrocytes or other mammalian cells has not been extensively investigated. Of seventeen mycoplasma serotypes investigated, four bind to putative sialic acid receptors (Manchee and Taylor-Robinson, 1969). *Mycoplasma pneumoniae* and *Mycoplasma gallisepticum* appear to attach to sialic acid receptors on the erythrocytes or tracheal epithelial cells. Pretreatment of these cells with sialic-acid-containing materials prevented this attachment (Sobeslavsky *et al.*, 1968).

D. Receptor for Hormones

Recently Hollenberg and Cuatrecasas (1975) demonstrated that the fat cell receptor for insulin is a glycoprotein, but that sialic acid does not appear to be directly involved in the binding of insulin to the membrane. Neuraminidase had no effect on the capacity of the cell to adsorb insulin (Cuatrecasas and Illiano, 1971), but it profoundly affected the rate of glucose oxidation. It was suggested (Hollenberg and Cuatrecasas, 1975) that neuraminidase may remove sialic acid units, perhaps on the receptor site of the glycoprotein, thus affecting a component involved in the transmission of the insulin-binding signal to the glucose transport site.

It is not clear whether or not sialic acid is involved in the interaction of the adrenocorticotropic hormone (ACTH) with the receptor at the surface of rat adrenal cells. Haksar *et al.* (1974), however, reported that neuraminidase treatment of isolated cells inhibited the cyclic AMP response to ACTH stimulation. These authors suggest that sialic acid may play a role in transmitting the signal from the ACTH–receptor interaction to the catalytic unit of adenyl cyclase.

Receptor sites for serotonin have been found wherever physiological reaction to it takes place. Neuraminidase action is followed by a loss of the inhibition of the activity of serotonin by lysergic acid (Gielen, 1968) and a decrease of the number of active receptor sites; a similar decrease was observed in the presence of inhibitors of sialic acid biosynthesis (Wesemann and Zilliken, 1968). The selective destruction of the receptor sites by neuraminidase in the presence of EDTA and the reactivations by addition of tissue lipids (Woolley and Gommi, 1964) led to the identification of gangliosides as receptors in the cell membrane. This was confirmed by reactivation with purified gangliosides (Woolley and Gommi, 1965).

E. Receptor for Antibodies

Mammalian cell surfaces may contain antigens of various types. Among these are histocompatibility antigens (Tanigaki and Pressman, 1974), tumor-specific transplantation antigens (Klein, 1967), and blood-group-specific antigens (Watkins, 1967). All known cell-surface antigens of these types are glycoproteins. Antigen specificity may reside either in the peptide moiety, as appears to be the case with the histocompatibility antigens (Nathenson *et al.*, 1970), or in the glycoprotein side chains, as for the blood group antigens (Watkins, 1967). In addition, new antigenic carbohydrate structures may be exposed on the cell surface by enzymic action, as described in the section on masking, or possibly under *in vivo* conditions. An example of new antigen formation is the Thomsen effect. The Friedenreich antigen is formed at the surface of the human erythrocyte by the action of dilute acid or neuraminidase. A galactose residue is exposed, which is not only immunogenic but is susceptible to attack by a naturally occurring γ_M-type antibody, to cause agglutination (Uhlenbruck *et al.*, 1969).

Despite the many known examples of antigenic carbohydrate structures at the cell surface, the only known surface antigens in mammalian cells in which sialic acid plays the dominant haptenic role are the M and N blood group antigens of the human erythrocyte (Landsteiner and Levine, 1928). Both the M and N antigenic structures possess terminal N-acetylneuraminic acid residues (Springer *et al.*, 1972), and removal of these with neuraminidase destroys either M or N activity. Springer *et al.* (1972) have suggested that the N antigen is the immediate precursor of the M antigen, and that a terminal β-D-galactopyranosyl residue, susceptible to β-galactosidase action, may be present in the N antigen structure. The detailed carbohydrate structures involved in the M and N specificities have not, as yet, been rigorously established (cf. Chapter 2), although oligosaccharide structures have been proposed by Springer and Desai (1974). These structures are characterized by terminal N-acetylneuraminosyl residues linked by α-(2→3) bonds to β-D-galactopyranosyl residues. These chains are presumably bound to 2-acetamido-2-deoxy-α-D-galactopyranosyl residues linked by O-glycosyl bonds to the serine or threonine residues of the peptide backbone.

Lisowska and Duk (1975a,b) have recently presented evidence suggesting that, although antigenic activity may indeed be dependent upon the sialic-acid-containing carbohydrate chains, differences in specificity between the M and N activities may reside in genetically controlled peptide sequences.

F. Receptor for Circulating Glycoproteins

In investigations by Morell *et al.* (1968), later by Ashwell and Morell (1974), and by others (Pricer and Ashwell, 1971; Morell *et al.*, 1971), it was demonstrated that all serum glycoproteins, with only one known exception, transferrin, require the presence of bound sialic acid for continuous circulation. By the use of glycoproteins labeled with either, or both, ^{3}H and ^{64}Cu, it was shown that the sialic-acid-free glycoproteins in the serum were soon bound to isolated parenchymal cell membranes, whereas intact glycoproteins under similar conditions were not bound. In a review of these studies, Ashwell and Morell (1974) reported that binding of these macromolecules requires the presence of both calcium ions and sialic acid receptors. Thus, it appears that hepatic uptake of serum glycoproteins requires both the absence of sialic acid as terminal residues in the serum glycoproteins and the presence of sialic acid residues at the surface of liver cells.

G. Receptor for Tetanus Toxin

Gangliosides were identified as receptors for the tetanus toxin in nervous tissues (van Heyningen, 1959). Gangliosides fix many times their own weight of toxin, and the binding constant increases with the number of N-acetylneuraminic acid residues. The presence of the sialic acid residues and that of free carboxylic groups is essential for toxin fixation (van Heyningen and Miller, 1961; van Heyningen, 1963), and the lack of receptor activity of hexosamine-free gangliosides from horse erythrocytes has been attributed to the presence of N-glycolyl residues or of different fatty acids (van Heyningen, 1959).

The role of sialic acid in the fixation of the toxin was further illustrated by the release of the toxin fixed on brain structures by neuraminidase treatment (Kryzhanovskii *et al.*, 1972, 1974). This fixation is greatly increased by the formation of complexes with cerebrosides or sphingomyelin (Mellanby and van Heyningen, 1965), which do not themselves bind to the toxin. Investigation with electrical and optical techniques, of model membranes containing gangliosides, have shown that the gangliosides are able to act as receptors in the membrane, that the bound toxin maintains its structural integrity, and that the interaction is confined to the surface without modification of the bilayer structure (Clowes *et al.*, 1972). The lack of relationship between the lethal activity of the toxin and the total ganglioside content has suggested that only a small fraction of the specific loci are involved in the essential functions inhibited by the toxin (Burton and Balfour, 1962).

V. SIALIC ACID IN NORMAL AND MALIGNANT OR TRANSFORMED CELLS

In view of the complexity of the macromolecular structures at the surfaces of mammalian cells, it is not surprising that comparisons of the quantity of sialic acid at the cell surface of malignant or transformed cells (cf. Chapter 8) with that at the surface of normal cells would appear contradictory. It has been reported, for example, that in the course of changing growth conditions, carbohydrate-containing macromolecules may be either added (Miller *et al.*, 1975; Smets and Broeckhuysen-Davies, 1972) or deleted (Miller *et al.*, 1975; Smets and Broeckhuysen-Davies, 1972; Cikes *et al.*, 1973). Thus, at least for some surface components, the environment in which a cell line grows is crucial. The large, surface glycoprotein, epiglycanin, of the TA3-Ha mammary carcinoma cell of the strain A mouse is an example. This glycoprotein may represent more than 1% of the dry weight of the cells when grown in the ascites form (Codington *et al.*, 1975*b*). Miller *et al.* (1975) recently reported, however, that after several weeks in suspension culture, epiglycanin was barely detectable. Consistent with the loss of this sialic-acid-rich component was the concurrent loss of surface sialic acid in cultured TA3-Ha cells, which was only a fraction of that of the cells grown in the ascites form. If the cultured form of the TA3-Ha cell was again grown intraperitoneally in the mouse, a high concentration of epiglycanin could be detected at the cell surface.

Other apparently contradictory examples have been reported. Ohta *et al.* (1968) observed that virally transformed 3T3 or BHK cells possess, in culture, less surface sialic acid than their nontransformed cell counterparts, a result consistent with the report of Grimes (1973) that virally transformed cells generally contain, in culture, less sialyltransferase than nontransformed cells. On the other hand, Kalant *et al.* (1964) reported in an investigation of tumors grown *in vivo* that diaminoazobenzene-induced rat hepatoma cells consistently possess more surface sialic acid than normal rat liver cells. van Beek *et al.* (1973) recently corroborated a previous report of Warren *et al.* (1972) that transformed cells grown *in vitro* consistently possess a glycoprotein fraction having a sialic acid content higher than that of the corresponding normal cells. In a series of investigations of glycolipids in normal and transformed cells, Hakomori and coworkers (Hakomori and Murakami, 1968; Hakomori, 1975) and others (Mora *et al.*, 1971) have observed in transformed cells a number of examples of the deletion of carbohydrate components. Notable among these components was sialic acid (Hakomori and Murakami, 1968).

An understanding of the significance of quantitative differences in surface sialic acid between different types of cells, between transformed and nontransformed cells, and between cells grown under different conditions, must await further investigation of the physicochemical properties and biological functions of the surface macromolecules, both glycoproteins and glycolipids, to which the sialic acid residues are attached.

VI. ROLE OF SIALIC ACID IN CELL-TO-CELL INTERACTION

The interaction between surface components plays a major role in the adhesion of cells. This process is of considerable importance during embryonic development and in the formation of tumors and their dissemination (Abercrombie and Ambrose, 1962; Weiss and Mayhew, 1967). In recent years, carbohydrate structures that contain sialic acid residues have been implicated in cell-to-cell or cell-to-substrate interaction. These sialic acid residues may play a direct role or an indirect one by masking residues of D-galactose and D-galactosamine to which they are generally linked.

Klenk (1958b) and Gottschalk (1958) were the first to suggest that the negative charges at the surface of the erythrocyte were due to sialic acid residues, and this hypothesis was confirmed by cell electrophoresis before and after treatment with neuraminidase. Generally, the role of sialic acid in the phenomena of adhesion or aggregation has been implicated only indirectly and is based on the removal of sialic acid from the cell surface by neuraminidase. Unfortunately, it is not well established whether all the sialic acid residues located at the surface of the cell are susceptible to the action of the enzyme, and neuraminidases from different sources have different specificities (see Chapter 10). It has also been shown that the carboxylic groups of the sialic acid residues are oriented outward into the electrokinetic surface of the cell in general (Weiss, 1967). Consequently, neuraminidase does not act at the plane of the cell but more deeply in the outer layer, which may limit its action. An additional complication resulting from the use of neuraminidase is the attachment of the exogenously added neuraminidase to the surface of cells and its subsequent action on membrane components or extracellular substrates (McQuiddy and Lilien, 1973; Sachtleben et al., 1973). Finally, it should always be kept in mind that the definition of attachment of cells is an operational one that varies with the methods used by each investigator (see discussion in Roseman, 1974; and Roseman et al., 1974), which may explain the contradictory observations reported in the past.

Various mechanisms have been proposed for the attachment of cells to an inactive support, such as glass or plastic (adhesion); to themselves before or after a degradative process (aggregation); to antibodies, other proteins, or charged polymers (agglutination); or to specific components of blood (clotting). These mechanisms may involve various chemical forces: ionic, covalent, or hydrogen bonding; van der Waals forces; bonding between proteins, electrostatic attractions, or surface energy and contact angles.

In the earliest theory of cell adhesion or aggregation, Tyler (1947) and P. Weiss (1947) proposed an antigen–antibody interaction as the cause of attraction. The requirement of destruction of the antigen or antibody site for detachment (Roseman, 1974) and the probable absence of antibodies at the surface of cells (Roth, 1973) makes this concept unlikely. Curtis (1960; 1962) based his theory of cell adhesion or aggregation on the Verwey and Overbeck (1948) theory applied to the adhesion of two parallel membranes at about 150 Å of distance. This theory is based on the existence of van der Waals forces between membranes if the constitutions of these membranes are similar. It is evident that the sialic acid residues, which have a flexible chain of alcohol groups (C_7 to C_9), could play an important role in this type of interaction.

Abercrombie and Ambrose (1962) stressed the importance of the highly negative charge of nonaggregative erythrocytes, thus suggesting that the main repulsive forces were probably of electrostatic nature, and that calcium ions could be responsible for attraction (see also Coman, 1961; Weiss and Mayhew, 1967). This theory attributes a major role to the carboxylic group of sialic acid, both in the attraction and in the repulsion of cells, a role that could also be fulfilled by other acidic groups of the plasma-membrane proteins or of adhering proteoglycans. Another theory attributes a major role to sialic acid in cell-to-glass adhesion or cell-to-cell aggregation by relating the attraction to the rigidity of the cell surface, which is maintained by the numerous negative charges (L. Weiss, 1965); the author himself admits that his theory is highly speculative (L. Weiss, 1968). The most recent attempt in developing a general theory extends the concepts presented earlier by Tyler (1947) and P. Weiss (1947). Roseman (1970; see also Roth, 1973) suggests that the early events in cell-to-cell interaction may be explained by an interaction between carbohydrate residues and glycosyl transferases located at the surface of the cell. Roseman and his associates (Orr and Roseman, 1969; Roth *et al.*, 1971; Walther *et al.*, 1973) have devised methods for the quantitative determination of cellular aggregation and, thus, were able to support their theory by experimental evidence (see reviews in Roth, 1973; and in Roseman, 1974). According to this new concept, sialic acid residues could have a direct interaction with neuraminidases or they may mask galactosyl or galactosaminyl residues from interacting with sialyltransferases. (For a rebuttal of the role of galactosyltransferase in cellular adhesion, see Deppert et al., 1974.)

In the following paragraphs, various interactions between cells or between inanimate supports or polymers and cells will be reviewed and their relevance to the various proposed mechanisms discussed.

A. Cellular Adhesion

Weiss (1961, 1963) observed that neuraminidase acts similarly to trypsin in detaching cultured cells from glass and proposed that neuraminidase treatment alters the organization of the membrane and increases deformability (Weiss, 1965). On the other hand, after comparing normal epidermal cells and cancer cells in their attachment to glass or to each other, Coman and associates suggested different mechanisms for the cell-to-glass and the cell-to-cell interactions (Coman, 1961; Berwick and Coman, 1962). Curtis (1962), however, disagreed with this suggestion and affirmed that both mechanisms were the same, since a similar gap of 100–200 Å exists between cells or between cells and glass, and both interactions are similarly dependent on the pH and on the presence or absence of divalent cations; cells attached either to other cells or to glass react similarly with enzymes, with the exception of neuraminidase, and the adhesions are independent of temperature changes. The latter observation was not confirmed by Roth (1968).

Preliminary treatment of Chinese hamster ovarian cells (Kraemer, 1966) or 3T3 cells (Kolodny, 1972) with neuraminidase had no effect on the attachment to or spreading over glass, a result that is consistent with our present knowledge of the rapid resynthesis of sialic acid residues at the surface of mammalian cells (see Section III). Walther et al. (1973) concluded that the kinetics of adhesions were different, after quantitatively determining the adhesion to glass, to plastic, or to homologous cell layers of numerous cell types, including embryonic cells and established cell lines of neural, epithelial, and fibroblastic morphology.

B. Intercellular Aggregation

Cellular aggregation is closely involved in the differentiation and cancer processes since, in both cases, cellular aggregates release cells that migrate and adhere again to each other or to different cells. Because of our limited knowledge of the chemical structures at the surface of cells, most observations on the role of sialic acid residues in intercellular aggregates have been obtained indirectly by treatment of normal, transformed, or cancer cells with neuraminidase. No general mechanism directly implicating sialic acid residues has been developed as yet since the removal of these residues by neuraminidase or as glycopeptides by trypsin gave results varying with each type of cell. For example, Kemp (1968; 1970) found that treatment of embryonic chick muscle cells with crystalline trypsin or with neuraminidase decreased the aggregation of the cells. This led him to propose a modification of the theory of L. Weiss (1964), who had implicated the structural rigidity of the microvilli as a result of studies of the sarcoma 37 cell (Weiss, 1965). Kemp suggested that the nonadhesiveness of cancer cells and of normal, neuraminidase-treated tissue cells depends on a low and a high deformability, respectively, whereas the adhesiveness of tissue cells depends on a normal deformability.

On the other hand, Deman and Bruyneel (1973) stressed the role of the electrostatic charges of sialic acid residues in explaining the increase of aggregation of HeLa cells after neuraminidase treatment, in agreement with the lyophobic colloid theory of Curtis (1967). These observations were similar to the ones made earlier by Cook et al. (1963), who found a decrease in the electrophoretic mobility of the ascites form of sarcoma 37 cells after neuraminidase treatment, whereas the cells derived from the solid form remained unaffected. Identical results had also been obtained by Wallach and de Perez Esandi (1964) in the study of Ehrlich ascites carcinoma cells, MC_1M_{SS} solid sarcoma cells, and MC_1M_{AA} ascites sarcoma cells derived from MC_1M_{SS} cells; and by Kojima and Maekawa (1970; 1972) in the study of island-forming and non-island-forming rat hepatoma cells. In this latter study, no reduction of the surface charge of the island-forming cells could be observed upon neuraminidase treatment, although some sialic acid residues were released both from adhering and nonadhering cells. The difficulty in interpreting the results of neuraminidase treatment due to the complexity of the cell surface components is illustrated by this last-mentioned study (Kojima and Maekawa, 1972); treatment with neuraminidase did not alter the electrophoretic mobility of island-forming cells, unless they had been treated previously with hyaluronidase and chondroitinase; these results led Kojima and Maekawa (1972) to suggest the presence of various layers of polysaccharides at the surface of the cell. Further work with HeLa cells (Deman et al., 1974) showed opposite effects for neuraminidase and trypsin, the former increasing, the latter decreasing aggregation. Calcium ions promoted the aggregation of HeLa cells, whereas magnesium ions had no effect. Neither calcium nor magnesium ions, nor neuraminidase treatment had any effect upon the aggregation of human erythrocytes, which suggests that trypsin removes a material responsible for aggregation, whereas neuraminidase counteracts this effect (Deman *et al.*, 1974). It is of interest that the aggregation of growing HeLa cells is only slightly increased by neuraminidase treatment, whereas a strong increase resulted from treatment of nongrowing cells in dense cultures (Deman and Bruyneel, 1975).

Comparison of the action of neuraminidase on erythrocytes, where no effect was observed although all the sialic acid residues are located at the surface of the cell, with that on HeLa cells, where only 0.4% of the 6.0×10^8 residues of sialic acid per cell contribute to the charge of the surface, presents the paradox of the enzyme promoting aggregation by acting on residues located in cryptic positions. This unexpected result was explained by Deman *et al.* (1974) on the basis of the rigidity of the cell surface (Weiss, 1965; Curtis, 1967), which seems more applicable here than Roseman's (1970) theory.

On the other hand, both Vicker and Edwards (1972), and Lloyd and Cook (1974) explained their results in terms of Roseman's (1970) theory. The first-mentioned authors were able to obtain suspensions of baby hamster kidney (BHK) fibroblasts after a very short treatment with a low concentration of trypsin; on treatment with a low concentration of neuraminidase, the cells showed increased aggregation. Since about one-third of the charges at the surface of the cell had been removed, the aggregation effect could be explained on the basis of Curtis' (1967) theory. After transformation with polyoma virus, however, very little response to the enzyme was observed, although the charge density of the transformed cell was higher than that of the untransformed cell. Thus, it is more likely that residues such as β-D-galactosyl or N-acetyl-α-D-galactosaminyl, which are uncovered by neuraminidase treatment, are responsible for aggregation. Similarly, Lloyd and Cook (1974) found that treatment of 16C malignant, rat dermal fibroblasts with neuraminidase increased aggregation and that this effect could be reversed by addition of sialic-acid-free, bovine submaxillary mucin, or sialic-acid-free, galactose-free fetuin. According to Roseman's (1970) theory, the carbohydrate residues adjacent to the sialic acid residues are responsible for aggregation, and appearance of these residues may play a modulatory role in cellular interaction (Lloyd and Cook, 1974).

All the evidence presented so far indicates that only sialic acid residues linked to glycoproteins are involved in cell aggregation. Past studies on glycolipids have shown that only the concentration of the neutral glycolipids of normal cells is increased with an increase in cell density, whereas the concentration of neutral glycolipids of transformed cells remains unchanged under similar growth conditions. Recently, however, Yogeeswaran and Hakomori (1975) reported an increased content of "band 4" gangliosides of 3T3 cells at the early stage of cell-to-cell contact. This increase was related to the suppression of membrane-bound sialidase activity. No similar phenomenon could be observed with transformed cells. These results suggest, for cell-surface sialidase, an important role in cell interaction and in the regulation of ganglioside metabolism during cell growth.

In a reinvestigation of the action of neuraminidase on neural cells and chick-embryo muscle cells, McQuiddy and Lilien (1971) were unable to support the earlier observations of Kemp (1968, 1970), probably because of differences in the enzymic treatment: no increase of aggregation resulted from neuraminidase treatment, and the authors discount any role of sialic acid in the aggregation process.

The role of sialic acid in the interaction between cells of different origin has been far less studied than in the interaction between identical cells. Cormack (1970) observed a decreased attachment of neuraminidase-treated Walker 25 tumor cells to the mesothelial membrane of rat, but the interpretation of this experiment performed *in vivo* is complicated by the resynthesis of sialic acid residues at the surface of the treated cell. Weiss and Cudney (1971) found no effect of neuraminidase on the immunolysis of mastocytoma P815 cells by sensitized spleen cells, and it is doubtful that sialic acid residues play a direct role in the interaction between different cells. It may play an indirect one, however, for example, by masking receptor sites at the surface of either human peripheral blood lymphocytes, or sheep red blood cells, as shown by the increased stability of the rosettes formed by these cells upon neuraminidase treatment (Galili and Schlesinger, 1974).

C. Agglutination

1. Hemagglutination. The role of sialic acid residues in the M and N antigens located at the surface of the erythrocyte has been discussed in Section IV, E. No effect of neuraminidase treatment on the agglutination of erythrocytes with various antisera could be detected by Bird and Wingham (1970), probably because of the lack of sensitivity of the manual technique used (Greenwalt and Steane, 1973a). Greenwalt and Steane (1973a) were able to detect with an AutoAnalyzer continuous-flow technique an increase in agglutination with anti- A, -B, -H, -D, and -c sera after neuraminidase treatment, but with a different relationship for each antibody. These results suggested that a mechanism more complicated than the reduction of the zeta potential was implicated, maybe the decrease of ordered water near the site of antigen–antibody reaction (Good and Wood, 1971a,b; 1972). It is of interest that the same authors (Greenwalt and Steane, 1973b) observed a similar increase of agglutinability with M- and N-antisera under similar conditions, although sialic acid has been proposed to be part of the M,N determinant (see IV, E). This unexpected result was explained by a possible loss of sialic acid residues around the M- and N-antigenic sites.

Various basic polymers are known to agglutinate erythrocytes, and this phenomenon has been used in order to better understand hemagglutination. Khan and Zinneman (1970) found that neuraminidase-treated cells did not precipitate with protamine sulfate, while intact cell agglutination was inhibited by heparin. Voigtmann and Uhlenbruck (1970) correlated the decrease of agglutination by Polybrene (hexadimethrine bromide) with the decrease of sialic acid residues. Bovine erythrocytes, however, could not be precipitated by Polybrene, which suggests that a structural component may be required for agglutination. Similar results were obtained by Greenwalt and Steane (1973c) with Polybrene and protamine, but these authors observed no decrease of agglutination with polylysine after neuraminidase treatment; they explain this result on the basis of disturbance of the water molecules vicinal to the site of interaction.

Polyagglutinability involving all the blood cells of one patient has been shown to be related to a partial deficiency of sialic acid residues at the surface of the erythrocyte (Lalezari and Al-Mondhiry, 1973).

2. Platelet Agglutination and Aggregation. The "adhesion" of platelets to each other is an essential step in the formation of a hemostatic plug or an intravascular thrombolic deposit. Many widely differing materials cause "aggregation" of platelets, such as thrombin, collagen fibers, fatty acid, immune complexes, 5-hydroxytryptamine, adrenaline, etc., probably all mediated through the action of adenosine di- or triphosphate (Haslam, 1967). These various examples of agglutination will be discussed only where sialic acid residues have been shown clearly to be implicated.

Hovig (1965) was the first to observe that neuraminidase-treated platelets "aggregated" spontaneously when resuspended in citrated plasma. Incubation of washed human platelets with cytidine N-acetylneuraminic acid monophosphate in the presence of sialyl transferase increased the number of sialic acid residues at the surface of the platelets by 19–38% (Mester, 1971) with a corresponding increased "aggregation" with 5-hydroxytryptamine and a decreased "aggregation" with adenosine diphosphate. This difference in behavior of the two aggregating agents is probably due to different receptor sites (Mester *et al.*, 1972). Similar effects of neuraminidase on the platelet "aggregation" induced by adenosine diphosphate, norepinephrine, collagen, or serotonin were reported by Davis *et al.* (1972). Marked thrombocytopenia was observed after injection of neuraminidase into mice (Gasic *et al.*, 1968), rats (Choi *et al.*, 1972), and rabbits (Gröttum and Jeremic, 1973), and a low content of platelet sialic acid was observed by Gröttum and Solum (1969) in two cases of congenital thrombocytopenia with giant platelets.

Neuraminidase treatment of bovine factor VIII produced strong aggregation, which occurred in two waves. When galactose oxidase was added to the incubation mixture, no aggregation occurred. This observation suggests that the agglutination of platelets by bovine factor VIII results from a substrate–enzyme complex, in agreement with Roseman's (1970) theory (Vermylen *et al.*, 1973). Neuraminidase induced an increase in agglutination only in normal subjects and hemophilia patients, but not in von Willebrand patients, thus showing that the platelet-agglutinating property of human factor VIII is related to the content of factor VIII-related antigen and independent of factor VIII-procoagulant activity (Vermylen *et al.*, 1974). *N*-Acetylneuraminic acid and 2-acetamido-2-deoxy-D-mannose were shown to correct the prolonged clotting time of factor VIII-deficient blood (Rubin *et al.*, 1970), an effect difficult to interpret with our present knowledge of glycoprotein or glycolipid biosynthesis. The fact that synthetic *N*-acetylneuraminic acid was less active than the material obtained from natural sources suggests the presence of other active components.

On the basis of a concentration of 2.4×10^6 molecules of sialic acid per cell, a high sialyltransferase activity, and a "measurable" neuraminidase activity at the surface of the human blood platelet, Bosmann (1972) proposed that platelet aggregation could result from an interaction, at the surface of the cell, between the sialyltransferases and the galactosyl residues liberated by the action of the neuraminidase. This application of Roseman's (1970) theory to platelet aggregation would require a dynamic equilibrium of transfers and removals of sialic acid residues, with continuous formation, degradation, and reformation of the enzyme–substrate complexes.

3. Other Blood Cells. In their study of blood cells in cheek-pouch venules Atherton and Born (1973) found that neuraminidase treatment caused near complete adhesion of the granulocytes to the vascular endothelium, but addition of *N*-acetylneuraminic acid had no effect, thus establishing a clear difference between the adhesion of platelets and that of granulocytes. Increase of sialic acid concentration in the plasma by neuraminidase treatment or addition of *N*-acetylneuraminic acid increased the time of the first appearance of thrombi in cheek-pouch venules, but neither neuraminidase nor *N*-acetylneuraminic acid modified the platelet count (Atherton and Born, 1973).

Small lymphocytes leave the blood circulation through an endothelial cell that normally occurs only in the postcapillary venules of lymph nodes. Pretreatment of the small lymphocytes with neuraminidase or trypsin greatly diminished their rate of migration (Born, 1974).

VII. PHYSIOLOGICAL ROLE OF SIALIC ACID RESIDUES

A. Transport of Ions, Amino Acids, and Proteins

Conflicting results were obtained in the study of the influence of sialic acid on the flow of ions through the cell membrane: Glick and Githens (1965) observed a sharp response of the K^+ ions to the removal of sialic acid with neuraminidase in L1210 leukemia cells, whereas the transport of Na^+ ions was only slightly inhibited; the transport of K^+ ions was inhibited regardless of the direction of flow. In Ehrlich ascites cells, however, removal of sialic acid residues with neuraminidase did not alter the content of Na^+ and K^+ ions in the cells. Only a very small reduction in the unidirectional fluxes of K^+ ions was observed after neuraminidase treatment. These observations led Weiss and Levinson (1969) to conclude that anionic sites on the cell membrane were not of major importance in regulating the intracellular concentration of Na^+ and K^+ ions or the unidirectional, transmembrane flux of K^+ ions.

Sialic acid residues were found to be important in the transport of amino acids and proteins in cancer cells, since neuraminidase treatment of HeLa cells decreased the net accumulation of α-aminoisobutyric acid without altering the rate of efflux of preloaded cells (Brown and Michael, 1969), and the same treatment of L1210 leukemia cells inhibited the outward flow of proteins without influencing lysis or the release of nucleosides or sugars. Some relative specificity in the release of proteins was shown by disk–gel electrophoresis (Glick et al., 1966).

B. Phagocytosis

In an earlier investigation, Fisher and Ginsberg (1956a,b) had reported the inhibition of glycolysis and phagocytosis after treating polymorphonuclear leukocytes of guinea pig with neuraminidase. Recent reinvestigation (Noseworthy et al., 1972) of these phenomena indicates that suppression of both glycolysis and phagocytosis may be due to an inhibitor of glycolysis present in the crude preparations of neuraminidase used earlier, and not to the removal of sialic acid residues. These results were confirmed by Constantopoulos and Najjar (1973) with intact and neuraminidase-treated polymorphonuclear phagocytes in the absence of tuftsin, leukokinin, and serum. These authors found, however, that sialyl residues were absolutely required for the stimulation of phagocytosis by the three materials just mentioned.

C. Anaphylactic Shock, Hypercapnia, and Brain Excitability

The slow-reacting substance of anaphylaxis was isolated from the lung tissue of guinea pigs subjected to anaphylactic shock, and was shown to be composed of gangliosides (Smith, 1966); however, the role of sialic acid residues was not ascertained.

Gangliosides are generally considered as stable components of neuronal membranes. They were found to contain less sialic acid, however, when isolated from patients who were cyanotic before death, from cerebral tissue adjacent to a variety of intracerebral lesions, or from the brain of asphyxiated cats. Hypercapnia and the resulting respiratory acidosis are believed to be responsible for the release of an activator of brain glycosidases, which would cause the loss of about one-third of the 10^{11} neuraminic acid residues per neuron (Lowden and Wolfe, 1964).

The role of sialic acid residues in the brain tissue was illustrated by the restoration of excitability to applied electrical pulses of cerebral tissue that had lost its reactivity by exposure to cold. This restoration was obtained by addition of sialic-acid-containing glycoproteins or gangliosides (McIlwain, 1960).

D. Lymphocyte Stimulation

In 1972, Novogrodsky and Katchalski observed that lymphocytes isolated from different sources underwent blastogenesis after mild treatment with periodate, and that a preliminary treatment of mouse spleen lymphocytes with neuraminidase reduced the effect of the periodate oxidation. Since the periodate-induced blastogenesis is eliminated by reduction, it is probable that the aldehyde groups formed by oxidation of the side chain of the sialyl residues are responsible for the physiological effect. Novogrodsky and Katchalski (1972) demonstrated that the sites of activity induced by Concanavalin A binding are different from the sites induced by periodate oxidation. Blastogenesis was also induced by oxidation with galactose oxidase of the terminal galactosyl residue exposed by neuraminidase treatment (Novogrodsky and Katchalski, 1973a), showing that the presence of any aldehyde group at the extremity of the carbohydrate chain was responsible for the physiological effect. Neuraminidase treatment also released additional sites for binding soya bean agglutinin, which resulted in an increased transformation of mouse spleen lymphocytes (Novogrodsky and Katchalski, 1973b).

E. Sperm Capacitation

Capacitation is the transformation of the sperm of some animal species necessary for the penetration of the spermatozoa through the zona pellucida of the egg for fertilization. The cumulus cells of the oviduct of the hamster collected shortly after ovulation were shown to attach and to capacitate the spermatozoa. This property of the cells was abolished by treatment with neuraminidase, showing that sialic acid plays an important role in this effect (Gwatkin et al., 1972).

VIII. CONCLUSIONS

As illustrated in the preceding pages, the various physiological roles of sialic acid residues of the cell surface have been explained in the past in terms of direct and indirect effects. The direct effects result either from the chemical properties of the carboxylic group or from the rigidity of the membrane due to the charged groups. The indirect effects are attributed to the masking of a neighboring galactose or galactosamine residue by direct glycosidic linkage or to the masking of sites located farther along the chain or on neighboring molecules by bulk effect.

It is of interest that the two major residues located at the extremities of carbohydrate chains of glycoproteins (where they may have the most influence and which seem to be complementary to each other), sialic acid and L-fucose (see Dische, 1965), have the most complex biosynthetic pathways of all the carbohydrate components of glycoproteins and glycolipids. If such complex biosynthetic mechanisms have resisted changes through evolution, it is probably because the endproducts have very specific roles to play, more specific than simply the masking of receptor sites. The influence of the charges of the carbohydrate group at the end of a long carbohydrate chain cannot be minimized, but it is difficult to attribute any specific role to this charge that may not be duplicated by numerous other components of proteins or other charged polymers.

One part of the sialic acid molecule that has received little consideration is the polyhydroxylated side chain C_7 to C_9. We would like to suggest that this side chain plays a more important role than envisaged previously, since it is the result of the rather complex biochemical pathway that builds sialic acid.

This side chain seems to be the only example of a covalently bound, flexible, polyhydroxylated structure at the surface of the cell. Although very little is known about the interactions of polyhydroxylated structures, the well-known interactions between starch or cellulose molecules suggest that such interactions do exist. Additional evidence for the possible interaction of the side chain of sialic acid is found in the numerous biological activities resulting from its periodate oxidation. It is known that periodate oxidation may act at various locations of a carbohydrate or protein polymer, but the very low concentration (*ca.* 0.002–0.5 mM), the very short time of application (generally a few minutes), and the resistance of the cells to this treatment suggest that it is the side chain of sialic acid residues located at the cell surface rather than the –S–S– protein bonds located deep inside the membrane that are attacked (see Liao *et al.*, 1973). Periodate oxidation of cells was shown to inhibit the mating of *E. coli* U-12 (Sneath and Lederberg, 1961), to block reaggregation of dissociated embryonic cells (Moscona, 1962), and to act directly on the cytoplasmic cell surface and not with the jelly coat of eggs for activation and fertilization (Runnström *et al.*, 1959). Other effects, such as lymphocyte stimulation, have been previously discussed (see Section VII, D).

The ability of the side chain of a sialic acid residue to interact with the protein part of other molecules has been shown in the polymerization of periodate-oxidized α_1-acid glycoprotein (Hughes and Jeanloz, 1966), and it is conceivable that this undegraded side chain may have more subtle effects through hydrogen bonding. Such effects could conceivably be of paramount importance during embryonic development and differentiation, and could be involved in many as yet unknown complex cellular interactions which must characterize the living organism. The side chain is generally assumed to be in the equatorial position, when the sialyl residue is in the $^2C_5$ chair conformation, but the presence of the bulky carboxylic group and the lack of a substituent at C_3 may certainly decrease the stability of this conformation and many skew and boat conformations are possible between the two extreme $^2C_5$ and $^5C_2$ chair conformations. These would allow a great freedom of movement to the side chain. Since any attractive effect on the carboxylic group would be reflected in the conformation of the molecule, and consequently on the position of the side chain in space, subtle influences could thus be transmitted along the sialyl residue structure. Although no direct experimental evidence has been presented as yet for this effect, the unique presence of this polyhydroxylated side chain at the surface of animal cells certainly does deserve attention.

IX. REFERENCES

Abercrombie, M., and Ambrose, E. J., 1962, The surface properties of cancer cells: A review, *Cancer Res.* **22**:525.
Allen, A. K., Neuberger, A., and Sharon, N., 1973, The purification, composition and specificity of wheat-germ agglutinin, *Biochem. J.* **131**:155.
Apffel, C. A., and Peters, J. H., 1970, Regulation of antigenic expression, *J. Theor. Biol.* **26**:47.
Ashwell, G., and Morell, A. G., 1971, Galactose: A cryptic determinant of glycoprotein catabolism, *in: Glycoproteins of Blood Cells and Plasma* (G. A. Jamieson and T. J. Greenwalt, eds.), pp. 173–189, Lippincott, Philadelphia.
Ashwell, G., and Morell, A. G., 1974, The role of surface carbohydrates in the hepatic recognition and transport of circulating glycoproteins, *Adv. Enzymol.* **41**:99.
Atherton, A., and Born, G. V., 1973, Effects of neuraminidase and N-acetyl neuraminic acid on the adhesion of circulating granulocytes and platelets in venules, *J. Physiol. (Lond.)* **234**:66P.
Balazs, E. A., and Jacobson, B., 1966, Interaction of amino sugars and amino sugar-containing macromolecules with viruses, cells, and tissues, *in: The Amino Sugars* (E. A. Balazs and R. W. Jeanloz, eds.), Vol. IIB, pp. 361–395, Academic Press, New York.
Bekesi, J. G., St. Arneault, G., and Holland, J. F., 1971, Increase of leukemia L1210 immunogenicity by *Vibrio cholerae* neuraminidase treatment, *Cancer Res.* **31**:2130.
Berwick, L., and Coman, D. R., 1962, Some chemical factors in cellular adhesion and stickiness, *Cancer Res.* **22**:982.
Bhavanandan, V. P., and Davidson, E. A., 1975, Personal communication.
Bird, G. W. G., and Wingham, J., 1970, N-Acetylneuraminic (sialic) acid and human blood group antigen structure, Vox. Sang. **18**:240.
Blix, G., and Jeanloz, R. W., 1969, Sialic acids and muramic acid, *in: The Amino Sugars* (R. W. Jeanloz, ed.), Vol. IA, pp. 213–265, Academic Press, New York.
Born, G. V. R., 1974, Research on the mechanisms of the intravascular adhesion of circulating cells, *in: Platelets and Thrombosis* (S. Sherry and A. Scriabine, eds.), pp. 113–126 University Park Press, Baltimore.
Bosmann, H. B., 1972, Platelet adhesiveness and aggregation. II. Surface sialic acid, glycoprotein: N-acetylneuraminic acid transferase, and neuraminidase of human blood platelets, *Biochim. Biophys. Acta* **279**:456.
Bray, B. A., and Alexander, B., 1969, Progressive cleavage of sialic acid from platelets and their electrophoretic mobility, *Blood* **34**:523.
Brecht, H., Ammerling, U., and Rott, R., 1971, Undisturbed release of influenza virus in the presence of univalent antineuraminidase antibodies, *Virology* **46**:337.
Brown, D. M., and Michael, A. F., 1969, Effect of neuraminidase on the accumulation of alpha-aminoisobutyric acid in HeLa cells, *Proc. Soc. Exp. Biol. Med.* **131**:568.
Burger, M. M., 1969, A difference in the architecture of the surface membrane of normal and virally transformed cells, *Proc. Natl. Acad. Sci. U. S.* **62**:994.
Burnet, F. M., 1951, Microproteins in relation to virus action, *Physiol. Rev.* **31**:131.

Burnet, F. M., and Anderson, S. G., 1947, The "T" antigen of guinea pig and human red cells, *Austr. J. Exp. Biol. Med. Sci.* **25:**213.

Burton, R. M., and Balfour, Y. M., 1962, Tetanus toxin activity and ganglioside content of rat brain, *Biochem. Pharmacol.* **11:**974.

Choi, S.-I., Simone, J. V., and Journey, C. J., 1972, Neuraminidase-induced thrombocytopenia in rats, *Br. J. Haematol.* **22:**93.

Choppin, P. W., and Schied, A., 1974, Identification of biological activities of paramyxovirus glycoproteins. Activation of cell fusion, hemolysis, and infectivity by proteolytic cleavage of an inactive precursor protein of Sendai virus, *Virology* **57:**475.

Cikes, M., Friberg, S., Jr., and Klein, G., 1973, Progressive loss of H-2 antigens with concomitant increase of cell-surface antigen(s) determined by Moloney leukemia virus in cultured mucine lymphomas, *J. Natl. Cancer Inst.* **50:**347.

Clowes, A. W., Cherry, R. J., and Chapman, D., 1972, Physical effects of tetanus toxin on model membranes containing ganglioside, *J. Mol. Biol.* **67:**49.

Codington, J. F., 1975, Masking of cell-surface antigens on cancer cells, in: *Cellular Membranes and Tumor Cell Behavior 28th Annual Symposium,* M. D. Anderson Hospital and Tumor Institute, Houston, pp. 399–419, William and Wilkins, Baltimore.

Codington, J. F., Sanford, B. H., and Jeanloz, R. W., 1970, Glycoprotein coat of the TA3 cell. I. Removal of carbohydrate and protein material from viable cells, *J. Natl. Cancer Inst.* **45:**637.

Codington, J. F., Sanford, B. H., and Jeanloz, R. W., 1972, Glycoprotein coat of the TA3 cell. Isolation and partial characterization of a sialic acid containing glycoprotein fraction, *Biochemistry* **11:**2559.

Codington, J. F., Sanford, B. H., and Jeanloz, R. W., 1973, Cell surface glycoproteins of two sublines of the TA3 tumor, *J. Natl. Cancer Inst.* **51:**585.

Codington, J. F., Tuttle, B., and Jeanloz, R. W., 1974, Methods for the removal, isolation and characterization of glycoprotein fragments from the TA3 tumor cell surface, *Colloq. Int. Cent. Natl. Rech. Sci.* **221:**957.

Codington, J. F., Linsley, K. B., Jeanloz, R. W., Irimura, T., and Osawa, T., 1975a, Immunochemical and chemical investigations of the structure of glycoprotein fragments obtained from epiglycanin, a glycoprotein at the surface of the TA3-Ha cancer cell, *Carbohyd. Res.* **40:**171.

Codington, J. F., Cooper, A. G., Brown, M. C., and Jeanloz, R. W., 1975b, Evidence that the major cell surface glycoprotein of the TA3-Ha carcinoma contains the *Vicia graminea* receptor sites, *Biochemistry* **14:**855.

Coman, D. R., 1961, Adhesiveness and stickiness: Two independent properties of the cell surface, *Cancer Res.* **21:**1436.

Constantopoulos, A., and Najjar, V. A., 1973, The requirement of membrane sialic acid in the stimulation of phagocytosis by the natural tetrapeptide, tuftsin, *J. Biol. Chem.* **248:**3819.

Cook, G. M. W., and Stoddart, R. W., 1973, *Surface Carbohydrates of the Eukaryotic Cell,* Academic Press, New York.

Cook, G. M. W., Heard, D. H., and Seaman, F., 1960, A sialomucopeptide liberated by trypsin from the human erythrocyte, *Nature (Lond.)* **188:**1011.

Cook, G. M. W., Heard, D. H., and Seaman, F., 1961, Sialic acids and the electrokinetic charge of the human erythrocyte, *Nature (Lond.)* **191:**44.

Cook, G. M. W., Heard, D. H., and Seaman, G. V. F., 1962, The electrokinetic characterization of the Ehrlich ascites carcinoma cell, *Exp. Cell Res.* **28:**27.

Cook, G. M. W., Seaman, G. V. F., and Weiss, L., 1963, Physicochemical differences between ascitic and solid forms of Sarcoma 37 cells, *Cancer Res.* **23**:1813.
Cormack, D., 1970, Effect of enzymatic removal of sialic acid on the adherence of Walker 256 tumour cells to mesothelial membrane, *Cancer Res.* **30**:1459.
Cuatrecasas, P., and Illiano, G., 1971, Membrane sialic acid and the mechanism of insulin action in adipose tissue cells, *J. Biol. Chem.* **246**:4938.
Currie, G. A., 1967, Masking of antigens on the Landschütz ascites tumor, *Lancet* **1967(2)**:1336.
Currie, G. A., and Bagshawe, K. D., 1967, The masking of antigens on trophoblast and cancer cells, *Lancet* **1967(1)**:708.
Currie, G. A., and Bagshawe, K. D., 1968a, The effect of neuraminidase on the immunogenicity of the Landschütz ascites tumor: Site and mode of action, *Br. J. Cancer* **22**:588.
Currie, G. A., and Bagshawe, K. D., 1968b, The role of sialic acid in antigenic expression: Further studies of the Landschütz ascites tumor, *Br. J. Cancer* **22**:843.
Curtis, A. S. G., 1960, Cell contacts: Some physical considerations, *Am. Nat.* **94**:37.
Curtis, A. S. G., 1962, Cell contact and adhesion, *Biol. Rev.* **37**:82.
Curtis, A. S. G., 1967, *The Cell Surface: Its Molecular Role in Morphogenesis,* Logos Press, Academic Press, London.
Davis, J. W., Yue, K. T. N., and Phillips, P. E., 1972, The effect of neuraminidase on platelet aggregation induced by ADP, norepinephrine, collagen or serotonin, *Thromb. Diath. Haemorrh.* **28**:221.
Deman, J. J., and Bruyneel, E., 1973, A method for the quantitative measurement of cell aggregation, *Exp. Cell Res.* **81**:351.
Deman, J. J., and Bruyneel, E. A., 1975, Intercellular adhesiveness and neuraminidase effect following release from density inhibition of cell growth, *Biochem. Biophys. Res. Commun.* **62**:895.
Deman, J. J., Bruyneel, E. A., and Mareel, M. M., 1974, A study on the mechanism of intercellular adhesion. Effects of neuraminidase, calcium, and trypsin on the aggregation of suspended HeLa cells, *J. Cell. Biol.* **60**:641.
Deppert, W., Werchau, H., and Walter, G., 1974, Differentiation between intracellular and cell surface glycosyl transferases. Galactosyl transferase activity in intact cells and in cell homogenate, *Proc. Natl. Acad. Sci. U.S.* **71**:3065.
Dische, Z., 1965, Amino- sugar-containing compounds in mucuses and in mucous membranes, in: *The Amino Sugars* (E. A. Balazs and R. W. Jeanloz, eds.), Vol. IIA, pp. 115–140, Academic Press, New York.
Drzeniek, R., and Gauhe, A., 1970, Differences in substrate specificity of myxovirus neuraminidases, *Biochem. Biophys. Res. Commun.* **38**:651.
Eylar, E. H., Madoff, M. A., Brady, O. V., and Oncley, J. L., 1962, The contribution of sialic acid to the surface charge of the erythrocyte, *J. Biol. Chem.* **237**:1992.
Fazekas de St. Groth, S., 1949, Modification of virus receptors by metaperiodate; properties of IO_4-treated red-cells, *Austr. J. Exp. Biol. Med. Sci.* **27**:65.
Fisher, T. N., and Ginsberg, H. S., 1956a, The reaction of influenza viruses with guinea pig polymorphonuclear leucocytes. II. The reduction of white blood cell glycolysis by influenza viruses and receptor-destroying enzyme (RDE), *Virology* **2**:637.
Fisher, T. N., and Ginsberg, H. S., 1956b, The reaction of influenza viruses with guinea pig polymorphonuclear leucocytes. III. Studies on the mechanism by which influenza viruses inhibit phagocytosis, *Virology* **2**:656.
Forrester, J. A., Ambrose, E. J., and Stocker, M. G. P., 1964, Microelectrophoresis of normal and transformed clones of hamster kidney fibroblasts, *Nature (Lond.)* **201**:945.

Fresen, K. O., and Ubendorfer, A., 1973, Physicochemical membrane changes in Ehrlich ascites tumor cells infected with oncolytic influenza virus, *Arch. Gesamte Virusforsch.* **41**:267.

Friberg, S., 1972, Comparison of an immunoresistant and an immunosusceptible ascites subline from murine tumor TA3. I. Transplantability, morphology, and some physicochemical characteristics, *J. Natl. Cancer Inst.* **48**:1463.

Friedenreich, V., 1930, *The Thomsen Hemagglutination Phenomenon. Production of a Specific Receptor Quality in Red Cell Corpuscles by Bacterial Activity*, Levin and Munksgaard, Copenhagen.

Fuhrmann, G. F., Granzer, E., Kübler, W., Rueff, F., and Ruhenstroth-Bauer, G., 1962, Neuraminsäurenbedingte Strukturunterschiede der Zellmembranen normaler und maligner Leberzellen, *Z. Naturforsch.* **17b**:610.

Galili, U., and Schlesinger, M., 1974, The formation of stable E rosettes after neuraminidase treatment of either human peripheral blood lymphocytes or of sheep red blood cells, *J. Immunol.* **112**:1628.

Gasic, G., and Gasic, T., 1963, Removal of PAS positive surface sugars in tumor cells by glycosidases, *Proc. Soc. Exp. Biol. Med.* **114**:660.

Gasic, G. J., Gasic, T. B., and Stewart, C. C., 1968, Antimetastatic effects associated with platelet reduction, *Proc. Natl. Acad. Sci. U.S.* **61**:46.

Gesner, B. M., and Woodruff, J. J., 1969, Factors affecting the distribution of lymphocytes, in: *Cellular Recognition* (R. T. Smith and R. A. Good, eds.), pp. 79–90, Appleton-Century-Crofts, New York.

Gielen, W., 1968, Vorkommen und biologische Bedeutung der Neuraminsäure, *Naturwissenschaften* **55**:104.

Glick, J. L., and Githens, S., III, 1965, The role of sialic acid in potassium transport of L1210 leukaemia cells, *Nature (Lond.)* **208**:88.

Glick, J. L., Goldberg, A. R., and Pardee, A. B., 1966, The role of sialic acid in the release of proteins from L1210 leukemia cells. *Cancer Res.* **26**:1774.

Glick, M. C., 1974, Chemical components of surface membranes related to biological properties, in: *Biology and Chemistry of Eucaryotic Cell Surface* (E. Y. C. Lee and E. E. Smith, eds.), pp. 213–240, Academic Press, New York.

Good, W., and Wood, J. E., 1971a, The hydrational effect of alkali metal and halide ions on the Rh-anti-Rh system, *Immunology* **90**:37.

Good, W., and Wood, J. E., 1971b, The hydrational effect of alkaline-earth chlorides and selected non-electrolytes on the Rh-anti-Rh system, *Immunology* **21**:617.

Good, W., and Wood, J. E., 1972, Hydrational aspects of A-anti-A and B-anti-B interactions, *Immunology* **23**:423.

Gottschalk, A., 1958, General discussion, in: *Chemistry and Biology of Mucopolysaccharides* (G. E. W. Wolstenholme and M. O'Connor, eds.), pp. 306–313, Little Brown, Boston.

Greenaway, P. J., and LeVine, D., 1973, Binding of N-acetyl-neuraminic acid by wheat germ agglutinin, *Nature (Lond.) New Biol.* **241**:191.

Greenwalt, T. J., and Steane, E. A., 1973a, Quantitative haemagglutination. 4. Effect of neuraminidase treatment on agglutination by blood group antibodies, *Br. J. Haematol.* **25**:207.

Greenwalt, T. J., and Steane, E. A., 1973b, Quantitative haemagglutination. 5. Influence of in vivo ageing and neuraminidase treatment on the M and N antigens of human red cells, *Br. J. Haematol.* **25**:217.

Greenwalt, T. J., and Steane, E. A., 1973c, Quantitative haemagglutination. 6. Relationship of sialic acid content of red cells and aggregation by polybrene, protamine and poly-L-lysine, *Br. J. Haematol.* **25**:227.

Grimes, W. J., 1973, Glycosyltransferase and sialic acid levels of normal and transformed cells, *Biochemistry* **12**:990.
Gröttum, K. A., and Jeremic, M., 1973, Neuraminidase injections in rabbits. Reduced platelet surface charge, aggregation and thrombocytopenia, *Thromb. Diath. Haemorrh.* **29**:461.
Gröttum, K. A., and Solum, N. O., 1969, Congenital thrombocytopenia with giant platelets. A defect in the platelet membrane, *Br. J. Haematol.* **16**:277.
Grothaus, E. A., Flye, M. W., Yunis, E., and Amos, D. B., 1971, Human lymphocyte antigen reactivity modified by neuraminidase, *Science* **173**:542.
Gwatkin, R. B. L., Andersen, O. F., and Hutchinson, C. F., 1972, Capacitation of hamster spermatozoa *in vitro*: The role of cumulus components, *J. Reprod. Fertil.* **30**:389.
Hakomori, S.-I., 1975, Structures and organization of cell surface glycolipids. Dependency on cell growth and malignant transformation, *Biochim. Biophys. Acta* **417**:55.
Hakomori, S.-I., and Murakami, W. T., 1968, Glycolipids of hamster fibroblasts and derived malignant transformed cell lines, *Proc. Natl. Acad. Sci. U.S.* **59**:254.
Haksar, A., Maudsley, D. V., Kimmel, G. L., Peron, F. G., Robidoux, W. R., Jr., and Gahnon, G., 1974, Adrenocorticotropin stimulation of cyclic adenosine 3'-5'-monophosphate formation in isolated rat adrenal cells. The role of membrane sialic acid, *Biochim. Biophys. Acta* **362**:356.
Haslam, R. J., 1967, Mechanisms of blood platelet aggregation, in: *Physiology of Hemostasis and Thrombosis* (S. A. Johnson and W. H. Seegers, eds.), pp. 88–112, C. C. Thomas, Springfield, Ill.
Henle, W., Henle, G., Groupé, V., and Chambers, L. A., 1944, Studies on complement fixation with the viruses of influenza, *J. Immunol.* **48**:163.
Hirst, G. K., 1941, The agglutination of red cells by allantoic fluid of chick embryos infected with influenza virus, *Science* **94**:22.
Hirst, G. K., 1942, Adsorption of influenza hemagglutinins and virus by red blood cells, *J. Exp. Med.* **76**:195.
Hirst, G. K., 1943, Adsorption of influenza virus on cells of the respiratory tract, *J. Exp. Med.* **78**:99.
Hirst, G. K., 1945, Annual Report of the Rockefeller Found, Int. Health Div., p. 50, quoted in Hirst (1948).
Hirst, G. K., 1948, The nature of the virus receptors of red cells. I. Evidence on the chemical nature of the virus receptors of red cells and of the existence of a closely analogous substance in normal serum, *J. Exp. Med.* **87**:301.
Holland, J. F., St. Arneault, G., and Bekesi, G., 1972, Combined chemo- and immunotherapy of transplantable and spontaneous murine leukemia, *Abstr. Am. Assoc. Cancer Res.* **1972**:83.
Hollenberg, M. D., and Cuatrecasas, P., 1975, Insulin: Interaction with membrane receptors and relationship to cyclic purine nucleotides and cell growth, *Fed. Proc.* **34**:1556.
Hovig, T., 1965. The effect of various enzymes on the ultrastructure, aggregation and clot retraction ability of rabbit blood platelets, *Thromb. Diath. Haemorrh.* **13**:184.
Huang, R. T. C., Rott, R., and Klenk, H. D., 1973, On the receptor of influenza viruses. 1. Artificial receptor for influenza virus, *Z. Naturforsch.* **28c**:342.
Hughes, R. C., and Jeanloz, R. W., 1966, Sequential periodate oxidation of α_1-acid glycoprotein, *Biochemistry* **5**:253.
Hughes, R. C., Sanford, B. H., and Jeanloz, R. W., 1972, Regeneration of the surface glycoproteins of a transplantable mouse tumor cell after treatment with neuraminidase, *Proc. Natl. Acad. Sci. U.S.* **69**:942.

Hughes, R. C., Palmer, P. D., and Sanford, B. H., 1973, Factors involved in the cytotoxicity of normal guinea pig serum for cells of murine tumor TA3 sublines treated with neuraminidase, *J. Immunol.* **111**:1071.

Kalant, H., Mons, W., and Guttman, M., 1964, Sialic acid content of normal rat liver and of DAB-induced hepatomata, *Can. J. Physiol. Pharmacol.* **42**:25.

Kemp, R. B., 1968, Effect of the removal of cell surface sialic acids on cell aggregation *in vitro, Nature (Lond.)* **218**:1255.

Kemp, R. B., 1970, The effect of neuraminidase (3:2:1:18) on the aggregation of cells dissociated from embryonic chick muscle tissue, *J. Cell Sci.* **6**:751.

Khan, M. Y., and Zinneman, H. H., 1970, The role of sialic acid in hemagglutination, *Am. J. Clin. Pathol.* **54**:715.

Klein, G., 1967, Tumor antigens, *in: The Specificity of Cell Surfaces* (B. D. Davis and L. Warren, eds.), pp. 165–180, Prentice-Hall, Englewood Cliffs, N. J.

Klenk, E., 1958*a*, Neuraminic acid, *in: Chemistry and Biology of Micropolysaccharides* (G. E. W. Wolstenholme and M. O'Connor, eds.), pp. 296–305, Little Brown, Boston.

Klenk, E., 1958*b*, General discussion, *in: Chemistry and Biology of Mucopolysaccharides* (G. E. W. Wolstenholme and M. O'Connor, eds.), pp. 306–313, Little Brown, Boston.

Klenk, E., and Lempfrid, H., 1957, Über die Natur der Zellreceptoren für das Influenzavirus, *Hoppe-Seyl. Z.* **307**:278.

Klenk, E., and Stoffel, W., 1956, Zur Kenntnis der Zellreceptoren für das Influenzavirus. Über das Vorkommen von Neuraminsäure im Eiweiss des Erthrocytenstromas, *Hoppe-Seyl. Z.* **303**:78.

Kojima, K., and Maekawa, A., 1970, Difference in electrokinetic charge of cells between two cell types of ascites hepatoma after removal of sialic acid, *Cancer Res.* **30**:2858.

Kojima, K., and Maekawa, A., 1972, A difference in the architecture of surface membrane between two cell types of rat ascites hepatoma, *Cancer Res.* **32**:847.

Kolodny, G. M., 1972, Effect of various inhibitors on re-adhesion of trypsinized cells in culture, *Exp. Cell Res.* **70**:196.

Kraemer, P. M., 1966, Regeneration of sialic acid on the surface of Chinese hamster cells in culture. I. General characteristics of the replacement process, *J. Cell Physiol.* **68**:85.

Kraemer, P. M., 1967, Configuration change of surface sialic acid during mitosis, *J. Cell Biol.* **33**:197.

Kryzhanovskii, G. N., and Sakharova, O. P., 1972, Effect of neuraminidase on the protagon-tetanus toxin complex, *Byull. Eksp. Biol. Med.* **73**:36.

Kryzhanovskii, G. N., Rozanov, A. Ya., and Bondarchuk, G. N., 1973, *In vitro* release of tetanotoxin, fixed by neurostructures, under the effect of neuraminidase, *Byull. Eksp. Biol. Med.* **76**:26.

Lalezari, P., and Al-Mondhiry, H., 1973, Sialic acid deficiency of human red blood cells associated with persistent red cell, leucocyte, and platelet polyagglutinability, *Br. J. Haematol.* **25**:399.

Landsteiner, L., and Levine, P., 1928, On individual differences in human blood, *J. Exp. Med.* **47**:757.

Lee, A., 1968, Effect of neuraminidase on the phagocytosis of heterologous red cells by mouse peritoneal macrophages, *Proc. Soc. Exp. Biol. Med.* **128**:891.

Lee, E. Y. C., and Smith, E. E. (eds.), 1974, *Biology and Chemistry of Eucaryotic Cell Surfaces,* Academic Press, New York.

LeVine, D., Kaplan, M. J., and Greenaway, P. J., 1972, The purification and characterization of wheat-germ agglutinin, *Biochem. J.* **129**:847.

Levinson, B., Peper, D., and Belyavin, G., 1969, Substituted sialic acid prosthetic groups as determinants of viral hemagglutination, *J. Virol.* **3**:477.

Liao, T.-H., Gallop, P. M., and Blumenfeld, O. O., 1973, Modification of sialyl residues of sialoglycoprotein(s) of the human erythrocyte surface, *J. Biol. Chem.* **248**:8247.

Lisowska, E., and Duk, M., 1975a, Modification of amino groups of human-erythrocyte glycoproteins and the new concept on the structural basis of M and N blood-group specificity, *Eur. J. Biochem.* **54**:469.

Lisowska, E., and Duk, M., 1975b, Effect of modification of amino groups of human erythrocytes on M, N, and N_{vg} blood group specificities, *Vox Sang.* **28**:392.

Lloyd, C. W., and Cook, G. M. W., 1974, On the mechanism of the increased aggregation by neuraminidase of 16C malignant rat dermal fibroblasts *in vitro*, *J. Cell Sci.* **15**:575.

Lowden, J. A., and Wolfe, L. S., 1964, Studies on brain gangliosides. IV. The effect of hypercapnia on gangliosides *in vivo*, *Can. J. Biochem.* **42**:1703.

Lukert, P. D., 1972, Chemical characterization of avian infectious bronchitis virus receptor sites, *Am. J. Vet. Res.* **33**:987.

Madoff, M. A., Ebbe, S., and Baldini, M., 1964, Sialic acid of human blood platelets, *J. Clin. Invest.* **43**:870.

Manchee, R. J., and Taylor-Robinson, D., 1969, Utilization of neuraminic acid receptors by mycoplasmas, *J. Bacteriol.* **98**:914.

Marchalonis, J. J., and Edelman, G. M., 1968, Isolation and characterization of a hemagglutinin from *Limulus polyphemus*, *J. Mol. Biol.* **32**:453.

McClelland, L., and Hare, R., 1941, The adsorption of influenza virus by red cells and a new *in vitro* method of measuring antibodies for influenza virus, *Can. J. Publ. Health* **32**:530.

McCrea, J. F., 1954, Studies on influenza virus receptor substance and receptor-substance analogues. II. Isolation and purification of a mucoprotein receptor substance from human erythrocyte stroma treated with pentane, *Yale J. Biol. Med.* **26**:191.

McIlwain, H., 1960, Characterization of constituents of blood plasma and of the brain which restore excitability to isolated cerebral tissue, *Biochem. J.* **76**:16P.

McQuiddy, P., and Lilien, J. E., 1971, Sialic acid and cell aggregation, *J. Cell Sci.* **9**:823.

McQuiddy, P., and Lilien, J. E., 1973, The binding of exogenously added neuraminidase to cells and tissues in culture, *Biochim. Biophys. Acta* **291**:774.

Meindl, P., Bodo, G., Palese, P., Schulman, J., and Tuppy, H., 1974, Inhibition of neuraminidase activity by derivatives of 2-deoxy-2,3-dehydro-*N*-acetylneuraminic acid, *Virology* **58**:457.

Mellanby, J., and van Heyningen, W. E., 1965, Fixation of tetanus toxin by subcellular fractions of brain, *J. Neurochem.* **12**:77.

Mester, L., 1971, Role of sialoglycoproteins in platelet aggregation, in: *Platelet Aggregation* (J. Caen, ed.), pp. 131–135, Masson, Paris.

Mester, L., Szabados, L., Born, G. V., and Michal, F., 1972, Changes in the aggregation of platelets enriched in sialic acid, *Nature (Lond.) New Biol.* **236**:213.

Miller, D. K., Cooper, A. G., Brown, M. C., and Jeanloz, R. W., 1975, Reversible loss, in suspension culture, of a major cell-surface glycoprotein of the TA3-Ha mouse tumor, *J. Natl. Cancer Inst.* **55**:1249.

Mora, P. T., Cumar, F. A., and Brady, R. O., 1971, A common biochemical change in SV40 and polyoma virus transformed mouse cells coupled to control of cell growth in culture, *Virology* **46**:60.

Morell, A. G., Irving, R. A., Sternlieb, I., Scheinberg, I. H., and Ashwell, G., 1968, Physical and chemical studies on ceruloplasmin. V. Metabolic studies on sialic acid-free ceruloplasmin *in vivo*, *J. Biol. Chem.* **243**:155.

Morell, A. G., Gregoriadis, G., Scheinberg, I. H., Hickman, J., and Ashwell, G., 1971, The role of sialic acid in determining the survival of glycoproteins in the circulation, *J. Biol. Chem.* **246**:1461.

Moscona, A. A., 1962, Analysis of cell recombinations in experimental synthesis of tissues *in vitro*, *J. Cell Comp. Physiol.* **60**:Suppl. 1, 65.

Nathenson, S. G., Shimada, A., Yamane, K., Muramatsu, T., Cullen, S., Mann, D. L., Fahey, J. L., and Graff, R., 1970, Biochemical properties of papain-solubilized murine and human histocompatibility alloantigens, *Fed. Proc.* **29**:2026.

Nicolson, G. L., 1974, The interactions of lectins with animal cell surfaces, *Intern. Rev. Cytol.* **39**:89.

Noseworthy, J., Jr., Korchak, H., and Karnovsky, M. L., 1972, Phagocytosis and the sialic acid of the surface of polymorphonuclear leukocytes, *J. Cell Physiol.* **79**:91.

Novogrodsky, A., and Katchalski, E., 1972, Membrane site modified on induction of the transformation of lymphocytes by periodate, *Proc. Natl. Acad. Sci. U.S.* **69**:3207.

Novogrodsky, A., and Katchalski, E., 1973*a*, Induction of lymphocyte transformation by sequential treatment with neuraminidase and galactose oxidase, *Proc. Natl. Acad. Sci. U.S.* **70**:1824.

Novogrodsky, A., and Katchalski, E., 1973*b*, Transformation of neuraminidase-treated lymphocytes by soybean agglutinin, *Proc. Natl. Acad. Sci. U.S.* **70**:2515.

Ohta, N., Pardee, A. B., McAuslan, B. R., and Burger, M. M., 1968, Sialic acid contents and controls of normal and malignant cells, *Biochim. Biophys. Acta* **158**:98.

Orr, C. W., and Roseman, S., 1969, Intercellular adhesion: I. A quantitative assay for measuring the rate of adhesion, *J. Membrane Biol.* **1**:109.

Palese, P., Schulman, J. L., Bodo, G., and Meindl, P., 1974*a*, Inhibition of influenza and parainfluenza virus replication in tissue culture by 2-deoxy-2,3-dehydro-*N*-trifluoroacetylneuraminic acid, *Virology* **59**:490.

Palese, P., Tobita, K., Ueda, M., and Compans, R. W., 1974*b*, Characterization of temperature sensitive influenza virus mutants defective in neuraminidase, *Virology* **61**:397.

Pardoe, G. I., and Uhlenbruck, G., 1970, Characteristics of antigenic determinants of intact cell surfaces, *J. Med. Lab. Technol.* **27**:249.

Patel, A. A., and Srinivasa Rao, S., 1966, Subcellular fixation of tetanus toxin in susceptible and resistant species, *Br. J. Pharmacol.* **26**:740.

Pricer, W. E., and Ashwell, G., 1971, The binding of desialylated glycoproteins by plasma membranes of rat liver, *J. Biol. Chem.* **246**:4825.

Rapin, A. M. C., and Burger, M. M., 1974, Tumor cell surfaces: General alterations detected by agglutinins, *Adv. Cancer Res.* **20**:1.

Rios, A., and Simmons, R. L., 1973, Immunospecific regression of various syngeneic mouse tumors in response to neuraminidase-treated tumor cells, *J. Natl. Cancer Inst.* **51**:637.

Rios, A., and Simmons, R. L., 1974, Active specific immunotherapy of minimal residual tumor: Excision plus neuraminidase-treated tumor cells, *Intern. J. Cancer* **13**:71.

Rogentine, G. N., Jr., and Plocinik, B. A., 1974, Carbohydrate inhibition studies of the naturally occurring human antibody to neuraminidase-treated human lymphocytes, *J. Immunol.* **113**:848.

Roseman, S., 1970, The synthesis of complex carbohydrates by multiglycosyltransferase systems and their potential function in intercellular adhesion, *Chem. Phys. Lipids* **5**:270.

Roseman, S., 1974, Complex carbohydrate in intercellular adhesion, *in: The Cell Surface in Development* (A. A. Moscona, ed.), pp. 255–271, John Wiley and Sons, New York.

Roseman, S., Rottmann, W., Walther, B., Öhman, R., and Umbreit, J., 1974, Measurement of cell–cell interactions, *Methods Enzymol.* **22:**597.
Rosenberg, S. A., and Einstein, A. B., Jr., 1972, Sialic acids on the plasma membrane of cultured human lymphoid cells. Chemical aspect and biosynthesis, *J. Cell Biol.* **53:**466.
Rosenberg, S. A., and Rogentine, G. N., Jr., 1972, Natural human antibodies to hidden membrane components, *Nature (Lond.) New Biol.* **239:**203.
Rosenberg, S. A., and Schwarz, S., 1974, Murine autoantibodies to a cryptic membrane antigen: Possible explanation for neuraminidase-induced increase in cell immunogenicity, *J. Natl. Cancer Inst.* **52:**1151.
Rosenberg, S. A., Plocinik, B. A., and Rogentine, G. N., Jr., 1972, Unmasking of human lymphoid cell heteroantigens by neuraminidase treatment, *J. Natl. Cancer Inst.* **48:**1271.
Roth, S., 1968, Studies on intercellular adhesive selectivity, *Dev. Biol.* **18:**602.
Roth, S., 1973, A molecular model for cell interactions, *Q. Rev. Biol.* **48:**541.
Roth, S., McGuire, E. J., and Roseman, S., 1971, An assay for intercellular adhesive specificity, *J. Cell Biol.* **51:**525.
Rubin, H., Fernbach, T., and Ritz, N. D., 1970, Corrective effect of sialic acid on the clotting of factor VIII deficient blood, *Thromb. Diath. Haemorrh.* **24:**152.
Runnström, J., Hagström, B. E., and Perlmann, P., 1959, Fertilization, in: *The Cell,* Vol. I (J. Brachet and A. E. Mirsky, eds.), pp. 327–397, Academic Press, New York.
Sachtleben, P., Gsell, R., and Mehrishi, J. N., 1973, Neuraminidase and anti-neuraminidase serum: Effect on the cell surface properties, *Vox Sang.* **25:**519.
Sanderson, A. R., Cresswell, P., and Welsh, K. I., 1971, Involvement of carbohydrate in the immunochemical determinant area of HL-A substances, *Nature (Lond.) New Biol.* **230:**8.
Sanford, B. H., 1967, An alteration in tumor histocompatibility induced by neuraminidase, *Transplantation* **5:**1273.
Sanford, B. H., and Codington, J. F., 1971, Further studies on the effect of neuraminidase on tumor cell transplantability, *Tissue Antigens* **1:**153.
Sanford, B. H., Codington, J. F., Jeanloz, R. W., and Palmer, P. D., 1973, Transplantability and antigenicity of two sublines of the TA3 tumor, *J. Immunol.* **110:**1233.
Schied, A., and Choppin, P. W., 1973, Isolation and purification of the envelope proteins of Newcastle Disease Virus, *J. Virol.* **11:**263.
Schied, A., and Choppin, P. W., 1974, The hemagglutinating and neuraminidase protein of a paramyxovirus: Interaction with neuraminic acid in affinity chromatography, *Virology* **62:**125.
Sethi, K. K., and Brandis, H., 1972, *In vitro* cytotoxicity of normal serum factor(s) on neuraminidase-treated Ehrlich ascites cells, *Z. Immun. Forsch.* **143:**426.
Sethi, K. K., and Brandis, H., 1973, Neuraminidase induced loss in transplantability of murine leukaemia 1210, induction of immunoprotection and the transfer of induced immunity to normal DBA/2 mice by serum and peritoneal cells, *Br. J. Cancer* **27:**106.
Sharon, N., and Lis, H., 1972, Lectins: Cell-agglutinating and sugar-specific proteins, *Science* **177:**949.
Simon-Reuss, I., Cook, G. M. V., Seaman, G. V. F., and Heard, D. H., 1964, Electrophoretic studies on some types of mammalian tissue cell, *Cancer Res.* **24:**2038.
Simmons, R. L., and Rios, A., 1971, Immunotherapy of cancer: Immunospecific rejection of tumors in recipients of neuraminidase-treated tumor cells plus BCG, *Science* **174:**591.
Simmons, R. L., and Rios, A., 1972, Modification of immunogenicity in experimental

immunotherapy and prophylaxis, *in: Membranes and Viruses in Immunopathology* (S. B. Day and R. A. Good, eds.), pp. 563–576, Academic Press, New York.

Simmons, R. L., and Rios, A., 1973, Differential effect of neuraminidase on the immunogenicity of viral associated and private antigens of mammary carcinomas, *J. Immunol.* **111**:1820.

Simmons, R. L., Rios, A., and Ray, P. K., 1970, Mechanism of neuraminidase induced antigen unmasking, *Surg. Forum* **21**:265.

Simmons, R. L., Rios, A., Lundgren, G., Ray, P. K., McKhan, C. F., and Haywood, C. R., 1971a, Immunospecific regression of methylcholanthrene fibrosarcoma with the use of neuraminidase, *Surgery* **70**:38.

Simmons, R. L., Rios, A., and Ray, P. K., 1971b, Effect of neuraminidase on growth of 3-methylcholanthrene-induced fibrosarcoma in normal and immunosuppressed syngeneic mice, *J. Natl. Cancer Inst.* **47**:1087.

Simmons, R. L., Lepschultz, M. L., Rios, A., and Ray, P. K., 1971c, Failure of neuraminidase to unmask histocompatibility antigens on trophoblasts, *Nature (Lond.) New Biol.* **231**:111.

Smets, L. A., and Broekhuysen-Davies, J., 1972, Shielding of antigens and concanavaline A agglutination sites by a surface coat of transplantable mouse lymphosarcoma cells, *Eur. J. Cancer* **8**:541.

Smith, W. G., 1966, Release of ganglioside from guinea-pig lung tissue during anaphylaxis, *Nature (Lond.)* **209**:1251.

Sneath, P. H. A., and Lederberg, J., 1961, Inhibition by periodate of mating in *Escherichiae coli* K-12, *Proc. Natl. Acad. Sci. U.S.* **47**:86.

Sobeslavsky, O., Prescott, B., and Chanock, R. M., 1968, Adsorption of *Mycoplasma pneumoniae* to neuraminic acid receptors of various cells and possible role in virulence, *J. Bacteriol.* **96**:695.

Springer, G. F., and Desai, P. R., 1974, Common precursors of human blood groups MN specificities, *Biochem. Biophys. Res. Commun.* **61**:470.

Springer, G. F., Tegtmeyer, H., and Huprikar, S. V., 1972, Anti-N reagents in elucidation of the genetical basis of human blood group MN specificities, *Vox Sang.* **22**:325.

Suttajit, M., and Winzler, R. J., 1971, Effect of modification of N-acetylneuraminic acid on the binding of glycoproteins to influenza virus and on susceptibility to cleavage by neuraminidase, *J. Biol. Chem.* **246**:3398.

Tanigaki, N., and Pressman, D., 1974, The basic structure and the antigenic characteristics of HL-A antigens, *Transplant. Rev.* **21**:15.

Thomsen, O., 1927, Ein vermehrungsfähiges Agens als Veränderer des isoagglutinatorischen Verhaltens der roten Blutkörperchen, eine bisher unbekannte Quelle der Fehlbestimmung, *Z. Immun. Forsch.* **52**:85.

Tiffany, T. M., and Blough, H. A., 1971, Attachment of Myxoviruses to artificial membranes. Electron microscopic studies, *Virology* **44**:18.

Tsvetkova, I. V., and Lipkind, M. A., 1973, Studies on the role of myxovirus neuraminidase in virus-cell receptor interaction by means of direct determination of sialic acid split from cells. 3. One-step growth kinetics of accumulation of the sialic acid liberated from NDV-infected chick embryo cells, *Arch. Gesamte Virusforsch.* **42**:125.

Tuppy, H., and Gottschalk, A., 1972, The structure of sialic acids and their quantitation, *in: Glycoproteins, Their Composition, Structure and Function* (A. Gottschalk, ed.), pp. 403–449, Elsevier Publishing Co., Amsterdam.

Tyler, A., 1947, An auto-antibody concept of cell structure, growth and differentiation, *Growth* **10(Suppl.)**:7.

Uhlenbruck, G., and Wintzer, G., 1970, Topochemical arrangement of neuraminic acid containing receptors within the cell membrane, in: *Blood and Tissue Antigens* (D. Aminoff, ed.), pp. 289–305, Academic Press, New York.
Uhlenbruck, G., Pardoe, G. I., and Bird, G. W. G., 1969, On the specificity of lectins with a broad agglutination spectrum. II. Studies on the nature of the T-antigen and the specific receptors for the lectin of *Arachis hypogaea, Z. Immun. Forsch.* **138:**423.
van Beek, W. P., Smets, L. A., and Emmelot, P., 1973, Increased sialic acid density in surface glycoprotein of transformed and malignant cells—A general phenomenon? *Cancer Res.* **33:**2913.
van Heyningen, W. E., 1959, Tentative identification of the tetanus toxin receptors in nervous tissue, *J. Gen. Microbiol.* **20:**310.
van Heyningen, W. E., 1963, The fixation of tetanus toxin by ganglioside, *Biochem. Pharmacol.* **12:**437.
van Heyningen, W. E., and Miller, P. A., 1961, The fixation of tetanus toxin by ganglioside, *J. Gen. Microbiol.* **24:**107.
Vermylen, J., Donati, M. B., de Gateano, G., and Verstraete, M., 1973, Aggregation of human platelets by bovine or human factor VIII: Role of carbohydrate side chains, *Nature (London) New Biol.* **244:**167.
Vermylen, J., de Gateano, G., Donati, M. B., and Verstraete, M., 1974, Platelet-aggregating activity in neuraminidase-treated human cryoprecipitates: Its correlation with Factor-VIII-related antigen, *Br. J. Haematol.* **96:**645.
Verwey, E. J. W., and Overbeck, J. T. G., 1968, *Theory of the Stability of Lyophobic Colloids*, Elsevier Publishing Co., Amsterdam.
Vicker, M. G., and Edwards, J. G., 1972, The effect of neuraminidase on the aggregation of BHK21 cells and BHK21 cells transformed by polyoma virus, *J. Cell Sci.* **10:**759.
Voigtmann, R., and Uhlenbruck, G., 1970, Untersuchungen über den Mechanismus der Erythrozytenaggregation, *Thromb. Diath. Haemorrh.* **24:**530.
Wallach, D. H. F., and Eylar, E. H., 1961, Sialic acid in the cellular membranes of Ehrlich ascites-carcinoma cells, *Biochim. Biophys. Acta* **52:**594.
Wallach, D. H. F., and de Perez Esandi, M. V., 1964, Sialic acid and the electrophoretic mobility of three tumor cell types, *Biochim. Biophys. Acta* **83:**363.
Walther, B. T., Öhman, R., and Roseman, S., 1973, A quantitative assay for intercellular adhesion, *Proc. Natl. Acad. Sci. U.S.* **70:**1569.
Warren, L., and Glick, M. C., 1968, Membranes of animal cells. II. The metabolism and turnover of the surface membrane, *J. Cell Biol.* **37:**729.
Warren, L., Fuhrer, J. P., and Buck, C. A., 1972, Surface glycoproteins of normal and transformed cells: A difference determined by sialic acid and a growth-dependent sialyl transferase, *Proc. Natl. Acad. Sci. U.S.* **69:**1838.
Watkins, E., Jr., Ogata, Y., Anderson, L. L., Watkins, E., III, and Waters, M. F., 1971, Activation of host lymphocytes cultured with cancer cells treated with neuraminidase, *Nature (Lond.) New Biol.* **231:**83.
Watkins, W. M., 1967, Blood-group substances, in: *The Specificity of Cell Surfaces* (B. D. Davis and L. Warren, eds.), pp. 257–279, Prentice-Hall, Englewood Cliffs, N. J.
Weiss, L., 1961, Sialic acid as a structural component of some mammalian tissue cell surfaces, *Nature (Lond.)* **191:**1108.
Weiss, L., 1963, Studies on cellular adhesion in tissue-culture. V. Some effects of enzymes on cell-detachment, *Exp. Cell Res.* **30:**509.
Weiss, L., 1964, Cellular locomotive pressure in relation to initial cell contacts, *J. Theor. Biol.* **6:**275.

Weiss, L., 1965, Studies on cell deformability. Effect of surface charge, *J. Cell Biol.* **26:**735.

Weiss, L., 1967, *The Cell Periphery, Metastasis and Other Contact Phenomena*, North-Holland, Amsterdam.

Weiss, L., 1968, Studies on cellular adhesion in tissue culture. IX. Electrophoretic mobility and contact phenomena, *Exp. Cell Res.* **51:**609.

Weiss, L., and Cudney, T. L., 1971, Some effects of neuraminidase on the *in vitro* interactions between spleen and mastocytoma (P815) cells, *Int. J. Cancer* **7:**187.

Weiss, L., and Hauschka, T. S., 1970, Malignancy, electrophoretic mobilities and sialic acids at the electrokinetic surface of TA3 cells, *Int. J. Cancer* **6:**270.

Weiss, L., and Levinson, C., 1969, Cell electrophoretic mobility and cationic flux, *J. Cell Physiol.* **73:**31.

Weiss, L., and Mayhew, E., 1967, Ribonucleic acid within the cellular peripheral zone and the binding of calcium to ionogenic sites, *J. Cell Physiol.* **69:**281.

Weiss, P., 1947, Problem of specificity in growth and development, *Yale J. Biol. Med.* **19:**235.

Wesemann, W., and Zilliken, F., 1968, Rezeptoren der Neurotransmitter. IV. Serotoninrezeptor und Neuraminsäurestoffwechsel der glatten Muskulatur, *Hoppe-Seyl. Z.* **349:**823.

Woodruff, J. J., 1974, Role of lymphocyte surface determinants in lymph node homing, *Cell. Immunol.* **13:**378.

Woodruff, J. J., and Gesner, B. M., 1969, The effect of neuraminidase on the fate of transfused lymphocytes, *J. Exp. Med.* **129:**551.

Woolley, D. W., and Gommi, B. W., 1964, Serotonin receptor: V, Selective destruction by neuraminidase plus EDTA and reactivation with tissue lipids, *Nature (Lond.)* **202:**1074.

Woolley, D. W., and Gommi, B. W., 1965, Serotonin receptors, VII. Activities of various pure gangliosides as receptors, *Proc. Natl. Acad. Sci. U.S.* **53:**959.

Yamakawa, T., Matsumoto, M., Suzuki, S., and Iida, T., 1956, The chemistry of the lipids of post-hemolytic residue or stroma of erythrocytes. VI. Sphingolipids of erythrocytes with respect to blood group activities. *J. Biochem. (Tokyo)* **43:**41.

Yogeeswaran, G., and Hakomori, S., 1975, Cell contact-dependent ganglioside changes in mouse 3T3 fibroblasts and a suppressed sialidase activity on cell contact, *Biochemistry* **14:**2151.

Zakstelskaya, Y., Molibog, E. V., Yakhno, M. A., Evstigneeva, N. A., Isachenko, V. A., Privalova, I. M., and Khorlin, A. Ya., 1972, Use of synthetic inhibitors of neuraminidase and hemagglutinin for the study of the functional role of active subunits of membranes of myxo- and paramyxoviruses, *Vopr. Virusol.* **17:**223.

Chapter 8

The Altered Metabolism of Sialic-Acid-Containing Compounds in Tumorigenic-Virus-Transformed Cells

Peter H. Fishman and Roscoe O. Brady

I. INTRODUCTION

At this point in time, the quest for information regarding phenomena regulating cell growth has focused to a considerable extent on investigations of alterations in the chemical composition of the surfaces of neoplastic cells. Relatively few discoveries have occurred so far in this regard, but the recently observed alteration of ganglioside composition and the underlying metabolic reactions responsible for this change in tumorigenic-virus-transformed cells may prove to be an important contribution in this area. We shall review the nature of these changes in tumorigenic virus transformation in this chapter. We shall also touch upon some differences in glycoprotein metabolism which have recently been observed in these cells, and we shall make some conclusions regarding the significance of the differences between normal and tumorigenic cells.

PETER H. FISHMAN AND ROSCOE O. BRADY • Developmental and Metabolic Neurology Branch, National Institute of Neurological Diseases and Stroke, National Institutes of Health, Bethesda, Maryland 20014.

II. EXPERIMENTAL PROCEDURES

A. Cells and Cell Culture

Our interest in complex lipids in neoplastic cells was initiated through an investigation of glycolipid metabolism in human leukocytes. It was observed that white blood cells from normal individuals contain several sphingolipid hydrolases. The activity of these enzymes in leukocytes parallels that in systemic organs and tissues. For example, the activity of glucocerebrosidase (reaction 1) is diminished in patients with

$$\text{Ceramide-Glucose} + H_2O \xrightarrow{\text{Glucocerebroside-}\beta\text{-Glucosidase}} \text{Ceramide} + \text{Glucose} \quad (1)$$

Gaucher's disease (Brady *et al.*, 1965), and this attenuation is accurately reflected in glucocerebrosidase activity in extracts of leukocytes obtained from these patients (Kampine *et al.*, 1967a). Because such a cellular change depends on the physiological state, it was considered of interest to examine glucocerebrosidase activity in leukemic leukocytes. The specific activity of this enzyme is increased three to four times in myelogenous leukemic leukocytes over that in normal white blood cells (Kampine *et al.*, 1967b). This finding prompted us to investigate sphingolipid composition and the activity of sphingolipid hydrolases in human tumors. Generally, the activity of these catabolic enzymes was unusually high in tumors, compared with the control tissue specimens which were available. The pattern of neutral sphingolipids in tumor tissue was not remarkably different from that in the controls.

The results obtained in these experiments were always subject to some doubt concerning their reproducibility and their significance since there was usually a heterogeneity of cell types, hemorrhage, and necrosis in the tumor tissue specimens. Therefore, we sought a better controlled experimental situation and decided to investigate these phenomena in mammalian cells established in tissue culture where the genotypic and phenotypic properties of such cells were known. A number of clonal cell lines which exhibited low tumorigenicity and contact inhibited growth in culture were available, as were highly malignant transformed clonal lines derived from them. The first group of cells we examined were the inbred mouse embryo AL/N lines (Takemota *et al.*, 1968). Normal contact-inhibited, spontaneously transformed, and simian virus 40 (SV40) and polyoma virus (Py) tumorigenic-transformed cell lines were employed. Other mouse lines investigated included the Swiss 3T3 and Balb/c 3T3 and the respective SV40- and Py-transformed derivatives as well as flat revertants selected from the transformed lines. The origin of these cell lines is given in detail elsewhere (Brady et al., 1973a). In addition, some experiments with cultured cells of other species have been done; these include baby hamster kidney (BHK), chick embryo fibroblast (CEF), newborn rat kidney (NRK), and human diploid fibroblast (WI38) and derivative lines transformed by various oncogenic viruses. Tissue culture conditions were essentially those conventionally employed for the respective cell lines except that labeled precursors of glycolipids and glycoproteins such as [^{3}H]N-acetylmannosamine, [^{14}C]glucosamine, and [^{3}H]- and [^{14}C]fucose were added to the culture medium when appropriate.

B. Isolation, Identification, and Quantification of Gangliosides

Much of the work described in this chapter deals with alterations of ganglioside composition and metabolism as a consequence of transformation of cells by tumorigenic viruses. Gangliosides were isolated by conventional procedures which include extracting the cells with chloroform–methanol (2:1, v/v), mixing with an aqueous solvent system (Folch et al., 1957), dialyzing the upper phase, and purifying the gangliosides by passage over a small column of Sephadex G-25 superfine grade (Cumar et al., 1970). Gangliosides were separated by thin-layer chromatography on silica gel (Mora et al., 1969) or by a minor modification of this method (Fishman et al., 1973). The individual components were detected and quantitated by several methods. (1) The thin-layer plates were exposed to iodine vapor and the appropriate areas marked and scraped from the plates. The gangliosides were then quantitated for their sialic acid content by reacting with resorcinol reagent or thiobarbituric acid. (2) The plates were sprayed with resorcinol reagent, the positive areas scraped from the plates and extracted with butylacetate–butanol (85:15, v/v) and the absorbance determined spectrophotometrically at 580 nm (Mora et al., 1969). (3) The gangliosides were visualized with resorcinol spray and quantitated directly by scanning the plates at 580 nm with a Zeiss chromatogram spectrophotometer (Brady et al., 1969). (4) The migration of the individual gangliosides was determined with rhodamine spray, the areas scraped from the plate, and the gangliosides eluted with chloroform–methanol–water (10:10:1, v/v/v). Following methanolysis, the gangliosides were quantitated as the trimethylsilyl derivatives of their monosaccharides by gas–liquid chromatography (Dijong et al., 1971).

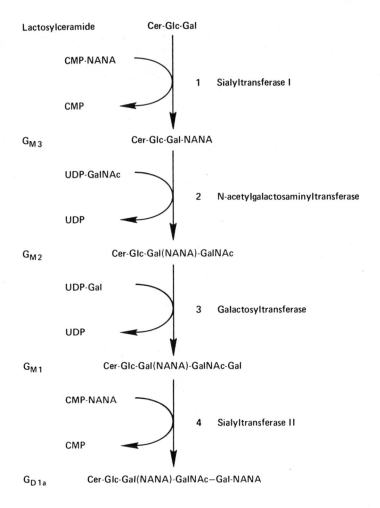

FIGURE 1. Pathway of biosynthesis of the common gangliosides of normal mouse cell lines.

C. Assay of Enzymes Involved in Glycolipid Metabolism

The reactions involved in the biosynthesis of gangliosides which are present in normal mouse cell lines are summarized in schematic form in Figure 1. For further details, see Chapter 4. The activity of the respective individual enzymes is generally determined as indicated in Table I. Occasionally, minor modifications such as adjustment of pH are required, depending on the various cell lines under investigation.

TABLE I. Reaction Mixtures and Incubation Conditions for Glycosyltransferase Assays[a]

Glycosyltransferase[b]	Reaction mixture[c]
Sialyltransferase I	50 nmoles lactosylceramide, 10^5 cpm CMP-NANA-^{14}C, 5 μmoles Na cacodylate (pH 6.35), 750 μg cutscum, 100 μg cardiolipin, 150 μg of cell protein in final volume of 50 μl
N-Acetylgalactosaminyltransferase	50 nmoles G_{M3}, 10^5 cpm UDP-GalNAc-^{14}C, 2.5 μmoles Na cacodylate (pH 7.2), 200 μg Triton X-100, 1 μmole $MnCl_2$, 75 μg of cell protein in final volume of 50 μl
Galactosyltransferase	10 nmoles G_{M2}, 2×10^5 cpm UDP-Gal-^{14}C, 5 μmoles Na cacodylate (pH 7.2), 100 μg Triton CF-54, 50 μg Tween 80, 1 μmole $MnCl_2$, 150 μg of cell protein in final volume of 50 μl
Sialyltransferase II	50 nmoles G_{M1}, 10^5 cpm CMP-NANA-^{14}C, 5 μmoles Na cacodylate (pH 6.5), 100 μg Triton CF-54, 50 μg Tween 80, 150 μg cell protein in a final volume of 50 μl

[a] From Fishman et al., 1972.
[b] The reaction catalyzed by each enzyme is depicted in Figure 1.
[c] The reaction mixtures are usually incubated at 37°C for 2 hr. The reaction products are isolated and quantitated as previously described (Fishman et al., 1972).

The catabolism of gangliosides was also a primary concern in searching for an understanding of the alteration of ganglioside composition in tumorigenic-virus-transformed cells. The catabolism (cf. Chapter 5) of the major gangliosides G_{M3} and G_{D1a} (cf. Chapter 2, Table I) is initiated by the enzymatic hydrolysis of the terminal molecule of sialic acid from the respective compounds [reactions (2) and (3)]:

Cer-Glc-Gal-NANA (G_{M3}) + H_2O

$\xrightarrow{\text{Neuraminidase}}$ Cer-Glc-Gal + NANA   (2)

Cer-Glc-Gal-(NANA)-GalNAc-Gal-NANA (G_{D1a}) + H_2O

$\xrightarrow{\text{Neuraminidase}}$ Cer-Glc-Gal-(NANA)-GalNAc-Gal + NANA

G_{D1a} and G_{M3} were specifically labeled in the N-acetylneuraminic acid moieties through biosynthesis *in vivo* using [^{3}H]N-acetylmannosamine as the labeled precursor (Kolodny *et al.*, 1970). The catabolism of these substances was determined by measuring the quantity of [^{3}H]NANA released using appropriate conditions of incubation (Cumar *et al.*, 1970).

III. GANGLIOSIDE METABOLISM IN CULTURED MOUSE CELL LINES

A. Distribution of Gangliosides in Normal and Virally Transformed Cells

Examination of neutral glycolipid composition of normal and tumorigenic cells showed no conspicuous differences. The idea occurred to us that it might be worthwhile to investigate the pattern of gangliosides in these cells since gangliosides are known to be highly concentrated in membranous elements of cells. For detailed information, see Chapter 3. A family of gangliosides ranging from G_{M3} to G_{D1a} was found in the control, contact-inhibited, NAL/N mouse cells. The ganglioside components usually appear as doublets on thin-layer chromatograms, presumably because of the presence of both N-acetylneuraminyl and the slower migrating N-glycolylneuraminyl derivatives of the respective gangliosides. It is also possible that differences in the chain length of the fatty acid moiety of the ceramide portion could affect the mobility of gangliosides. When the ganglioside composition of an SV40-transformed derivative cell line was investigated, the only ganglioside found was G_{M3}. It was of immediate interest to examine the pattern of gangliosides in spontaneously transformed non-contact-inhibited tumorigenic cells. Such derivative lines of the AL/N strain were available, and it was found that the ganglioside composition of these cells was essentially similar to the control NAL/N cell line. A typical thin-layer chromatogram of gangliosides in these and SV40-transformed cells is shown in Figure 2. We then examined the ganglioside patterns in many other control, spontaneously transformed and tumorigenic-virus-transformed derivative lines. Normal Balb/c and 3T3 Swiss mouse cells have essentially the same gangliosides as the control NAL/N cells. The SV40 or polyma virus (Py) transformants of these cells showed the same simplification of ganglioside pattern as that observed in the SV40-

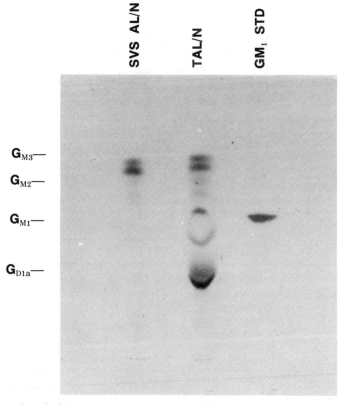

FIGURE 2. Ganglioside patterns of mouse AL/N cell lines. TAL/N is a spontaneously transformed tumorigenic cell line with an essentially normal ganglioside complement. SVS AL/N is a simian virus 40 transformant (Mora *et al.*, 1969).

transformed AL/N cells (Table II). These investigations have been expanded, and a similar modification in ganglioside composition has been found in DNA-virus-transformed cell lines with few exceptions.

The changes in ganglioside composition detected by quantitative analysis were also dramatically reflected by the pattern of labeling of gangliosides when appropriate precursors were added to the culture media (Figure 3). Only the smallest ganglioside, G_{M3}, was labeled in SV40-virus-transformed AL/N cells, whereas the entire family of gangliosides became radioactive in the controls when [^{3}H]N-acetylmannosamine or [^{14}C]glucosamine were used as precursors (Brady and Mora, 1970).

B. Enzymatic Studies

1. Catabolism of Gangliosides. The lack of gangliosides larger than G_{M3} in the tumorigenic DNA-virus-transformed cells conceivably could be due to excessively rapid catabolism of the larger gangliosides or an inability to synthesize the higher homologs. The first possibility was eliminated through investigations of the catabolism of G_{M3} and G_{D1a}. The rate of hydrolysis of these components was essentially the same in the control and virus-transformed cells (Cumar *et al.*, 1970).

2. Biosynthesis of Gangliosides. Since an abnormality of ganglioside catabolism in tumorigenic-virus-transformed cells was excluded, we

TABLE II. Distribution of Gangliosides in AL/N and Swiss Mouse Cell Lines and in Polyoma- and SV40-Virus-Transformed Derivative Lines[a]

Cell type	Growth in culture	Gangliosides, nmoles/mg of protein				
		G_{D1a}	G_{M1}	G_{M2}	G_{M3}	Total
NAL/N	Contact-inhibited	1.8	1.5	0.8	0.6	4.7
SVS AL/N[b]	High-density	0.16	0.22	0.1	1.9	2.4
Py AL/N[c]	High-density	0.1	0.15	0.2	1.8	2.3
Swiss 3T3	Contact-inhibited	2.4	2.6	1.8	4.0	10.8
Py11[c]	High-density	0.1	0.2	0.05	3.2	3.5
SV101[b]	High-density	0.05	0.1	0.05	3.5	3.7
SV Py 3T3[d]	High-density	0.6	0.8	0.40	4.8	6.6

[a] From Brady *et al.*, 1973.
[b] SV40-transformed.
[c] Py-transformed.
[d] SV40- and Py-transformed.

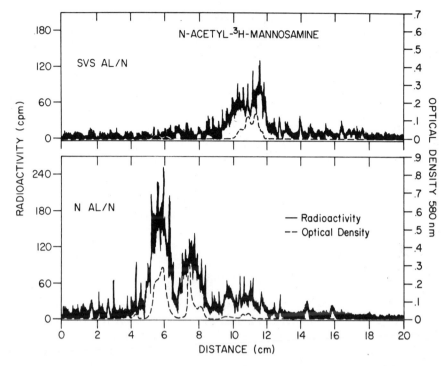

FIGURE 3. Scans of absorbance at 580 nm and radioactivity of thin-layer chromatograms of gangliosides extracted from NAL/N (control) and SVS AL/N cells grown in the presence of N-acetyl-[^{3}H]mannanosamine in the tissue culture medium. Reproduced with permission from Brady and Mora, 1970.

directed our attention to an examination of ganglioside synthesis in controls and virus-transformed derivative cell lines. Ganglioside formation occurs by the sequential addition of hexoses and N-acetylneuraminic acid to the elongating oligosaccharide chain. See Chapter 4. Since our analytical and labeling studies indicated a drastic diminution of gangliosides larger than G_{M3} in the virus-transformed cells, we considered it reasonable to investigate the conversion of G_{M3} to the next higher homolog, ganglioside G_{M2} (reaction 2 in Figure 1), which is catalyzed by the enzyme hematoside:UDP-GalNAc N-acetylgalactosaminyltransferase. The activity of this aminosugar transferase is drastically reduced in SV40- and Py-transformed mouse cell lines (Table III). Thus, the decreased activity of this enzyme may be the cause of the altered ganglioside pattern in the tumorigenic DNA-virus-transformed cells.

TABLE III. Uridine Diphosphate N-Acetylgalactosamine: Hematoside N-Acetylgalactosaminyltransferase Activity in Normal and Transformed Mouse Cell Lines[a]

Cell line	State	Aminosugar transferase activity, nmoles/mg protein/hr
Swiss 3T3	Normal	0.48
SV101	SV40-transformed	0.06
Py11	Py-transformed	0.08
NAL/N	Normal	0.76
TAL/N	Spontaneously transformed	0.87
SVS AL/N	SV40-transformed	0.13
Py AL/N	Py-transformed	0.11

[a] The enzyme activity was assayed in whole-cell homogenates under optimum conditions as given in Table I.

C. Effect of Growth and Culture Conditions on Ganglioside Metabolism

1. Growth Rates and Cell Density. Virally transformed cells are altered in several important phenotypic characteristics. For the most part, they are smaller and more rounded than the parent cells. Their doubling time in culture is shorter. Also, in contrast to contact-inhibited cells, which stop growing when a monolayer of cells covers the substratum, transformed cells will pile up on top of one another and continue to grow until a much higher cell density is reached. It was very important to examine whether the alteration in ganglioside synthesis was caused by the insertion of the viral genome into the genetic apparatus of the transformed cell or whether it was the result of some ancillary aspect of change in growth in culture. We therefore examined the synthesis and composition of gangliosides in normal and transformed cells while they were sparsely distributed in the culture flask, at the time of confluency, and in the postconfluent state. There was no significant change in N-acetylgalactosaminyltransferase activity or other enzymes involved in ganglioside biosynthesis throughout these various stages of growth (Figure 4). Thus, the decreased activity of this critical enzyme for ganglioside synthesis is not caused by variation of these parameters in culture.

2. Culture Conditions and Selection. The effects of other culture parameters have recently been reviewed (Mora, 1973). In general, variation of culture media, pH, and serum had little or no effect on ganglioside metabolism. Contact-inhibited cells grown under crowded conditions for a number of passages whereby the cells became less

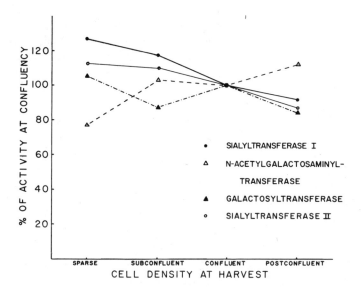

FIGURE 4. Effect of cell density on glycosyltransferase activities. (Reproduced with permission from Fishman et al., 1972.)

contact-inhibited and reached high saturation densities resulted in no observable changes in ganglioside pattern. Extensive cultivation (over 200 passages) of normal contact-inhibited or virally transformed cell lines also had little or no effect on the ganglioside composition and glycosyltransferase activities of these cell lines. In contrast, there were observable changes of these biochemical properties when spontaneously transformed mouse cell lines were so subcultured. (See also Section V, C.)

Selective pressures were also applied to the malignant TAL/N and SVS-AL/N cell lines by transplantation of the cells into syngeneic mice and reestablishing the tumors in culture (Mora, 1973). Even after three such serial transfers, the SVS-AL/N cell lines continued to express an altered ganglioside content and low aminosugar transferase activity. In general, the TAL/N derivative lines maintained their complement of higher gangliosides and high enzyme activity. Occasionally a tumor would give rise to a cell culture with an attenuated ganglioside pattern and low N-acetylgalactosaminyltransferase activity (unpublished results of P. H. Fishman and P. T. Mora). This phenomenon is under further investigation.

Another aspect of culture conditions is clonal selection. Subclones of a NIL hamster cell line were reported to have extensive differences in glycolipid composition (Sakiyama et al., 1972). However, another group working with subclones of this same cell line found little variation in glycolipid pattern (Critchley and MacPherson, 1973). Clonal variations of the ganglioside composition of SV40-transformed Swiss 3T3 cells has recently been reported (Yogeeswaren et al., 1972). The possibility of clonal variation is relevant to the interpretation of the altered ganglioside metabolism in virally transformed mouse cells (see Section V, D).

3. **Interaction Between Normal and Transformed Cells.** We also undertook an investigation to determine whether a transmissible factor was produced by the transformed cells and caused the diminution of hematoside: UDP-GalNAc N-acetylgalactosaminyltransferase activity. Normal and virus-transformed cells were grown in culture flasks using a low barrier to keep the two types of cells separated but which allowed the culture fluid to contact the cells in both compartments. There was no evidence of a factor diffusing from the virus-transformed cells which lessened aminosugar transferase activity in the control cells (Mora et al., 1971). Similarly, there was no indication of a correction factor emanating from the control cells which restored the activity of this enzyme in the virus-transformed cells.

When control and virus-transformed cells were grown in intimate admixture, there was a 30% decrease in the predicted activity of the aminosugar transferase based on an analysis of the respective types of cells at the time of harvest. It is not yet clear whether this diminution is significant. Therefore, no decision can be made at this time regarding the possibility that an inhibitor or repressor of this enzyme might be passed from transformed cells to normal cells which are in contact with each other. What is quite clear is the fact that there is no accumulation of an inhibitor of N-acetylgalactosaminyltransferase in the transformed cells since the activity of this enzyme in mixed homogenates of normal and transformed cells is exactly that calculated from equivalent amounts of homogenates of the respective cells alone. (Cumar et al., 1970; Mora et al., 1971, 1973).

D. **Sialic-Acid-Containing Glycolipids in Transformed Cells Obtained from Other Species**

1. **Hematoside in Hamster Cells.** Hakomori and Murakami (1968) reported that the levels of hematoside, the only ganglioside in BHK, was reduced in spontaneous and Py-transformed BHK cells. Similar results were observed with BHK cells transformed by two strains of Rous sarcoma virus (RSV) (Hakomori et al., 1968). The reduced hematoside content in at least the Py-BHK cells was attributed to decreased activity of lactosylceramide: CMP-NANA sialyltransferase, reaction 1 in Figure

1 (Den et al., 1971). Experiments in our own laboratory on normal and Py-BHK cells obtained from Dr. Walter Eckhart of the Salk Institute did not show a difference in the G_{M3} content of these cell lines. The quantity of hematoside and level of sialyltransferase activity were similar in both cell lines and not affected by cell density in culture (Table IV). Further studies in another polyoma transformant as well as two lines transformed by temperature-sensitive mutants of the virus gave similar results.

The glycolipids of another established hamster cell line, NIL, have also been extensively studied. The only sialylglycolipid in these cells is hematoside. There was approximately a 50% increase in the incorporation of [1-^{14}C]palmitate into this glycolipid when the cells increased in density but transformation by hamster sarcoma virus or an adeno 7/SV40 hybrid had no effect (Robbins and Macpherson, 1971). When labeled palmitate incorporation into G_{M3} was compared among six NIL clonal lines, there was a tenfold variation (Sakiyama et al., 1972). Two clones showed cell density-dependent increases in incorporation, two no effect, and two showed a decrease. The clones exhibiting the strongest density-dependent increase in palmitate incorporation into hematoside were transformed by hamster sarcoma or polyoma viruses. The transformed subclones lost this cell-density-dependent effect. One disturbing aspect of these experiments is the method utilized, i.e., incorporation of radioactivity from a precursor into glycolipid products as an indicator of the quantity of material present. In fact, when the hematoside content was determined analytically in a clone that did not show a density effect for this glycolipid by the radioisotope technique, there was 50% less hematoside in cells from a confluent culture than from a growing culture

TABLE IV. Hematoside and Sialyltransferase Levels in Normal and Polyoma-Transformed BHK Cells

Cell line	Density at harvest	Hematoside,[a] nmoles/mg protein	Sialyltransferase,[b] cpm/hr/mg protein
BHK	Sparse	3.89	3994
Py-BHK	Sparse	3.10	4511
BHK	Confluent	3.88	7411
Py-BHK	Confluent	3.41	6558

[a] Hematoside was extracted from the cells, isolated by thin-layer chromatography, visualized with resorcinol spray, and quantitated by direct densitometry as described in Section II.
[b] Lactosylceramide: CMP-NANA sialyltransferase activity was assayed on whole cell homogenates as described in Section II.

(Sakiyama *et al.*, 1972). Similar discrepancies between the two methods have been observed by others (Hakomori *et al.*, 1971).

2. Gangliosides in Chick Embryo Fibroblasts. Chick embryo fibroblasts (CEF) also contain sialylglycolipids; these include hematoside, disialohematoside (G_{D3}), and two uncharacterized compounds (Hakomori *et al.*, 1971). There was some increase in these compounds when the cells became confluent. When the cells were transformed by RSV, the G_{M3} and G_{D3} content decreased significantly. The time course of these changes was investigated by infecting the cells with excessive virus. Under these conditions most of the culture became transformed within two days. G_{D3} levels decreased during transformation, but G_{M3} content changed only after the cells became transformed. Infection of CEF cells with a Rous-associated leukosis type virus (RAV) which is nontransforming had no effect on the sialoglycolipids.

In contrast to these studies, CEF cells transformed by either RSV or a temperature-sensitive mutant of RSV showed no significant changes in hematoside or uncharacterized "higher gangliosides" (Warren *et al.*, 1972*a*). Unfortunately, the radioisotope technique of measuring labeled palmitate incorporation was used to estimate glycolipid content. Experiments in our laboratory, nevertheless, confirm these latter findings. CEF cells transformed with either a wild-type Rous sarcoma virus of the Brian high titer strain or a temperature-sensitive mutant (Bader and Brown, 1971) were obtained from Dr. John Bader, National Cancer Institute, and assayed for lactosylceramide: CMP-NANA sialyltransferase activity. Both of the transformed lines had almost twice the enzyme activity of the untransformed cells, and there was no modulation of this sialyltransferase when the temperature-sensitive transformants were grown at either the permissive or nonpermissive temperature (unpublished results of P. H. Fishman, J. P. Bader, and P. T. Mora).

3. Hematoside and Sialyltransferase in Newborn Rat Kidney Cells. Another transformed cell system that we have investigated has shown a positive correlation between expression of the transforming virus and glycolipid content. Newborn rat kidney cells (NRK) transformed by either the Kirsten isolate of murine sarcoma virus (Ki-MSV) or a temperature-sensitive mutant were analyzed for glycolipid composition. The major sialylglycolipid in NRK cells is G_{M3}. When the cells were grown at the permissive temperature for transformation (31°C), there was substantially less hematoside in both transformed lines than in the control cells (unpublished observation of P. H. Fishman, S. A. Aaronson, and R. O. Brady). At the nonpermissive temperature (39°), the cells transformed by the mutant had more hematoside than the control cells. Lactosylceramide: CMP-NANA sialyltransferase activity was determined in these cells. There was over twice as much enzyme activity in control cells grown at 31°C than in either transformed cell line. When grown at 39°C, the temperature-sensitive line had 50% more activity than the controls, whereas the wild-type transformant had 70% less than the control cells. These results are of great potential interest and may provide a model system for delineating the molecular basis of viral transformation. Once one has a specific biochemical marker that reflects the transformed state and can modulate simultaneously both of these parameters, one can design the pertinent experiments to trace the molecular processes involved.

IV. SIALIC ACID AND GLYCOPROTEINS IN TRANSFORMED CELLS

Most cells in culture have a substantial amount of sialic acid, the bulk of which is associated with glycoproteins (Weinstein et al., 1970) and is primarily localized on the cell surface (Glick et al., 1971), cf. Chapter 3. Because the plasma membrane of cells is of prime interest in the transformation process, considerable research has been directed towards investigating changes in sialic acid and sialoglycoproteins after transformation.

A. Sialic Acid and Sialyltransferase Activity in Transformed Cells

Total sialic acid content was found to be lower in several SV40-transformed clones of Swiss 3T3 mouse cells (Ohta et al., 1968). The plasma membrane as well as other membrane fractions of similar cell lines had reduced amounts of sialic acid (Wu et al., 1969). Isolated plasma membranes from Py-BHK cells had 60% less sialic acid than membranes from normal BHK cells (Makita and Seyama, 1971). These changes appear to be associated with reduced sialyltransferase activities in the transformed cells. Grimes (1970) determined the levels of sialic acid and sialyltransferase activity in normal and SV40-transformed Swiss and Balb 3T3 mouse cells. There was a close correlation between sialic acid content and sialyltransferase activity in each cell line even though enzyme levels were measured with exogenous acceptors such as desialylated fetuin or bovine submaxillary mucin. This correlation has been extended to a number of other cell lines (Grimes, 1973).

There is also an inverse correlation between the sialic content and saturation density of various Swiss 3T3 lines (Culp et al., 1971). Untransformed and flat revertant cell lines had the highest content of sialic acid, while spontaneous and SV40 transformants had the lowest amounts. Since contact-inhibited cells grow to low saturation densities (number of cells per unit area of substratum), these results suggest some relationship between sialic acid levels and social behavior of cells in culture. This observation is by no means universal, as Py-transformed Swiss 3T3 cells do not fit this pattern and have normal levels of sialic acid and sialyltransferase (Grimes, 1973). Our own studies on ganglioside sialyltransferases indicate that there are many other exceptions which include virally transformed hamster and chick cells (see Section IV,A) and mouse cells (Section V,C and Table VII).

B. Membrane Glycoproteins

Membrane glycoproteins have been isolated from normal and transformed cells and separated by chromatographic and electrophoretic procedures. Because cellular glycoproteins represent a very heterogeneous class of compounds, double-isotope biosynthetic techniques are normally employed for comparative studies. Thus, transformed and control cell lines are grown in separate culture media containing different isotopic forms of the same sugar such as [^{14}C]L-fucose and [^{3}H]L-fucose to label the glycoproteins in the respective types of cells. The cells or cellular fractions are solubilized, admixed, and analyzed. Differences in glycoproteins are detected by changes in the ^{14}C/^{3}H ratio.

Membrane fractions from normal and SV40-transformed Swiss 3T3 cells grown in the presence of glucosamine were solubilized and compared by gel filtration chromatography (Meezan et al., 1969). There were significant differences in the glycoprotein profiles of the two cell lines in all of the membrane fractions examined (surface, nuclei, mitochondrial, and microsomal). Labeled glycopeptides, obtained by digesting the membrane fractions with pronase, were cochromatographed on Sephadex G-50. Glycopeptides from the transformed cell membranes were enriched in higher-molecular-weight material. However, these results could not be confirmed when membrane glycoproteins from the same cell lines were compared by coelectrophoresis on SDS-polyacrylamide gels, or when the pronase-derived glycopeptides were cochromatographed on BioGel P-10 (Sakiyama and Burge, 1972). The discrepancy between the two studies may in part be due to the fact that the Swiss 3T3 cells had been subcultured for 15 additional passages and were less contact-inhibited.

Glucosamine-labeled plasma membranes obtained from normal, SV40-, and Py-transformed Swiss 3T3 cells have been compared by coelectrophoresis on SDS-polyacrylamide gels (Sheinin and Onodera,

1972). The glycoprotein patterns of the transformed lines were different from those of normal cells, and there were also differences among the three SV40-transformed clones examined. Work in our laboratory indicates that there are significant differences between the membrane glycoproteins of normal and transformed cells. Using double-isotope labeling techniques and SDS-gel electrophoresis of solubilized whole cells, SV40-transformed Swiss 3T3 and AL/N cells show an enrichment in high-molecular-weight (approximately 160,000 dalton) glycoproteins when compared to the contact-inhibited parental lines (unpublished results of P. H. Fishman and R. H. Quarles).

C. Glycopeptides of Transformed Cells

Warren et al. (1973) have compared the glycopeptides of a number of transformed cells and found a consistent difference. Fucose-labeled glycopeptides were prepared by treating the intact cells with trypsin and further digesting the released surface material with pronase. When the glycopeptides are cochromatographed on Sephadex G-50 columns, glycopeptides from transformed cells have more earlier eluting radioactive material than those from normal cells. These differences have been found in RSV- and Py-transformed BHK, MSV- and RSV-transformed Balb 3T3, and RSV-transformed CEF cells (Buck et al., 1970, 1971a).

The growth state of the cells plays a role in these changes. Glycopeptides from growing BHK- or RSV-BHK cells are enriched in higher-molecular-weight material when compared to glycopeptides from confluent cultures (Buck et al., 1971b). The differences in these surface glycopeptides from normal and transformed cells are most apparent when the cells are rapidly growing. Experiments with normal and SV40-transformed Swiss 3T3 cells confirm these results. SDS-gel electrophoresis of whole cell extracts show that fucose-labeled glycoproteins of sparse growing cultures are enriched in higher-molecular-weight components when compared to confluent cultures (unpublished observations of P. H. Fishman and R. H. Quarles).

D. Role of Sialic Acid and Sialyltransferase

Following treatment of the fucose-labeled glycopeptides derived from the surfaces of RSV-BHK cells with neuraminidase, the radioactive products appeared similar to glycopeptides from normal cells (Warren et al., 1972b). These results suggested that increased sialic acid content may account for the higher molecular weight of the glycopeptides from transformed cells. When sialyltransferase activity was assayed with desialylated glycopeptide as acceptor, extracts of transformed cells had over three times the sialyltransferase activity found in normal cells extracts. Enzyme activities in the two cell types were similar when untreated glycopeptide material from normal cells or desialylated fetuin or bovine mucin was the acceptor. Growing cells had more of this specific sialyltransferase than cells from confluent cultures (Warren et al., 1972b).

These changes in sialoglycopeptides appear to be associated with viral transformation. The change observed in RSV-transformed CEF is not seen in RAV-infected cells (Warren et al., 1972a). More important are experiments with CEF cells transformed by a temperature-sensitive (TS) mutant of RSV (Martin, 1970). Glycopeptides from TS-CEF cells grown at the permissive temperature (36°C) were similar to those from RSV-CEF cells, whereas TS-CEF cells grown at the nonpermissive temperature (41°C) contained glycopeptides like those from control CEF cells (Warren et al., 1972a). When assayed for the specific sialyltransferase, TS-CEF cells grown at 36°C had five times the activity as the same cells grown at 41°C (Warren et al., 1973).

Sialyltransferase activities in normal and virally transformed mouse 3T3 cells have also been measured with specific endogenous acceptors (Bosmann, 1972a). The acceptors were prepared by incubating the intact cells with neuraminidase and then trypsin. The desialylated material released from the cells was used to measure sialyltransferase activity in crude detergent extracts of the cells. Cells transformed by MSV, RSV, or Py had more acceptor and transferase activity than normal 3T3 cells. Surface sialyltransferase activity as measured by the ability of intact cells to transfer sialic acid-^{14}C from CMP-NANA-^{14}C to undefined surface acceptors was also higher in the transformed lines (Bosmann, 1972b).

E. Comments

Although it is tempting to connect changes in sialic acid, sialoglycoproteins, and sialyltransferases with expression of the transformed state of cells in culture, unexplained discrepancies among the experimental evidence prevents any conclusions from being drawn. Although sialic acid and sialyltransferase levels are reduced in many transformed cells, so are neutral and aminosugars (Wu et al., 1969; Makita and Seyama, 1971) and other glycosyltransferases (Bosmann and Eylar, 1968; Grimes, 1970, 1973). And how are these reduced levels of sialic acid and sialyltransferase reconciled with the increased amounts of sialoglycopeptides and sialyltransferase activity found by Warren? Admittedly, Warren and coworkers may be dealing with only a small, distinct population of glycopeptides and a specific sialyltransferase. However, it is important to note that this specific sialyltransferase is not confined to the cell surface (Warren et al., 1973) and the altered glycopeptides of transformed cells are present in other cell membrane fractions (Meezan et al., 1969; Keshgegian and Glick, 1973). Finally, these changes are influenced by the growth state of the cell and can be expressed by normal growing cells.

V. RELATIONSHIP BETWEEN VIRAL TRANSFORMATION AND ALTERED GANGLIOSIDE METABOLISM

The block in ganglioside biosynthesis observed in a variety of SV40- and Py-transformed mouse cells suggests that the alteration is related to some function common to both of these DNA tumor viruses (Mora et al., 1971). Several considerations have to be explored.

A. Productive Infection of Mouse Cells

If this altered ganglioside metabolism is related to some viral function, is it a function involved in transformation, productive infection, or both? Mouse cells are permissive for polyoma virus and are rarely transformed by it (cf. Eckhart, 1969). Upon viral infection, the cells undergo a series of discrete events which result in cell lysis and release of new virions (cf. Weil and Kara, 1970). These include the appearance of viral-specific T-antigen, induction of cellular and viral DNA synthesis which is maximum at 25–30 hr postinfection, synthesis of viral coat proteins and new virions (present at 50 hr.), and, finally, cell lysis and virus release (5–7 days after infection). Swiss 3T3 and TAL/N cells were treated with sufficient polyoma virus to ensure infection of all cells. Hematoside: UDP-GalN Ac N-acetylgalactosaminyl-transferase was then determined at two key times after infection (Table V). The activity of this enzyme was similar to that seen in mock-infected cells. The results indicate that the reduced aminosugar transferase activity in virally transformed cells is specifically related to some transforming function of the viruses and is not a consequence of lytic infection of the cell.

B. Ganglioside Metabolism in Flat Revertant Cell Lines

Since the altered ganglioside metabolism is found in virally transformed cell lines which have altered growth properties in culture and *in*

TABLE V. N-Acetylgalactosaminyltransferase Activity in Mouse Cells Lytically Infected and Transformed by Polyoma Virus[a]

Cell line	Time of harvest after exposure to virus (Transferase activity as percent of control)	
	27–28 hr	48–50 hr
TAL/N (control)	100	100
TAL/N (Py-infected)	129	131
Py AL/N (transformed)	0	1
Swiss 3T3 (control)	100	100
Swiss 3T3 (Py-infected)	107	80
Py11 (transformed)	4	12

[a] Data calculated from Mora et al., 1971.

vivo, is there a relationship between these properties? Phenotypically "flat revertant" cell lines have been selected by treating virally transformed cells with fluorodeoxyuridine (Pollack et al., 1968). A contact-inhibited flat SV40 derivative line obtained by this technique had the normal ganglioside pattern (Table VI) of the parent Swiss 3T3 cell line. Flat revertant polyoma cells obtained by this procedure had some phenotypic and karyotypic heterogeneity (Pollack et al., 1970), and the ganglioside composition of these cells was only partially restored to normal. The N-acetylgalactosaminyltransferase activity in these two cell lines paralleled the ganglioside compositions (Mora et al., 1971). These results indicated that there is a coordination in the expression of N-

TABLE VI. Distribution of Gangliosides in a Swiss 3T3 Cell Line, in SV40- and Py-Transformed Derivative Lines, and in Flat Sublines Derived from the Virus-Transformed Lines[a]

Cell lines	Gangliosides, nmoles/mg protein				
	G_{D1a}	G_{M1}	G_{M2}	G_{M3}	Total
Swiss 3T3	1.8	1.4	1.3	3.3	7.8
SV101	0.05	0.1	0.05	3.5	3.7
Py11	0.1	0.2	0.05	3.2	3.5
F1² SV101	2.6	0.8	1.4	2.9	7.7
F1 Py11	0.93	0.2	0.4	3.9	5.4

[a] From Mora et al., 1971.

acetylgalactosaminyltransferase activity and phenotypic growth properties of these mouse cell lines. Similar results were obtained with a flat revertant SV40-transformed Balb 3T3 line selected for serum factor dependence (Smith et al., 1971). Whereas this enzyme activity was low in the SV40 transformant, the level was similar to that in normal Balb 3T3 cells in the flat revertant (Fishman et al., 1973).

C. Specificity of the Altered Ganglioside Metabolism

Does viral transformation result in a specific or general reduction in ganglioside biosynthesis? The activities of the four glycosyltransferases involved in ganglioside synthesis (Figure 1) were determined in normal and virally transformed Swiss 3T3 and AL/N cells (Table VII). Sialyltransferase I activity was reduced only in the SVS-AL/N line; sialyltransferase II was diminished in SV101 but not Py11 and elevated in both transformed AL/N lines. Galactosyltransferase was elevated in the Swiss transformants and reduced in the AL/N transformed cells. The only consistent change was the reduction of N-acetylgalactosaminyltransferase activity (seven- to eightfold) in all four transformed cell lines.

The low galactosyltransferase activity found in the transformed AL/N cells may be related to another observation made on the spontaneously transformed TAL/N cell line. After prolonged cultivation of this line, the cells no longer contained the gangliosides G_{M1} and G_{D1a}. When the glycosyltransferases were assayed in early- and late-passage TAL/N cells, galactosyltransferase activity was almost absent from the late-passage cells; the other enzyme activities were not reduced (Brady et al., 1973). As both SVS-AL/N and Py-AL/N have also been in

TABLE VII. Glycosyltransferase Activities in Normal and Virus-Transformed Mouse Cell Lines[a]

	Cell line			
	Percent of Swiss 3T3		Percent of NAL/N	
Enzyme	SV-3T3	Py-3T3	SVS-AL/N	Py-AL/N
Sialyltransferase I	105	110	56	143
N-Acetylgalactosaminyltransferase	13	17	16	14
Galactosyltransferase	655	168	6	11
Sialyltransferase II	36	82	184	192

[a] Activities expressed as percent found in the normal contact-inhibited parental line. Data calculated from Fishman et al., 1972.

continuous culture for several years, the low galactosyltransferase activity in these cell lines may be due to similar phenomenon.

D. Generality of the Phenomenon

An extremely important question is whether the observed phenomenon of altered ganglioside metabolism is an obligatory consequence of tumorigenic-virus transformation. Cell transformation by oncogenic DNA viruses is an event of low efficiency. Mouse cells infected with SV40 undergo abortive infection, but only a few cells become permanently transformed (for review cf. Eckhart, 1969). The observed block in ganglioside biosynthesis may be due to viral transformation or could be the result of selection of cells with altered ganglioside metabolism. In addition, this change may not be a direct consequence of viral transformation but might have occurred during the prolonged cultivation of the cells as a result of secondary processes or selective pressures. The activity of N-acetylgalactosaminyltransferase varies considerably between different strains of mouse cells as well as different clones of the same strain. In addition, there are considerable changes in ganglioside biosynthesis when cells are established in culture (unpublished observations of P. H. Fishman and P. T. Mora) as well as when cells are cultivated for a prolonged time (see above). For these reasons, in all of our studies, we have directly compared established lines of virally transformed cells with the established parental line from which they were derived. With only two exceptions, the DNA-virus-transformed mouse cell lines are characterized by an absence of higher gangliosides and reduced aminosugar transferase activity. One exception was an SV40-transformed mouse cell line selected for resistance to 5-bromo-2-deoxyuridine (Dubbs et al., 1967). A second exception was a temperature-sensitive SV40-transformed Swiss 3T3 cell line (Renger and Basilico, 1972). This cell line is unusual in that the temperature-sensitive mutation is host cell related, as wild-type virus can be rescued from these cells. Similar changes in ganglioside composition have been observed in SV40- and Py-transformed Swiss 3T3 cells by Sheinen et al. (1971). More recently, however, this same group reported that two SV40-transformed clones of Swiss 3T3 have the more complex gangliosides found in normal mouse cells (Yogeeswaren et al., 1972). G_{D1a} was also found in an SV40-transformed 3T3 subclone and in an SV40-transformed primary mouse fibroblast (Diringer et al., 1972).

In order to reexamine this question, AL/N mouse embryo cells were infected with SV40 and a number of clones, both T-antigen-positive and -negative, were isolated and subcultured (unpublished results of L. Couvillion and P. T. Mora). When assayed for N-acetylgalactosaminyltransferase activity, all of the T-antigen-positive clones had at least fourfold less activity than the T-antigen-negative clones. Preliminary results with similar clones of spontaneously transformed TAL/N cells suggest that there is no correlation between aminosugar transferase activity and presence of tumor antigen. Since the parental TAL/N line is already malignantly transformed and does not exhibit an altered ganglioside metabolism, the possibility exists that transformation in these cells arose by at least one molecular event different from that observed in virally transformed cells. Because of this altered cellular function, these cells may not respond to the presence of an integrated tumor virus, and certain viral functions related to viral transformation may not be expressed. Such a mechanism would explain the normal ganglioside metabolism found in the temperature-sensitive SV40-transformed Swiss 3T3 cells (cited above). It is interesting to note that these cells are still independent of serum requirement at the nonpermissive temperature (Renger and Basilico, 1972). Similar distinctions have been made between spontaneously and virally transformed hamster cells with regard to their glycolipid metabolism (Critchley and Macpherson, 1973; Sakiyama and Robbins, 1973).

E. **Transformation of Mouse Cells by RNA Tumor Viruses and Other Agents**

The questions raised above can be answered more directly if a cell population can be quantitatively and rapidly transformed. This is possible with RNA tumor viruses which can transform cells virtually quantitatively in a relatively short period of time. Using a high input of murine sarcoma virus (MSV), mouse cells undergo a mass morphological transformation within two days after infection. Thus, morphological and biochemical changes can be temporally correlated. When Swiss 3T3 cells were infected with Moloney sarcoma virus, there was a drastic decrease in N-acetylgalactosaminyltransferase activity following transformation (Table VIII). The other ganglioside glycosyltransferases were either unaffected or elevated (Mora et al., 1973). Analysis of the ganglioside composition of the MSV transformed cells confirmed the enzyme studies; there was an 80% reduction in the amino sugar containing gangliosides (Figure 5). MSV is a defective virus which can transform a cell but cannot reproduce. Sarcoma virus replication requires a helper leukemia virus which is always present in MSV stocks. When Swiss 3T3 cells were infected with Moloney leukemia virus by itself, there was no effect on ganglioside biosynthesis (Mora et al.,

TABLE VIII. Levels of Ganglioside Glycosyltransferases during the Transformation of Swiss 3T3 Cells by Murine Sarcoma Virus

Days after infection	Glycosyltransferase, percent of control cells			
	NAN I	GalNAc	Gal	NAN II
2	91	80 (95)[a]	150 (240)	156
3	[b]	75	160	[b]
4	118	50 (53,53)[a]	325 (385,300)[a]	115 (85)
7	[b]	20	275	[b]

[a] Values in parentheses represent separate experiments. Data from Mora et al., 1973.
[b] Not determined.

1973). These experiments demonstrate that a mass transformation of mouse cells with an RNA tumor virus can directly and quantitatively result in a specific block in ganglioside biosynthesis. In this case, the reduction of aminosugar transferase activity was similar to that observed in DNA-virus-transformed mouse cells. This change is also a consequence of viral transformation and not productive infection.

However, the situation appears to be more complex. When a nonproductively transformed cell line transformed with the Kirsten isolate of MSV (Aaronson and Weaver, 1971) was analyzed for ganglioside pattern, an accumulation of G_{M2} was observed (Figure 6). This observation suggested a loss of galactosyltransferase activity (reaction (3) of Figure 1). Extracts of the cells were assayed for G_{M2}:UDP-Gal galactosyltransferase activity. This enzyme could not be detected in these cells (Table IX). Since we are dealing with a clonal isolate and not a mass transformation, other clones transformed by Kirsten sarcoma virus will have to be examined. However, the results suggest that transformation of mouse cells by different murine sarcoma viruses may lead to different specific blocks in ganglioside biosynthesis. A defect in the biosynthesis of G_{M1} also occurs in other transformed cells, including SV40-transformed human fibroblasts (Hakomori, 1970; P. H. Fishman, unpublished observations), spontaneously transformed mouse cells (see Section V,C), and chemically induced rat hepatoma cells (Brady et al., 1969). More recently, we have observed an absence of G_{M1} and G_{D1a} and reduced galactosyltransferase activity in Balb 3T3 cells transformed *in vitro* by chemical carcinogens and X-irradiation. These latter results indicate that the same cell line may exhibit a similar biochemical response to various transforming agents and that an altered ganglioside metabolism may be important to the transformation process.

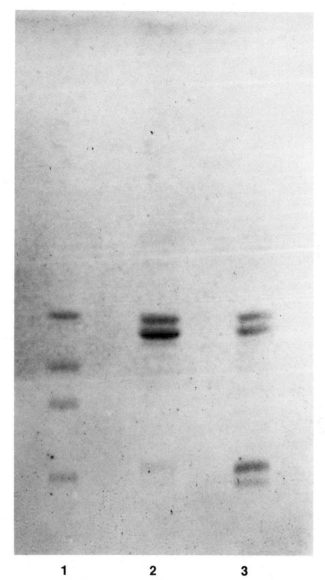

FIGURE 5. Thin-layer chromatogram of gangliosides from Mock and MSV infected Swiss 3T3 cells. Swiss 3T3 cells were infected with Moloney sarcoma virus (MSV) at a high multiplicity and analyzed for gangliosides six days post infection. Lane 1—authentic ganglioside standards in descending order: G_{M3}, G_{M2}, G_{M1}, and G_{D1a}. Lane 2—gangliosides from MSV-transformed cells. Lane 3—gangliosides from an equivalent amount of mock-infected cells. Reproduced with permission from Mora et al., 1973.

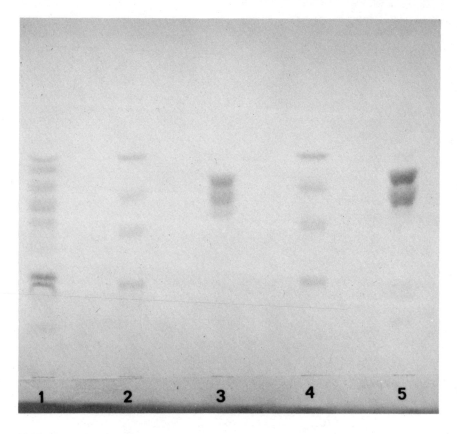

FIGURE 6. Thin-layer chromatogram of gangliosides from Balb 3T3 and KBalb cells. Lane 1—gangliosides from normal contact-inhibited Balb 3T3 cells, Lanes 2 and 4—authentic ganglioside standards (same as Figure 5). Lanes 3 and 5—gangliosides from two batches of KBalb, a nonproductively Kirsten sarcoma-virus-transformed subclone of Balb 3T3. Reproduced with permission from Fishman *et al.*, 1973.

VI. DISCUSSION

A. Molecular Basis of Altered Ganglioside Metabolism

By what mechanism can the insertion of a tumor virus genome into host cell DNA (Sambrook *et al.*, 1968) affect the activity of a host cell enzyme? Possibilities exist at the levels of DNA transcription, messenger RNA translation, and expression of enzyme activity. Unfortunately, our knowledge of the regulation of gene expression in mammalian cells

TABLE IX. Glycosyltransferase Activities in Normal and Ki-MSV-Transformed Balb 3T3 Cell Lines[a]

Enzyme	Activity in KBalb cells as percent of control Balb 3T3 cells
N-Acetylgalactosaminyltransferase	103
Galactosyltransferase	0
Sialyltransferase II	92

[a] Data calculated from Fishman et al., 1973.

is minute compared to our understanding of bacterial systems. However, several speculations can be made based on the data at hand.

1. The Integration Site Model. At the level of DNA, integration of viral DNA into host cell chromosomes may occur in the genome coding for the affected glycosyltransferase. Presence of this nonhost DNA may interfere with transcription of this gene or result in a frame-shift mutation whereby an inactive or partially active enzyme is synthesized. There are several lines of evidence that argue against this model. Viral DNA is also integrated during productive infection (Hirai and Defendi, 1972; Babiuk and Hudson, 1972) which does not affect ganglioside metabolism in mouse cells. However, it is possible that viral integration occurs at another site during lytic infection. The flat revertant cell lines also contain integrated viral genomes which can express at least one viral function (tumor antigen) but also have a normal complement of gangliosides. The increased chromosome number of these cells may result in an unaffected host gene coding for the glycosyltransferase. However, other data indicate that the number of integrated viral genomes in flat SV40 cells increases proportionally (A. Vogel, personal communication). Finally, our experiments with the temperature-sensitive MSV-NRK cells argues against this integration site model. Presumably, the viral DNA remains integrated at both temperatures, and at present our knowledge of DNA structure and transcription does not permit us to explain this anomaly. Nevertheless, this integration site model is attractive for several reasons. First, it would allow us to understand why there are exceptions to the general phenomenon of altered ganglioside metabolism in virally transformed cells. Studies with hybrids between virally transformed and untransformed cells indicate that the viral genome can be associated with more than one chromosome, possible in random fashion (Martin and Littlefield, 1968; Weiss et *al.*, 1968). Thus, integration of the viral genome into one particular host chromosome would be required for an altered ganglioside metabolism. Secondly, it would explain why the interaction of cells with different viruses results in different blocks in ganglioside biosynthesis. In these circumstances, viral integration may have occurred at different gene loci. Finally, it could explain the occasional alteration in activity of other glycosyltransferases in the ganglioside pathway. A similar situation is represented by the phenomenon of polarity found in bacteria, where a mutation in one gene of a polycistron alters the translation of subsequent genes (Epstein and Beckwith, 1968).

2. Viral Action at the Level of Translation. It is known that in eukaryotic cells induction and repression of enzyme synthesis can occur at the level of messenger RNA (Tomkins and Martin, 1970). Thus, the SV40, polyoma, and Moloney sarcoma viruses could contain the genetic information for a common repressor molecule which could interfere with the synthesis of aminosugar transferase. Both DNA viruses can code for 5–10 polypeptides (Eckhart, 1969), and the potential coding capacity of the large RNA tumor viruses is extensive.

This model is compatible with our findings with the temperature-sensitive MSV-NRK cell line. In these cells, the hypothetical repressor would be a heat-labile protein. Unfortunately, there are not yet any mouse cells transformed by a temperature-sensitive virus available to further explore this hypothesis. This model does not readily explain the flat revertant cell lines. Analogous to bacterial systems, one would expect such a repressor molecule to be dominant. However, the existence of these flat revertants indicates that host cell factors are involved in the expression of the transformed phenotype. Little is known about the regulation of glycolipid synthesis, but undoubtedly cellular control mechanisms do exist. It is conceivable that a viral gene product could interfere with the normal regulation and biosynthesis of these compounds. Such a mechanism could explain the elevated levels of other glycosyltransferases. Impaired feedback control could result in continual expression of these enzyme activities.

3. Viral Action at the Level of Enzyme Activity. A third possibility is that the activity of the affected glycosyltransferase is directly modulated by some viral gene product. When homogenates from normal and SV40-transformed or MSV-transformed mouse cells were admixed and assayed for hematoside: UDP-GalNAc N-acetylgalactosaminyltransferase activity, there was no evidence of such an inhibitor (Mora *et al.*, 1971, 1973). However, this enzyme is membrane-bound (Cumar *et al.*, 1970), as are the other glycosyltransferases. It is not unlikely that the activity and specificity of these membrane-bound glycosyltransferases are affected by other elements of the membrane in juxtaposition with them. An analogous situation is presented by the specificity imparted to galactosyltransferase by α-lactalbumin (Brew et al., 1968). The two proteins constitute the enzyme complex lactose synthetase. In the absence of α-lactalbumin, the galactosyltransferase has a wide specificity for acceptors other than glucose.

Isolation of the aminosugar transferase in soluble form may provide insight into this possibility and is being carried out. In addition, detailed kinetic analysis of this enzyme from normal and transformed cells may allow us to distinguish among some of these possibilities. Recent work by Pastan and coworkers demonstrated that the reduced levels of cyclic AMP in CEF cells transformed by a strain of RSV (Brian) were due to an altered K_m of the affected adenylcyclase (Anderson et al., 1973a). However, transformation by the Schmidt–Ruppin strain resulted in only a decreased V_{Max} of this enzyme (Anderson et al., 1973b). These findings point out the complexities involved in viral transformation and the difficulties in unraveling its molecular basis.

B. Significance

Finally, we might speculate on the significance of these alterations in ganglioside (and glycoprotein) metabolism upon viral transformation. Such changes may be relevant only if they explain the social behavior of transformed cells in culture and *in vivo*. Transformed cells in general exhibit a loss of growth control and numerous surface changes. What we are concerned with then is the transmission of membrane-mediated information both externally and internally and how this transmission has been altered by viral transformation. If we conceptualize the plasma membrane as a fluid mosaic (Singer and Nicholson, 1972), then the carbohydrate chains of glycolipids and glycoproteins represent polar entities floating above a hydrophobic environment. It has been shown that these groups are restricted to the external surface of the plasma membrane and to one surface of other membranes (Hirano et al., 1972). The informational potential of these groups is tremendous in various recognition phenomena (Roseman, 1970). Structural and topographical variation may influence cell behavior. Examples are cell-to-cell recognition and adhesiveness, contact inhibition of growth and movement, surface antigens and agglutination sites, substratum and serum requirements, and neoplastic processes.

Gangliosides may interact directly in such phenomena or indirectly by masking or unmasking other informational groups. Plasma membranes are enriched in gangliosides (Klenk and Choppin, 1970; Renkonen *et al.*, 1970), and the altered ganglioside pattern of transformed cells is also found in their plasma membranes (Sheinen *et al.*, 1971). One example of ganglioside involvement in the external flow of membrane information is the finding that ganglioside G_{M1} is the specific receptor for cholera toxin (Cuatrecasas, 1973). Interaction of the toxin with the G_{M1} receptor results in a stimulation of adenylcyclase and increased intracellular cyclic AMP levels (Cuatrecasas, 1973). Transformed cells lacking G_{M1} do not respond to the cholera toxin (unpublished observations of P. H. Fishman, M. D. Hollenberg, and P. Cuatrecasas). We would like to speculate that gangliosides act as receptors for other external factors found in serum or excreted from cells in culture. The absence of these receptors on the surfaces of transformed cells could explain their lack of response to such factors. One example is the ability of virally transformed cells to grow in serum factor-free media (Smith *et al.*, 1971).

The internal transmission of membrane-mediated information between plasma membrane and cell nucleus is also important to the regulation of growth control. The cell surface is not isolated from the nucleus, as there is a vast intracellular network of membrane structures which contains glycolipids, glycoproteins, and glycosyltransferases. Cells divide in a cycle consisting of four phases: S, during which chromosome replication occurs; G_2, the premitotic phase; M, the phase in which the cell divides; and G_1, which is between mitosis and DNA synthesis. Cells may also enter a resting phase, G_0.

There is now evidence accumulating that the metabolism and composition of complex carbohydrates changes during the cell cycle. The synthesis of glycoproteins was found to be maximal during the S phase in synchronized cultures of a mouse lymphoma line, while glycolipids appeared to be synthesized at the end of G_2 and during mitosis (Bosmann and Winston, 1970). There also appear to be effects of the cell cycle on surface glycoproteins. When glucosamine-labeled glycoproteins are removed from the surface of L mouse cells by trypsin and separated by ion exchange column chromatography, there are four classes of glycoproteins. In synchronized cultures, there is a preferential accumulation of one of these glycoprotein classes during mitosis (Brown, 1972). Fucose-labeled glycopeptides from the surface of BHK cells arrested in metaphase by vinblastin have also been analyzed (Glick and Buck, 1973). When compared to similar glycopeptides from nonmetaphase cells, the metaphase cells were enriched in the higher-molecular-weight glycopeptides characteristic of transformed cells (see Section IV,C,X). Similar results were obtained when glycopeptides from growing BHK cells were compared to nongrowing cells (Buck *et al.*, 1971). It appears that a glycopeptide alteration that is permanently expressed by virally transformed cells is also temporarily expressed by normal cells when they are rapidly growing and dividing.

Thus, if the final step in mitosis is the incorporation of a specific glycoprotein or glycolipid into the plasma membrane, this event may represent a switch point for entry of the cell into a new round of division (G_1) or a resting state (G_0). Activation of this switch may in turn be influenced by membrane-mediated information from outside the cell. Transformed cells may always be synthesing an "on" switch as suggested by their glycoprotein composition, or they may not be able to synthesize an "off" switch (inability to synthesize the more complex gangliosides). Until we know more about the function and regulation of these sialic-acid-containing compounds, we can only speculate.

VII. CONCLUDING REMARKS

Mouse cells contain a homologous series of gangliosides increasing in size from monosialyllactosylceramide (G_{M3}) to disialyltetrahexosylceramide (G_{D1a}). After transformation by oncogenic DNA and RNA viruses as well as by chemical carcinogens and X-irradiation, there is a reduction in the oligosaccharide chains of these compounds due to specific blocks in ganglioside biosynthesis. The decreased glycosyltransferase activities observed in transformed mouse cells appear to be associated with the transformation process. Although the molecular basis of these changes is unknown, and the relationship between these acidic membrane glycolipids and the behavior of cells in culture and *in vivo* is unclear, the discovery of an altered ganglioside metabolism in these transformed cells may be an important implement for cancer research.

ACKNOWLEDGMENTS

The authors wish to express their appreciation to Dr. Peter T. Mora of the National Cancer Institute for his close and continual collaboration in these studies. We thank Drs. Stuart Aaronson, John Bader, Robert Bassin, and George Todaro, National Cancer Institute; Robert Pollack and Arthur Vogel, Cold Spring Harbor Laboratory; Claudio Basilico, New York University; and Walter Eckhart, Salk Institute for generous donations of cell lines. The technical assistance of Mr. Roy Bradley, Miss Linda Couvillian, and Mrs. Vivian McFarland is acknowledged.

VIII. REFERENCES

Aaronson, S. A., and Weaver, C. A., 1971, Characterization of murine sarcoma virus (Kirsten) transformation of mouse and human cells, *J. Gen. Virol.* **13**:245.

Anderson, W. B., Johnson, G. S., and Pastan, I., 1973a, Transformation of chick-embryo fibroblasts by wild-type and temperature-sensitive Rous sarcoma virus alters adenylate cyclase activity, *Proc. Natl. Acad. Sci. U.S.* **70**:1055.

Anderson, W. B., Lovelace, E., and Pastan, I., 1973b, Adenylate cyclase activity is decreased in chick embryo fibroblasts transformed by wild type and temperature sensitive Schmidt–Ruppin Rous sarcoma virus, *Biochem. Biophys. Res. Commun.* **52**:1293.

Babiuk, L. A., and Hudson, J. B., 1973, Integration of polyoma virus DNA into mammalian genomes, *Biochem. Biophys. Res. Commun.* **47**:111.

Bader, J. P., and Brown, N. R., 1971, Induction of mutations in an RNA tumor virus by an analogue of a DNA precursor, *Nat. New Biol.* **234**:11.

Bosmann, H. B., 1972a, Sialyl transferase activity in normal and RNA- and DNA-virus transformed cells utilizing desialyzed, trypsinized cell plasma membrane external surface glycoproteins as exogenous acceptors, *Biochem. Biophys. Res. Commun.* **49**:1256.

Bosmann, H. B., 1972b, Cell surface glycosyltransferases and acceptors in normal and RNA- and DNA- virus transformed fibroblasts, *Biochem. Biophys. Res. Commun.* **48**:523.

Bosmann, H. B., and Eylar, E. H., 1968, Collagen-glucosyl transferase in fibroblasts transformed by oncogenic viruses, *Nature* **218**:582.

Bosmann, H. B., and Winston, R. A., 1970, Synthesis of glycoprotein, glycolipid, protein and lipid in synchronized L5178Y cells, *J. Cell. Biol.* **45**:23.

Brady, R. O., and Mora, P. T., 1970, Alteration in ganglioside pattern and synthesis in SV40 and polyoma virus transformed mouse cell lines, *Biochim. Biophys. Acta* **218**:308.

Brady, R. O., Kanfer, J. N., and Shapiro, D., 1965, Metabolism of glucocerebrosides II. Evidence of an enzymatic deficiency in Gaucher's disease, *Biochem. Biophys. Res. Commun.* **18**:221.

Brady, R. O., Borek, C., and Bradley, R. M., 1969, Composition and synthesis of gangliosides in rat hepatocyte and hepatoma cell lines, *J. Biol. Chem.* **244**:6552.

Brady, R. O., Fishman, P. H., and Mora, P. T., 1973a, Alterations of complex lipid metabolism in tumorigenic virus transformed cell lines, *Adv. Enzyme Reg.* **11**:231.

Brady, R. O., Fishman, P. H., and Mora, P. T., 1973b, Membrane components and enzymes in virally transformed cells, *Fed. Proc.* **32**:102.

Brew, K., Vanaman, T. C., and Hill, R. L., 1968, The role of α-lactalbumin and the A protein in lactose synthetase, a unique mechanism for the control of a biological reaction, *Proc. Natl. Acad. Sci. U.S.* **59**:491.

Brown, J. C., 1972, Cell surface glycoprotein I: Accumulation of a glycoprotein on the outer surface of mouse LS cells during mitosis, *J. Supermol. Struct.* **1**:1.

Buck, C. A., Glick, M. C., and Warren, L., 1970, A comparative study of glycoproteins from the surface of control and Rous sarcoma virus transformed hamster cells, *Biochemistry* **9**:4567.

Buck, C. A., Glick, M. C., and Warren, L., 1971a, Glycopeptides from the surface of control and virus-transformed cells, *Science* **172**:169.

Buck, C. A., Glick, M. C., and Warren, L., 1971b, Effect of growth on the glycoproteins from the surface of control and Rous sarcoma virus transformed hamster cells, *Biochemistry* **10:**2176.

Critchley, D. R., and Macpherson, I., 1973, Cell density dependent glycolipids in NIL 2 hamster cells, derived malignant and transformed cell lines, *Biochem. Biophys. Acta* **296:**145.

Cuatrecasas, P., 1973, Interaction of *Vibrio cholera* enterotoxin with cell membranes, *Biochemistry* **12:**3547.

Culp, L. A., Grimes, W. J., and Black, P. H., 1971, Contact-inhibited revertant cell lines isolated from SV40-transformed cells, *J. Cell. Biol.* **50:**682.

Cumar, F. A., Brady, R. O., Kolodny, E. H., McFarland, V. W., and Mora, P. T., 1970, Enzymatic block in the synthesis of gangliosides in DNA virus-transformed tumorigenic mouse cell lines. *Proc. Nat. Acad. Sci. U.S.* **67:**757.

Den, H., Schultz, A. M., Basu, M., and Roseman, S., 1971, Glycosyltransferase activities in normal and polyoma-transformed BHK cells, *J. Biol. Chem.* **246:**2721.

Dijong, I., Mora, P. T., and Brady, R. O., 1971, Gas chromatographic determination of gangliosides in mouse cell lines and in virally transformed derivative lines, *Biochemistry* **10:**4039.

Diringer, H., Strobel, G., and Koch, M. A., 1972, Glycolipids of mouse fibroblasts and virus transformed mouse cell lines, *Hoppe-Seyl. Z.* **353:**1759.

Dubbs, D. R., Kitt, S., de Torres, R. A., and Anken, M., 1967, Virogenic properties of bromodeoxyuridine-resistant Simian virus 40-transformed mouse kidney cells, *J. Virology* **1:**968.

Eckhart, W., 1969, Cell transformation by polyoma virus and SV40, *Nature* **224:**1069.

Epstein, W., and Beckwith, J. R., 1968, Regulation of gene expression, *Ann. Rev. Biochem.* **37:**411.

Fishman, P. H., McFarland, V. W., Mora, P. T., and Brady, R. O., 1972, Ganglioside biosynthesis in mouse cells: Glycosyltransferase activities in normal and virally transformed lines, *Biochem. Biophys. Res. Commun.* **48:**48.

Fishman, P. H., Brady, R. O., Bradley, R. M., Aaronson, S. A., and Todero, G. J., 1974, Absence of a specific ganglioside galactosyltransferase in murine sarcoma virus-transformed mouse cells, *Proc. Natl. Acad. Sci. U.S.* **71:**298.

Fishman, P. H., Brady, R. O., and Mora, P. T., 1973, Altered glycolipid metabolism related to viral transformation of established mouse cell lines, *in: Tumor Lipids: Biochemistry and Metabolism* (R. Wood, ed.), pp. 251–266, American Oil Chemists Society, Chicago.

Folch, J., Lees, M., and Stanley, G. H. S., 1957, A simple method for the isolation and purification of total lipids from animal tissues, *J. Biol. Chem.* **226:**497.

Glick, M. C., and Buck, C. A., 1973, Glycoproteins from the surface of metaphase cells, *Biochemistry* **12:**85.

Glick, M. C., Comstock, C. A., Cohen, M. A., and Warren, L., 1971, Membranes of animal cells. VIII. Distribution of sialic acid, hexosamines, and sialidase in the L cell, *Biochim. Biophys. Acta* **233:**247.

Grimes, W. J., 1970, Sialic acid transferases and sialic acid levels in normal and transformed cells, *Biochemistry* **9:**5083.

Grimes, W. J., 1973, Glycosyltransferase and sialic acid levels of normal and transformed cells, *Biochemistry* **12:**990.

Hakomori, S., 1970, Cell density-dependent changes of glycolipid concentrations in fibroblasts and loss of this response in virus-transformed cells, *Proc. Natl. Acad. Sci. U.S.* **67:**1741.

Hakomori, S., and Murakami, W. T., 1968, Glycolipids of hamster fibroblasts and derived malignant-transformed cells, *Proc. Nat. Acad. Sci. U.S.* **59**:254.

Hakomori, S., Saito, T., and Vogt, P. K., 1971, Transformation by Rous sarcoma virus: Effects on cellular glycolipids, *Virology* **44**:609.

Hakomori, S., Teather, C., and Andrews, H., 1968, Organizational difference of cell surface hematoside in normal and virally transformed cells, *Biochem. Biophys. Res. Commun.* **33**:563.

Hirai, K., and Defendi, V., 1972, Integration of SV40 DNA into the DNA of permissive monkey kidney cells, *J. Virol.* **9**:705.

Hirano, H., Parkhouse, B., Nicolson, G., Lennox, E. S., and Singer, S. J., 1972, Distribution of saccharide residues on membrane fragments from a myeloma-cell homogenate: its implications for membrane biogenesis, *Proc. Natl. Acad. Sci. U.S.* **69**:2945.

Kampine, J. P., Brady, R. O., Kanfer, J. N., Feld, M., and Shapior, D., 1967a, Diagnosis of Gaucher's disease and Niemann–Pick disease with small samples of venous blood, *Science* **155**:86.

Kampine, J. P., Brady, R. O., Yankee, R. A., Kanfer, J. N., Shapiro, D., and Gal, A. E., 1967b, Sphingolipid metabolism in leukemic leukocytes, *Cancer Res.* **27**:1312.

Keshgegian, A. A., and Glick, M. C., 1973, Glycoproteins associated with nuclei of cells before and after transformation by a ribonucleic acid virus, *Biochemistry* **12**:1221.

Klenk, H.-D., and Choppin, P. W., 1970, Glycosphingolipids of plasma membranes of cultured cells and an enveloped virus (SV5) found in these cells, *Proc. Natl. Acad. Sci. U.S.* **66**:57.

Kolodny, E. H., Brady, R. O., Quirk, J., and Kanfer, J. N., 1970, Preparation of radioactive Tay–Sachs ganglioside labeled in the sialic acid moiety, *J. Lipid Res.* **11**:144.

Makita, A., and Seyama, Y., 1971, Alterations of Forssman-antigenic reactivity and of monosaccharide composition in plasma membrane from polyoma-transformed hamster cells, *Biochim. Biophys. Acta* **241**:403.

Marin, G., and Littlefield, J. W., 1968, Selection of morphologically normal cell lines from polyoma-transformed BHK 21/13 hamster fibroblasts, *J. Virology* **2**:69.

Martin, G. S., 1970, Rous sarcoma virus: a function required for the maintenance of the transformed state, *Nature* **227**:1021.

Meezan, E., Wu, H. C., Black, P. H., and Robbins, P. W., 1969, Comparative studies on the carbohydrate-containing membrane components of normal and virus-transformed mouse fibroblasts, II. Separation of glycoproteins and glycopeptides by Sephadex chromatography, *Biochemistry* **8**:2518.

Mora, P. T., 1973, Cell growth regulation, cell selection and the function of the membrane glycolipids, in: *Membrane Mediated Information: Function and Biosynthesis of Membrane Lipids and Glycoproteins* (P. W. Kent, ed.), pp. 64–84, Oxford University Press, London.

Mora, P. T., Brady, R. O., Bradley, R. M., and McFarland, V. W., 1969, Gangliosides in DNA virus-transformed and spontaneously transformed tumorigenic mouse cell lines. *Proc. Nat. Acad. Sci. U.S.* **63**:1290.

Mora, P. T., Cumar, F. A., and Brady, R. O., 1971, A common biochemical change in SV40 and polyoma virus transformed mouse cells coupled to control of cell growth in culture, *Virology* **46**:60.

Mora, P. T., Fishman Bassin, R. H., Brady, R. O., and McFarland, V. W., 1973, Transformation of Swiss 3T3 cells by murine sarcoma virus is followed by decrease in a glycolipid glycosyltransferase, *Nature*, **245**:226.

Ohta, N., Pardee, A. B., McAuslan, B. R., and Burger, M. M., 1968, Sialic acid contents and controls of normal and malignant cells, *Biochem. Biophys. Acta* **158**:98.

Pollack, R. E., Green, H., and Todaro, G. J., 1968, Growth control in cultured cells: Selection of sublines with increased sensitivity to contact inhibition and decreasing tumor-producing ability, *Proc. Natl. Acad. Sci. U.S.* **60**:126

Pollack, R. E., Wolman, S., and Vogel, A., 1970, Reversion of virus-transformed cell lines: Hyperploidy accompanies retention of viral genes, *Nature* **228**:938.

Renger, H. C., and Basilico, C., 1972, Mutation causing temperature-sensitive expression of cell transformation by a tumor virus, *Proc. Natl. Acad. Sci. U.S.* **69**:109.

Renkonen, O., Gahmberg, C. G., Simons, K., and Kaarianinen, L., 1970, Enrichment of gangliosides in plasma membranes of hamster kidney fibroblasts, *Acta Chem. Scand.* **24**:733.

Robbins, P. W., and MacPherson, I. A., 1971, Control of glycolipid synthesis in cultured hamster cell line, *Nature* **229**:569.

Roseman, S., 1970, The synthesis of complex carbohydrates by multiglycosyltransferase systems and their potential function in intercellular adhesion, *Chem. Phys. Lipids* **5**:270.

Sakiyama, H., and Burge, B. W., 1972, Comparative studies of the carbohydrate-containing components of 3T3 and Simian virus 40 transformed 3T3 Mouse fibroblasts, *Biochemistry* **11**:1366.

Sakiyama, H., Gross, S. K., and Robbins, P. W., 1972, Glycolipid synthesis in normal and virus-transformed hamster cell lines, *Proc. Natl. Acad. Sci. U. S.* **69**:872.

Sakiyama, H., and Robbins, P. W., 1973, Glycolipid synthesis and tumorigenicity of clones isolated from the Nil-2 line of hamster embryo fibroblasts, *Red. Proc.* **32**:86.

Sambrook, J., Westphal, H., Srinivassan, P. R., and Dulbecco, R., 1968, the integrated state of viral DNA in SV40-transformed cells, *Proc. Natl. Acad. Sci. U.S.* **60**:1288.

Sheinin, R., and Onodera, K., 1972, Studies of the plasma membrane of normal and virus-transformed 3T3 mouse cells, *Biochim. Biophys. Acta* **274**:49.

Sheinin, R., Onodera, K., Yogeeswaran, G., and Murray, R. K., 1971, Studies of components of the surface of normal and virus-transformed mouse cells, in: *The Biology of Oncogenic Viruses 2nd LePetit Symposium* (L. G. Silvestri, ed.), pp. 274–285.

Singer, S. J., and Nicolson, G. L., 1972, The fluid mosaic model of the structure of cell membranes, *Science* **175**:720.

Smith, H. S., Sher, C. D., and Todaro, G. J., 1971, Induction of cell division in medium lacking serum growth factor by SV40, *Virology* **44**:359.

Takemoto, K. K., Ting, R. C. Y., Ozer, H. L., and Fabish, P., 1968, Establishment of a cell line from an inbred mouse strain for viral transformation studies: Simian virus 40 transformation and tumor production, *J. Natl. Cancer Inst.* **41**:1401.

Tomkins, G. M., and Martin, D. W., Jr., 1970, Hormones and gene expression, *Ann. Rev. Genet.* **4**:91.

Warren, L., Critchley, D., and MacPherson, I., 1972a, Surface glycoproteins and glycolipids of chicken embryo cells transformed by a temperature-sensitive mutant of Rous sarcoma virus, *Nature* **235**:275.

Warren, L., Fuhrer, J. P., and Buck, C. A., 1972b, Surface glycoproteins of normal and transformed cells: A difference determined by sialic acid and growth-dependent sialyl transferase, *Proc. Natl. Acad. Sci. U.S.* **69**:1838.

Warren, L., Fuhrer, J. P., and Buck, C. A., 1973, Surface glycoproteins of cells before and after transformation by oncogenic viruses, *Fed. Proc.* **32**:80.

Weil, R., and Kara, J., 1970, Polyoma "tumor antigen": An activator of chromosome replication?, *Proc. Natl. Acad. Sci. U.S.* **67**:1011.

Weinstein, D. B., Marsh, J. B., Glick, M. C., and Warren, L., 1970, Membranes of animal cells. VI. The glycolipids of the L cell and its surface membrane, *J. Biol. Chem.* **245**:328.

Weiss, M., Ephrussi, B., and Scaletta, L., 1968, Loss of T-antigen from somatic hybrids between mouse cells and SV-40 transformed human cells, *Proc. Natl. Acad. Sci. U.S.* **59**:1132.

Wu, H. C., Meezan, E., Black, P. H., and Robbins, P. W., 1969, Comparative studies on the carbohydrate-containing membrane components of normal and virus-transformed mouse fibroblasts. I. Glucosamine-labeling patterns in 3T3, spontaneously transformed 3T3 and SV40-transformed 3T3 cells, *Biochemistry* **8**:2509.

Yogeeswaran, G., Sheinin, R., Wherett, J. R., and Murray, R. K., 1972, Studies on the glycosphingolipids of normal and virally transformed 3T3 mouse fibroblasts, *J. Biol. Chem.* **247**:5146.

Chapter 9

Circulating Sialyl Compounds

Abraham Rosenberg and Cara-Lynne Schengrund

I. INTRODUCTION

A number of biologically active glycoproteins found in mammalian plasma contain sialic acid. The role(s) of sialic acid on circulating sialoglycoproteins has received much attention during the past 10–15 years. Areas of particular interest are (1) circulating sialoglycoproteins in disease and (2) the function of sialic acid in determining survival time of circulating sialoglycoproteins. Changes in the levels of plasma sialoglycoproteins have been reported in a variety of diseases (for review, Spiro, 1970; Winzler, 1971). Changes in sialic acid content of specific circulating sialoglycoproteins affect their rate of clearance from plasma (for review, Ashwell and Morell, 1975). Enzymatic, hormonal, and immunogenic activities of various sialoglycoproteins have been found to be related in some degree to their sialic acid content. This chapter will not attempt to review all the work that has been done on circulating sialoglycoproteins, but will restrict itself to aspects of the following: (a) a survey of circulating sialoglycoproteins, (b) changes in plasma sialic acid levels in disease, and (c) possible biological roles for sialic acid. Synthesis and catabolism of sialic-acid-containing compounds are reviewed in Chapters 4 and 5, respectively.

ABRAHAM ROSENBERG AND CARA-LYNNE SCHENGRUND • Department of Biological Chemistry, The Milton S. Hershey Medical Center, The Pennsylvania State University, Hershey, Pennsylvania 17033.

TABLE I. Plasma Sialoglycoproteins

Compound	Species source	Molecular weight	Sialic acid, residues/mole	Percent carbohydrate	Reference
Alkaline phosphatase	Human	—	—	—	Robinson and Pierce, 1964
Amylase	Marsupial mouse	—	2–3	—	Finnegan and Hope, 1970
α_1-Antitrypsin	Human	45,000	5^a	12.4	Winzler and Bocci, 1972
Antropinesterase	Rabbit	—	—	—	Margolis and Feigelson, 1964
Ceruloplasmin	Human	150,000	9	7	Jamieson, 1965, 1972
Cholinesterase	Horse	315,000	30^a	—	Heilbronn, 1962 Main et al., 1972
Fetuin	Fetal calf	48,400	13.6	22	Spiro, 1960 Spiro and Spiro, 1962
Fibrinogen	Bovine	330,000	7	2.3	Mester et al., 1963 Cunningham, 1971
α_1-Glycoprotein (3.5S)	Human	54,000	6^a	13.7	Schultze et al., 1955
Haptoglobulin	Bovine	85,000	15^a	22.7	Schultze, 1958 Winzler, 1960

Compound	Species	MW	Sialic acid residues per mole	% Carbohydrate	Reference
γ-Globulin	Rabbit	160,000	2	3.1	Nolan and Smith, 1962
Immunoglobulin G	Human	150,000	1	3[b]	Clamp and Johnson, 1972
Immunoglobulin A	Human	160,000	9	7[b]	Clamp and Johnson, 1972
Immunoglobulin M	Human	180,000	10	10.5[b]	Clamp and Johnson, 1972
		900,000	50	—	
Immunoglobulin D	Human	—	18	—	Clamp and Johnson, 1972
Immunoglobulin E	Human	—	18	—	Clamp and Johnson, 1972
α₂-Macroglobulin	Human	820,000	50	19.3	Winzler, 1960
					Winzler and Bocci, 1972
Orosomucoid	Human	41,000	16	41	Schultze et al., 1958
					Sjoholm, 1967
Oxytocinase	Human	90,000	4	5	Schultze et al., 1958
Transferrin					Blumberg and Warren, 1961
					Parker and Bern, 1962
Transferrin	Pig	—	2	—	Kristjansson and Cipera, 1963
Transferrin	Rabbit	70,000	2	—	Baker et al., 1968
					Scharmann et al., 1971
Transferrin	Bovine	75,000	5	—	Stratil and Spooner, 1971

[a] Sialic acid residues per mole were calculated from the percentage of sialic acid present.
[b] Per cent carbohydrate was calculated from the number of carbohydrate residues present per mole.

II. NORMAL PLASMA CONSTITUENTS

Table I lists nonhormonal plasma sialoglycoproteins, and when known, their molecular weight, number of sialic acid residues per mole, and carbohydrate content. Sialic acid content was estimated by either the Warren procedure (1959) or by treatment with microbial sialidase and measurement of sialic acid release and concomitant change in electrophoretic mobility or isoelectric point. Some of the physical and biological changes observed in the plasma sialoglycoproteins after removal of sialic acid residues are described below.

A. Circulating Sialoenzymes

Human serum alkaline phosphatases (EC 3.1.3.1) show a relationship between their type and the Lewis blood group (Arfors et al., 1963). The enzyme gives four bands on vertical starch gel electrophoresis. Incubation with sialidase alters the mobility of three of the four bands (Robinson and Pierce, 1964). Alkaline phosphatase which appears in the serum during pregnancy also contains sialic acid (Robinson et al., 1966) and is supposed to be of placental origin. Amylase (EC 3.2.1.1) found in serum is present in the form of isoenzymes having two or three sialic acid residues per molecule, while that from the pancreas lacks sialyl groups (Finnegan and Hope, 1970). It is possible that these groups inhibit uptake and destruction of the serum enzyme by the liver. Atropinesterase (EC 3.1.1.10), a nonspecific B-type esterase, is a sialoglycoprotein, as determined by change in electrophoretic mobility after sialidase treatment. Loss of the sialic acid residues has little effect on catalytic or immunogenic properties (Margolis and Feigelson, 1964). Human serum cholinesterase (EC 3.1.1.8) decreases in electrophoretic mobility after incubation with sialidase, but there is no change in enzymatic activity toward acetylcholine or butyrylcholine (Svensmark, 1961; Svensmark and Kristensen, 1962). Augustinsson and Ekedahl (1962) determined substrate specificity of human serum cholinesterase, its Michaelis constants, and the inhibitor constant for the esterase complex with prostigmine. No change from the normal was found in the sialidase-treated cholinesterase. The isoelectric point changes from 3 to 7 upon removal of the sialic acid residues (Svensmark and Kristensen, 1963).

Oxytocinase (EC 3.4.1), an α-aminopeptide amino acid hydrolase which inactivates oxytocin, is present in blood during pregnancy. It contains sialic acid (Tuppy et al., 1963) which, according to Sjöholm

(1967), is essential in maintaining the conformation but not the activity of the enzyme.

B. Serum Sialoglobulins

The sialic acid moieties of the α_1-acid glycoprotein orosomucoid, are not important antigenically (Athineos *et al.*, 1962). Orosomucoid production increases in individuals subjected to stress, but the sialic acid content is proportional to that of normal orosomucoid. However, patients with chronic disease, e.g., Hodgkin's and diabetes, have elevated α_1-acid glycoprotein with reduced sialic acid content. Schmid *et al.* (1964) suggest that abnormally large quantities of the protein are produced over a period of time, and compounds lacking sialic acid to varying degrees are produced. Fetuin (an α_1 glycoprotein) from fetal calf serum is acidic in behavior owing primarily to peripherally located sialic acid residues (Spiro, 1960). Oshiro and Eylar (1968) showed that microheterogeneity observed on starch gel electrophoresis was due to differences in sialic acid content and distribution. A possible function suggested for the sialyl residues is the maintenance of the secondary structure of the molecule (Green and Kay, 1963). Transferrins are β-globulins which contain sialic acid and carry iron. Again, several different protein bands are obtained upon electrophoresis attributable to different numbers of sialic acid residues per molecule (Parker and Bearn, 1962; Chen and Sutton, 1967). Kristjansson and Cipera (1963) have found that sialic acid is not involved in the iron-binding process.

It appears that, generally, sialic acid is *not* a part of an antigenic determinant (atropinesterase and orosomucoid) nor a functioning part of a catalytic or binding site (α_1-antitrypsin, atropinesterase, ceruloplasmin, and serum cholinesterase). However, sialic acid does seem to play a role in transport (α_1-antitrypsin). This effect is discussed more fully later in this chapter. Johnson *et al.* (1970) found that removal of sialic acid residues from the copper-containing protein ceruloplasmin had no effect on its kinetic properties as an oxidase. They concluded that the terminal sialic acid residues do not influence the structure and function of the "active sites" in ceruloplasmin.

C. Sialoglycoprotein Hormones

Several sialoglycoprotein hormones are present in relatively minor but essential quantities in the plasma. As early as 1948, it was reported that the biological activity of crude chorionic and pituitary gonadotropins was lost upon treatment with the receptor destroying enzyme (RDE) of *Vibrio cholerae* or influenza virus (Whitten, 1948), which suggests that sialic acid is essential for activity. Subsequent work has shown that sialic acid is not essential for immunological reactivity. The effect of sialic acid removal from several different hormones is given in Table II. It appears very likely that the inactivation of sialo hormones by sialic acid removal is a consequence of the enhanced uptake and destruction of the desialo compounds by the liver (for review, Ashwell and Morell, 1975). For example, studies on the effect of progressive desialylation of human chorionic gonadotropin (HCG) and its rate of clearance from plasma led Van Hall *et al.* (1971) to conclude that the biological inactivation of HCG observed was due to its rapid clearance from the plasma. This may be a general phenomenon and is discussed in more detail later in this chapter.

The presence of an inhibitor of follicle-stimulating hormone (FSH) in urine of the bonnet monkey, *Macaca radiata,* has been reported. It has been found to have sialidase activity which serves to biologically inactivate FSH (Sairam and Moudgal, 1971). Erythropoietin also loses biological activity *in vivo* after desialylation (Lowy *et al.,* 1960). Again, this effect has been traced to loss of stability during circulation and not to loss of action on target cells. The desialo compound is more active *in vitro* (Goldman *et al.,* 1974). Thyroid-stimulating hormone causes loss of sialic acid from the thyroid gland. Presumably this loss is due to release of thyroglobulin (Bates and Warren, 1963), which contains sialic acid. Enzymes for the addition of sialic acid to thyroglobulin molecules have been characterized in calf thyroid (Spiro and Spiro, 1968). Interestingly, thyroglobulin is not inactivated by removal of its sialyl residues.

Upon conversion to fibrin, fibrinogen loses 1.2 moles of sialic acid. Chandrasekhar *et al.* (1962) found that fibrinogen clotted faster after removal of sialic acid residues by sialidase treatment. Characterization of the sialic acid present in human fibrinogen showed only N-acetylneuraminic acid, while bovine fibrinogen contains about equal amounts of N-acetyl- and N-glycolylneuraminic acid (Chandrasekhar *et al.,* 1966). Later workers (Vermylen *et al.,* 1974) found that fibrinogen could not cause aggregation of human platelets following treatment with sialidase. However, the carbohydrate-containing compound, factor VIII, upon treatment with sialidase causes strong platelet aggregation. In von Willebrand's disease (characterized by a prolonged bleeding time and defective platelet adhesion to the lips of a wound) there is a decrease of antigenic factor VIII activity, and aggregation is very limited. Sialidase is believed to remove sialyl residues from factor VIII, exposing galactose residues which can react with sialyltransferase present in the

TABLE II. Effect of Sialidase Treatment on the Biological and Immunological Activity of Sialoglycohormones

Hormone	Function	Percent sialic acid	Effect of sialidase action on:		Reference
			Biological activity *in vivo*	Immunological activity	
Erythropoietin	Induces normal erythrocyte development	13–14	Inactivated	No effect	Lowy et al., 1960 Schooley and Garcia, 1971
Human pituitary luteinizing hormone	Acts on interstitial cells of both ovaries and testes.		Inactivated	Enhancement	Braunstein et al., 1971
Human chorionic gonadotropin	Maintaining growth of corpus luteum during pregnancy; stimulates Leydig tissue.	8.5	Inactivated	No effect	Barr and Collee, 1967 Van Hall et al., 1971 Kunii, 1971
Thyroglobulin	Stored form of thyroid hormones (thyroxine plus triiodothyronine)	1.2	No effect	No effect	Murthy et al., 1965 Spiro and Spiro, 1965 Shome et al., 1968
Follicle-stimulating hormone	Growth of graafian follicles; spermatogenesis	1.2	Inactivate 97% or more	Enhancement	Gottschalk et al., 1960 Vaitukaitis and Ross, 1971

platelet membrane, giving first platelet adhesion and then aggregation (Vermylen et al., 1974).

α_1-Antitrypsin, a sialic-acid-containing serum glycoprotein (Laurell, 1965), is of interest in that its deficiency is related to obstructive lung disease in adults and hepatitis in children. In the homozygous deficiency state (ZZ) α_1-antitrypsin, with a reduced sialic acid content, is at about 10% of the normal serum level. Bell and Carrell (1973) suggest that there is incomplete addition of sialic acid to the molecule, and therefore it accumulates intracellularly, with only small amounts leaving the cell via passive diffusion. In support of this hypothesis, asialo-α_1-antitrypsin was found to accumulate in large amounts on the rough endoplasmic reticulum in liver cells, contributing to their susceptibility to hepatitis, and constituting a major factor in childhood cirrhosis. Because of inability to leave the cell, insufficient levels of α_1-antitrypsin are available to protect the lung tissue from proteolysis in obstructive lung disease. The sialic acid residues of α_1-antitrypsin are not necessary for the binding of trypsin (Laurell, 1965), but they apparently are essential for its transport from the cell to the plasma (Bell and Carrell, 1973).

III. CIRCULATING SIALOGLYCOPROTEINS IN ABNORMAL PHYSIOLOGICAL STATES

A wide variety of diseases occur which are accompanied by changes in sialic acid levels in plasma. Various hypotheses have been put forward to explain the observed increases (Winzler, 1955; 1971). It has been suggested that increased serum sialoglycoprotein levels are associated with tissue proliferation rather than tissue destruction (Shetlar et al., 1949). Several questions remain unanswered: (1) why are more sialoglycoproteins synthesized in many diseases, or in those where there are reduced levels, what are the controls exerted; (2) do these compounds have a direct role in the disease process, or are they "artifacts," and what changes are manifested by the enzymes involved in their synthesis or breakdown; (3) if it were possible to restore the metabolism of these compounds to normal, would this combat the disease process; and (4) are sialic-acid-containing compounds active in viral infections?

Anderson (1965) showed a correlation between reduction of material recovered in the seromucoid fraction and loss of carbohydrate (sialic acid) from the seromucoid. Upon loss of carbohydrate, seromucoid components become less soluble in perchloric acid. Therefore, an apparent reduction in a given compound may be due to partial loss of carbohydrate with a concomitant decrease in recovery during isolation.

Studies with normal children (Sharma and Sur, 1967) showed that serum sialic acid levels were independent of age and sex. In children with various diseases—tubercular meningitis, rickets, and Hand–Schüller–Christian disease—the serum sialic acid levels increase.

Fetuin occurs in the serum of man and other animals following a variety of types of injury (Winzler, 1971), and may represent a recurrence of the stages observed in fetal growth. The mechanism is unknown.

A. Diabetes

Changes in serum levels of sialic acid have been found in patients with diabetes. De Moor *et al.* (1970) found increased levels of serum sialic acid in men of advanced age at the onset of diabetes mellitus. Shvartz and Paukman (1971) found elevated serum sialic acid in patients with diabetic micro- and macroangiopathies as compared with cases of uncomplicated diabetes. Lack of change in serum sialic acid in the ordinary diabetic was corroborated by Howard and Kelleher (1971). They found no change in plasma sialic acid concentration in normal or diabetic (mellitus) individuals during an oral glucose tolerance test. Chakrabarti *et al.* (1972) observed that the sialic acid level in the erythrocyte membrane of diabetic patients is lower than that of normal persons. The level in the untreated diabetic and the diabetic treated with oral hypoglycemic drugs did not show much difference. However, they observed a complete normalization of the sialic acid level in the erythrocyte membrane of diabetic patients treated with insulin. Sialic acid was shown by Cuatrecasas and Illiano (1971) to be involved in transport of glucose across the cell membrane. Part of the therapeutic mechanism of action of insulin may be to aid in restoration (exposure) of the surface membrane sialic acid residues, which have been implicated in the transport of glucose.

B. Inflammatory Reactions

Inflammatory reactions also may be accompanied by an increase in serum sialic acid levels. During inflammatory reactions the level of rabbit serum α_1-macroglobulin was observed to increase (Got *et al.*, 1967). In rats treated with croton oil or with CCl_4 to induce inflammation of the liver and liver injury, serum sialic acid levels increased (Jakab and Takács, 1970).

C. Infectious Psychoses

Agishev (1970) studied over 100 patients with acute infectious psychoses and found an increase in serum levels of haptoglobulins, ceruloplasmin, α-globulins, and sialic acid. When the psychotic condition was exacerbated, further increase was observed. He suggests the use of haptoglobin, ceruloplasmin, and sialic acid levels for diagnosing infectious psychoses. Engen (1971) found that dogs with canine distemper show a significant increase in serum sialic acid levels.

D. The Effect of Steroid Hormones

Changes in steroid hormones have also been shown to affect sialic acid levels in the blood. Pidemsky and Afonina (1970) found that adrenalectomy of mice was accompanied by a rise in sialic acid in blood and in liver homogenates. Malashevich (1970) observed that administration of corticosteroids and ACTH to normal rats resulted in a pronounced deficit of sialic acid in liver and myocardium. Analogous conditions in adrenalectomized animals resulted in an increase in the sialic acid level in all tissues under study after injection of prednisolone (brain, liver, and serum). After injection of hydrocortisone, sialic acid increased in the liver, and in the heart with ACTH administration. These results seem to indicate specificity of increase in sialic acid in the target organ for the particular hormone tested. Engen (1970) studied the effects of hydrocortisone and dexamethasone on serum sialic acid levels in dogs and found they increased significantly. Dexamethasone, a steroid with thirtyfold more glucocorticoid activity than hydrocortisone, gave a significant response at one-tenth the dosage of hydrocortisone.

E. Liver Disease

Patients having hepatic cirrhosis (Laennec's and biliary cirrhosis, primary and metastatic liver carcinoma, and hepatitis) were found to have partially desialylated thyroxine-binding globulin (TBG) (Marshall et al., 1974). Previously, it was found that the amount of thyroxine bound was not appreciably altered by removal of sialic acid residues (Blumberg and Warren, 1961). Asialo-TBG binds to liver cell plasma membranes and it is rapidly cleared by the liver in rats, rabbits, and humans. Marshall et al. (1974) studied the inhibitory effect of normal and cirrhotic serums on the binding of asialo-TBG to the liver cell plasma membrane and found that glycoproteins present in serum from cirrhotic patients had a marked elevation of inhibitory activity which probably reflects an increase in levels of asialoglycoproteins. The greatest inhibitory effect was by the macroglobulin and 7 S globulin fractions. They suggest that hepatic removal of asialoglycoproteins from the circulation may protect the animal from the production of autoantibodies to damaged proteins. Serum from patients with abnormal liver function due to congestive heart failure, pancreatitis, or extrinsic biliary tract obstruction had normal inhibitory activity.

F. Virus Inhibition of Hemagglutination

Pepper (1968) investigated the role of sialic acid in horse serum on the inhibition of hemagglutination by Asian influenza virus. Horse serum has N-acetyl, N-glycoyl, and the 4-O-acetylated derivatives of neuraminic acid (cf. Chapter 1). The biological activity of horse serum specifically toward the Az strain of influenza virus appears to be determined by the N,O-diacetylneuraminic acid component in α_2-macroglobulin.

G. Cancer

Singh et al. (1967) studied human malignancies without clinical evidence of metastases. All of the patients had elevated serum sialic acid levels. Patients with metastases had, on the average, higher sialic acid values than patients without metastases. However, the overlap in the range of values for the two groups was too great to use sialic acid content to determine the presence of metastases. Cabezas and Porto (1971) also found an increase in the mean values of sialic acid in cancer sera as compared with sera from normal adults. Rao and Sirsi (1970) found an increase of 36% in serum-bound sialic acid in rats bearing Yoshida ascites sarcomas in comparison to normal healthy rats. Free serum sialic acid in tumor-bearing rats was present at about half the concentration found in serum in normal rats.

Woodman (1974) developed a test to measure the "sialopolyanion" values of the serum after intravenous implantation into mice of leukemic cells or Lewis lung tumor cells. The test compares the dye-binding polyanions present in an aliquot of serum with the dye-binding polyanions present in an aliquot of serum treated with sialidase. The difference is presumably due to the sialopolyanions in the serum. He found a significant increase in serum sialopolyanions within 2 days after injection of the transformed cells, but the white blood cell count in leukemic mice increased only after 8 days and the Lewis lung tumor became palpable only after 5 days. Preliminary experiments with healthy subjects, with patients with nonmalignant disease, and with cancer patients gave similar results to those obtained for healthy, traumatized, or tumor-bearing mice. The amount of carbocyanine dye-binding polyanion is greater in sera from tumor-bearing animals than from traumatized ones.

H. Diet

Diet has been found to have an effect on serum sialic acid levels. Davis and Richmond (1968) kept rats on a protein-free diet until they lost about 25% of their original weight. Upon feeding protein, serum sialic acid values went up especially in the "2α-β globulin" and "2α-globulin" fractions. A 20% proportion of protein in the diet was optimal. However, refeeding did not completely restore the sialic acid level to normal. Taoka and Fillios (1971) found more rapid synthesis and attachment of sialic acid to glycoprotein(s) in animals fed low protein diets. They suggested that the sialoglycoproteins were needed for certain regulatory mechanisms, possibly translocation of compounds.

IV. ROLE OF SIALIC ACID IN CIRCULATING SIALOGLYCOCOMPOUNDS

As early as 1964, Perona *et al.*, observed that sialidase treatment of erythrocytes resulted in a decrease in half-life, as determined by loss of ^{51}Cr-labeled cells from the circulation. However, the mechanism of this reduction was not determined. See Chapter 7.

Morell *et al.* (1966) observed that a radioactive preparation of asialoceruloplasmin upon injection into rabbits was cleared from the circulation in a matter of minutes, vastly more rapidly than native ceruloplasmin, whose half-life is about 56 hr. In subsequent work (Morell *et al.*, 1968) they observed that the rapid disappearance of the asialoceruloplasmin from the serum was accompanied by an equally rapid accumulation of radioactivity within the liver, specifically in the lysosomal fraction of the parenchymal cells (Gregoriadis *et al.*, 1970).

Woodruff and Gesner (1968) studied the effect of sialidase treatment on ^{51}Cr-labeled rat lymphocytes. They found that sialidase treatment caused a change in the cells electrophoretic mobility and in agglutinability. Upon injection of the sialidase-treated ^{51}Cr-labeled lymphocytes into the rat, it was observed that the asialolymphocytes were rapidly taken up by the liver, with very few accumulating in lymphoid tissues. These results led them to conclude that the sialic acid components of the cell surface played an important role in determining the distribution of lymphocytes in the body.

Morell et al. (1971) reported the results of a study on the effect of desialylation on the circulatory lifetime of orosomucoid, fetuin, ceruloplasmin, haptoglobin, α_2-macroglobulin, thyroglobulin, lactoferrin, human chorionic gonadotropin, and follicle-stimulating hormone. They observed that all of the desialylated proteins mentioned were rapidly removed from the circulation and found in the liver. The only exception was transferrin. They were also able to show intracompetitive inhibition of hepatic uptake of the asialo compounds, and that the presence of fully sialylated compounds did not inhibit uptake of the asialo derivatives. Finally, they observed that the requirements for binding to the liver cells did not necessitate the presence of intact protein but were met by the asialoglycopeptides formed upon pronase treatment.

Since these initial observations, several reports have been made of experiments performed to try to clarify the role of sialic acid in hepatic uptake. Van Hall et al. (1971) found that upon progressive desialylation of HCG, the removal of 25% of the sialyl residues decreased the plasma half-life by 50%. Desialylation over the range 25–62% reduced the plasma half-life by a factor of 10. Therefore, when early workers observed a loss of hormonal biological activity upon desialylation they probably were observing a decrease in plasma half-life of the hormone. Nelsestuen and Suttie (1971) extended these studies by noting that the half-life of bovine prothrombin was reduced tenfold after desialylation. Pricer and Ashwell (1971) found that while the removal of sialic acid from the circulating sialoglycoprotein aided binding to the liver cell membrane, loss of sialic acid from the *surface* of the liver cell membrane resulted in loss of binding. Calcium was also needed for binding.

Regoeczi et al. (1974) were able to find a 15% and a 29% increase in the rate of removal of homologous asialotransferrin preparations from the plasma of rabbits and humans, respectively. However, they observed that when human asialotransferrin was injected into rabbits, it was removed 3.5 times more rapidly than the homologous asialotransferrin. Removal of normal human transferrin from rabbit plasma was not significantly different from the normal rabbit compound. In subsequent work (Regoeczi and Hatton, 1974) was found that progressively desialylated human transferrin molecules had plasma half-lives inversely proportional to the loss of sialic acid.

This phenomenon seems to apply to circulating cells as well. In addition to the aforementioned studies with lymphocytes, Gattegno et al. (1974) found that desialylated erythrocytes are also rapidly removed from the circulation. Based upon these observations, Rogers and Kornfeld (1971) developed the hypothesis that by coupling the proper glycopeptide to a test molecule, it should be possible to cause the molecule to be delivered selectively to hepatocytes. To test this idea, they covalently linked the glycopeptides of fetuin and α-G-immunoglobulin to lysozyme and to albumin, and treated the products with sialidase to obtain the asialo derivatives. The asialofetuin glycopeptide-protein compounds were rapidly removed from the circulation. However the α-G-glycopeptide-proteins were not. This indicates that there is specificity of uptake by the liver cells. They suggest that this technique offers a feasible method for directing the hepatic uptake of various proteins that otherwise would not be taken up by hepatocytes. In a current review, Ashwell and Morell (1975) suggest that this may be a potential means for replacement therapy in cases of genetic deficiency of lysosomal enzymes.

The aforementioned considerations suggest a function for the sialidase found associated with the plasma membranes of liver cells (Schengrund *et al.*, 1972; Visser and Emmelot, 1973) to desialylate circulating sialoglycoproteins for their subsequent uptake by the hepatocytes. The sialidase of erythrocyte membranes (Bosmann, 1974) may function in determining their half-life in the circulation.

That loss of sialic acid allows for rapid removal of glycoproteins and cells from the circulation, and the possible use of this finding in enzyme replacement therapy suggests a fruitful area for further intensive investigation.

ACKNOWLEDGMENT

This work was supported by National Institutes of Health, Grants NS08258 and CA14319.

V. REFERENCES

Agishev, V. G., 1970, Concerning the changes of the proteins in the blood serum in some infectious psychoses, *Psikhiar. S. S. Korsakova* **70**:1343–8.

Anderson, A. J., 1965, Factors affecting the amount and composition of the serum seromucoid fraction, *Nature* **208**:491–492.

Arfors, K. E., Beckman, L., and Lundin, L. G., 1963, Further studies on the association between human serum phosphatases and blood groups, *Acta Genet. Statist.* **13**:366–368.

Ashwell, G., and Morell, A. G., 1975, The role of surface carbohydrates in the hepatic recognition and transport of circulating glycoproteins, *Adv. Enzymol.* **41**:99–128.

Athineos, E., Thornton, M., and Winzler, R. J., 1962, Comparative antigenicity of native and "desialized" orosomucoid in rabbits, *Proc. Soc. Exp. Biol. Med.* **111**:353–356.

Augustinsson, K.-B., and Ekedahl, G., 1962, The properties of neuraminidase-treated serum cholinesterase, *Biochim. Biophys. Acta* **56**:392-393.
Baker, E., Shaw, D. C., and Morgan, E. H., 1968, Isolation and characterization of rabbit serum and milk transferrins. Evidence for difference in sialic acid content only, *Biochemistry* **7**:1371-1378.
Bates, R. W., and Warren, L., 1963, Studies on the sialic acid content of thyroid glands, *Endocrinology* **73**:1-4.
Barr, W. A., and Collee, J. G., 1967, Differences in the biological and immunological activities of human chorionic gonadotiopin after removal of sialic acid by enzymic hydrolysis, *J. Endocrinol.* **38**:395-399.
Bell, O. F., and Carrell, R. W., 1973, Basis of the defect in α-1-antitrypsin deficiency, *Nature* **243**:410-411.
Blumberg, B. S., and Warren, L., 1961, The effect of sialidase on transferrins and other serum proteins, *Biochim. Biophys. Acta* **50**:90-101.
Bocci, V., Viti, A., Russi, M., and Rita, G., 1971, Isoelectric fractionation of desialyzed interferon. *Experientia* **27**:1160-1161.
Bosmann, H. B., 1974, Red cell hydrolases. *Vox Sang.* **26**:497-512.
Braunstein, G. D., Reichert, Jr., L. E., Van Hall, E. V., Vaitukaitis, J. L., and Ross, G. T., 1971, The effects of desialylation on the biologic and immunologic activity of human pituitary luteinizing hormone, *Biochem. Biophys. Res. Commun.* **42**:962-967.
Cabezas, J. A., and Porto, E., 1970, Acidos Siálicos. XIII. Contenido en acido *N*-acetilneuramínico, hexosaminas y ácido pirúvico, y actividad *N*-acetil-β-glucosaminidásica de sueros normales y cancerosos, *Rev. Esp. Fisiol.* **26**:339-345.
Chakrabarti, T., Pawar, R. B., Shastri, M. G., Choudhary, D. R., Chakrabarti, C. H., and Dias, P. D., 1972, *N*-Acetyl neuraminic acid level in the erythrocyte membrane and serum of patients suffering from diabetes mellitus, *Ind. J. Med. Res.* **60**:1038-1042.
Chandrasekhar, N., Warren, L., Osbahr, A. J., and Laki, K., 1962, Role of sialic acid in fibrinogen. *Biochim. Biophys. Acta* **63**:337-339.
Chandrasekhar, N,, Osbahr, A. J., and Laki, K., 1966, Identification of sialic acids in human and bovine fibrinogen, *Biochem. Biophys. Res. Commun.* **23**:757-760.
Chen, S.-H., and Sutton, H. E., 1967, Bovine transferrins: Sialic acid and the complex phenotype, *Genetics* **56**:425-430.
Clamp, J. R., and Johnson, I., 1972, Immunoglobulins *in: Glycoproteins,* Part A (A. Gottschalk, ed.) p. 612-652, Elsevier Publishing Co., New York.
Cowan, N. J., and Robinson, G. B., 1972, The sequence of addition of terminal sugars to an immunoglobulin a myeloma protein, *Biochem. J.* **126**:751-754.
Cuatrecases, P., and Illiano, G., 1971, Membrane sialic acid and the mechanism of insulin action in adipose tissue cells, *J. Biol. Chem.* **246**:4938-4946.
Cunningham, L. W., 1971, *Microheterogeneity and Function of Glycoproteins in Glycoproteins of Blood Cells and Plasma* (G. A. Jamieson and T. J. Greenwalt, eds.), pp. 34-61, Lippincott Co., Philadelphia.
Davis, M. M., and Richmond, J. E., 1968, Effect of dietary protein on serum proteins. *Am. J. Physiol.* **215**:366-369.
De Moor, P., Bouillon, R., and Van Mieghem, W., 1970, Transcortin activity as related to the age at discovery of diabetes mellitus, *Clin. Chim. Acta* **30**:627-633.
Engen, R. L., 1970, Effect of steroids on serum sialic acid values in the canine, *Proc. Soc. Exp. Biol. Med.* **135**:778-780.
Engen, R. L., 1971, Serum sialic acid values in dogs with canine distemper, *Am. J. Vet. Res.* **32**:803-804.

Finnegan, D. J., and Hope, R. M., 1970, The role of sialic acid in the serum amylase isoenzyme pattern of the marsupial mouse *Sminthopsis crassicaudata. Austr. J. Exp. Biol. Med. Sci.* **48**:237–240.

Gattegno, L., Bladier, D., and Cornillot, P., 1974, The role of sialic acid in the determination of survival of rabbit erythrocytes in the circulation, *Carbohydrate Res.* **34**:361–369.

Goldwasser, E., Kung, C. K.-H., and Eliason, J., 1974, On the mechanism of erythropoietin-induced differentiation. XIII. The role of sialic acid in erythropoietin action, *J. Biol. Chem.* **249**:4202–4206.

Got, R., Cheftel, R.-I., Font, J., and Moretti, J., 1967, Étude de L' α_1-macroglobuline du serum de lapin. III. Biochimie de la copule glucidique, *Biochim. Biophys. Acta* **136**:320–330.

Gottschalk, A., Whitten, W. K., and Graham, E. R. B., 1960, Inactivation of follicle-stimulating hormone by enzymic release of sialic acid, *Biochim. Biophys. Acta* **38**:183–184.

Gregoriadis, G., Morell, A. G., Steinlieb, I., and Scheinberg, I. H., 1970, Catabolism of desialylated ceruloplasmin in the liver, *J. Biol. Chem.* **245**:5833–5837.

Green, W. A., and Kay, C. M., 1963, The influence of organic solvents and enzyme modification on the secondary structure of fetuin, *J. Biol. Chem.* **238**:3640–3644.

Heilbronn, E., 1962, Treatment of horse serum cholinesterase with sialidase, *Acta Chem. Scand.* **16**:516.

Heyward, J. T., Coleman, M. T., and Dowdle, W. R., 1972, Influenza antihemagglutinin and antinuraminidase activity of IgG and IgM in reference chicken antisera, *Proc. Soc. Exp. Biol. Med.* **140**:1289–1293.

Howard, C. B., and Kelleher, P. C., 1971, Plasma fucose and sialic acid concentrations during oral glucose tolerance tests in normal and diabetic (mellitus) humans, *Clin. Chim. Acta* **31**:75–80.

Itoh, C., Hashimoto, I., Onuma, Y., and Ishitoya, Y., 1968, Inhibitory effects of monoclonal immunoglobulins on anti-globulin reaction, *Tohoku J. Exp. Med.* **94**:307–313.

Jakab, J. F., and Takács, L., 1970, Effect of liver injury and of induced inflammation on the serum glycoprotein level, *Acta Med. Acad. Sci. Hung.* **27**:57–63.

Jamieson, G. A., 1963, Carbohydrate chains of human transferrin, *Fed. Proc.* **22**:538.

Jamieson, G. A., 1965, Studies on glycoproteins. 1. The carbohydrate portion of human ceruloplasmin, *J. Biol. Chem.* **240**:2019–2027.

Jamieson, G. A., 1971, *in: Glycoproteins of Blood Cells and Plasma* (G. A. Jamieson and T. J. Greenwalt, eds.), p. 68, Lippincott Co., Philadelphia.

Jamieson, G. A., 1972, *Ceruloplasmin, in Glycoproteins,* Part A (A. Gottschalk, ed.), Elsevier Publishing Co., pp. 676–685, New York.

Johnson, C. A., Lvstad, R. A., Walaas, E., and Walaas, O., 1970, The properties of neuraminidase-treated crystalline ceruloplasmin, *Experientia* **26**:134–135.

Kasavina, B. S., and Kolchinskaya, T. A., 1966, Studies of the human thyroid in some pathological conditions, *Clin. Chim. Acta* **13**:685–693.

Kristjansson, F. K., and Cipera, J. D., 1963, The effect of sialidase on pig transferrins, *Can. J. Biochem. Physiol.* **41**:2523–2527.

Kunii, H., 1971, Inactivation of human chorionic gonadotropin extracted from chorionic tissue and serum by neuraminidase, *Tohoku J. Exp. Med.* **105**:317–325.

Laurell, C.-B., 1965, Effect of neuraminidase, acetone, and chloroform on α_1-antitrypsins, *Scand. J. Clin. Lab. Invest.* **17**:297–298.

Lowy, P. H., Keighley, G., and Borsook, H., 1960, Inactivation of erythropoietin by neuraminidase and by mild substitution reactions, *Nature* **185**:102–103.

Main, A. R., Tarkan, E., Aull, J. L., and Soucie, W. G., 1972, Purification of horse serum cholinesterase by preparative polyacrylamide gel electrophoresis, *J. Biol. Chem.* **247**:566–571.

Malashevich, E. V., 1970, Dependence of the sialic acid content in tissues on functional state of the hypophysis-adrenal systems, *Ukr. Biokhim. Zh.* **42**:56–59.

Margolis, F., and Feigelson, P., 1964, Atropinesterase, a sialoprotein, *Biochim. Biophys. Acta* **89**:357–360.

Marshall, J. S., Green, A. M., Pensky, J., Williams, S., Zinn, A., and Carlson, D. M., 1974, Measurement of circulating desialylated glycoproteins and correlation with hepatocellular damage, *J. Clin. Invest.* **54**:555–562.

Mester, L., Moczar, E., and Laki, K., 1963, Structure et rôle de la partie glucidique du fibrinogène et de la fibrine: Sur les liaisons des composeś glucidiques, *Séance* **256**:307–308.

Morell, A. G., Van Den Hamer, C. J. A., Scheinberg, I. H., and Ashwell, G. A., 1966, Preparation of radioactive sialic acid-free ceruloplasmin labelled with tritium on terminal D-galactose residues, 1966, *J. Biol. Chem.* **241**:3745–3749.

Morell, A. G., Irvine, R. A., Steinlieb, I., Scheinberg, I. H., and Ashwell, G., 1968, Physical and chemical studies on ceruloplasmin, *J. Biol. Chem.* **243**:155–159.

Morell, A. G., Gregoriadis, G., Scheinberg, I. H., Hickman, J., and Ashwell, G., 1971, The role of sialic acid in determining the survival of glycoproteins in the circulation, *J. Biol. Chem.* **246**:1461–1467.

Murthy, P. V. H., Raghupathy, E., and Chaikoff, I. L., 1965, Studies on thyroid proteins. I. Isolation and properties of a glycopeptide from sheep thyroglobulin, *Biochem.* **4**:611–618.

Nelsestuen, G. L., and Suttie, J. W., 1971, Properties of asialo and aglycoprothrombin, *Biochem. Biophys. Res. Commun.* **45**:198–203.

Niedermeier, W. Schrohenloher, R. E., and Hurst, M., 1972, The localization of oligosaccharides in a human IgM protein, *J. Immunol.* **108**:346–351.

Nolan, C., and Smith, E. L., 1962, Glycopeptides. II. Isolation and properties of glycopeptides from rabbit γ-globulin, *J. Biol. Chem.* **237**:446–452.

Oshiro, Y., and Eylar, E. H., 1968, Physical and chemical studies on glycoproteins. III. The microheterogeneity of fetuin, a fetal calf serum glycoprotein, *Arch. Biochem. Biophys.* **127**:476–489.

Parker, W. C., and Bearn, A. G., 1962, Studies on the transferrins of adult serum, cord serum, and cerebrospinal fluid: The effect of neuraminidase, *J. Exp. Med.* **115**:83–105.

Pepper, D. S., 1968, The sialic acids of horse serum with special reference to their virus inhibitor properties, *Biochim. Biophys. Acta* **156**:317–326.

Perona, G., Cortesi, S., Xodo, P., Scandellari, C., Ghiotto, G., and de Sandre, G., 1964, Variations of *in vivo* survival, acetylcholinesterase activity and sensitivity to acid lysis in human erythrocytes treated with proteolytic enzymes and neuraminidase, *Acta Istopocia* **4**:287–295.

Pidemsky, E. L., and Afonina, T. D., 1970, The effect of some antiphlogistic preparations on the content of mucopolysaccharides in the blood and liver homogenates, *Patol. Fizioli Eksp. Ter.* **14**:74–76.

Pricer, W. E., and Ashwell, G., 1971, The binding of desialylated glycoproteins by plasma membranes of rat liver, *J. Biol. Chem.* **246**:4825–4833.

Rao, V. S., and Sirsi, M., 1970, Serum sialic acid in rats bearing yoshida ascites sarcoma, *Ind. J. Biochem.* **7**:184–186.

Regoeczi, E., and Hatton, M. W. C., 1974, Studies of the metabolism of asialotransferrins: The mechanism for the hypercatabolism of human asialotransferrin in the rabbit, *Can. J. Biochem.* **52**:645–651.

Regoeczi, E., Hatton, M. W. C., and Wong, K.-L., 1974, Studies of the metabolism of asialotransferrins: Potentiation of the catabolism of human asialotransferrin in the rabbit, *Can. J. Biochem.* **52**:155–161.

Robinson, J. C., and Pierce, J. E., 1964, Studies on inherited variants of blood proteins. III. Sequential action of neuraminidase and galactose oxidase on transferrin $B_{1-2}b_2$, *Arch. Biochem. Biophys.* **106**:348–352.

Robinson, J. C., and Pierce, J. E., 1964, Differential action of neuraminidase on human serum alkaline phosphatases, *Nature* **204**:472–473.

Robinson, J. C., Pierce, J. E., and Blumberg, B. S., 1966, The serum alkaline phosphatase of pregnancy, *Am. J. Obst. Gynec.* **94**:559–565.

Rogers, J. C., and Kornfeld, S., 1971, Hepatic uptake of proteins coupled to fetuin glycopeptide, *Biochem. Biophys. Res. Commun.*, **45**:622–629.

Rule, A. H., and Boyd, W. C., 1964, Relationships between blood group agglutinogens: Role of sialic acids, *Transfusion* **4**:449–456.

Sairam, M. R., and Moudgal, N. R., 1971, On the mechanism of action of the monkey urinary follicle stimulating hormone inhibitor—Its sialidase activity, *Ind. J. Biochem. Biophys.* **8**:141–146.

Scharmann, W., Brückler, J., and Blobel, H., 1971, Wirkung bakterieller Neuraminidasen auf Transferrin vom Menschen, Rind und Kaninchen, *Biochim. Biophys. Acta* **229**:136–142.

Schengrund, C.-L., Jensen, D., and Rosenberg, A., 1972, Localization of sialidase in the plasma membrane of rat liver cells, *J. Biol. Chem.*, **247**:2742–2746.

Schmid, K., Burke, J. F., Debray-Sachs, M., and Tokita, K., 1964, Sialic acid-deficient α_1-acid glycoprotein produced in certain pathological states, *Nature* **204**:75–76.

Schooley, J. C., and Garcia, J. F., 1971, The destruction by neuraminidase of the biological activity of erythropoietin when complexed with antierythropoietin, *Proc. Soc. Exp. Biol. Med.* **138**:66–68.

Schooley, J. C., and Mahlmann, L. J., 1971, Inhibition of the biologic activity of erythropoietin by neuraminidase *in vivo*, *J. Lab. Clin. Med.* **78**:765–770.

Schultze, H. E., 1958, Über Glykoproteine. *Deut. Med. Wochschr.* **83**:1742–1752.

Schultze, H. E., Gollner, I., Heide, K., Schonenberger, M., and Schwick, G. Z., 1955, Zur Kenntnis der α-Globuline des menschlichen Normalserums, *Z. Naturforsch.* **106**:463–473.

Schultze, H. E., Schmidtberger, R., and Haupt, H., 1958, Untersuchungen uber die gebundenen Kohlenhydrate in isolierten Plasmaproteiden, *Biochem. Z.* **329**:490–507.

Sharma, N. C., and Sur, B. K., 1967, Serum fucose and sialic acid levels in indian children and adults under normal and pathological conditions, *Ind. J. Med. Res.* **55**:380–384.

Shetlar, M. R., Foster, J. V., Kelly, K. H., Shetlar, C. L., Bryan, R. S., and Everett, M. R., 1949, The serum polysaccharide level in malignancy and in other pathological conditions, *Cancer Res.* **9**:515–519.

Shome, B., Parlow, A. F., Ramirez, V. D., Elrick, H., and Pierce, J. G., 1968, Human and porcine thyrotropins: A comparison of electrophoretic and immunological properties with the bovine hormone, *Arch. Biochem. Biophys.* **103**:444–455.

Shownkeen, R. C., Thomas, M. B., and Hartree, A. A., 1973, Sialic acid and tryptophan content of subunits of human pituitary luteinizing hormone, *J. Endocrinol.* **59**:201–202.

Shvartz, L. S., and Paukman, L. I., 1971, Diabetic angiopathies and mucopolysaccharide metabolism, *Probl. Endocrinol.* **17**:37–41.

Singh, R., Sur, B. K., Agarwal, S. N., and Ramraju, B., 1967, Serum sialic acid in malignancy, *Ind. J. Med. Res.* **55**:270–273.

Sjöholm, I., 1967, Biochemical studies on oxytocin and oxytocinase, *Acta Pharm. Suedcica* **4**:81–96.
Spiegelberg, H. L., and Weigle, W. O., 1966, Studies on the catabolism of γ subunits and chains, *J. Immunol.* **95**:1034–1040.
Spiro, R. G., 1960, Studies on fetuin, a glycoprotein of fetal serum I. Isolation, chemical composition, and physicochemical properties, *J. Biol. Chem.* **235**:2860–2869.
Spiro, R. G., 1970, Structure metabolism and biology of glycoproteins, *Ann. Rev. Biochem.* **39**:599–638.
Spiro, M. J., and Spiro, R. G., 1962, Composition of the peptide portion of fetuin, *J. Biol. Chem.* **237**:1507–1510.
Spiro, R. G., and Spiro, M. J., 1965, The carbohydrate composition of the thyroglobulins from several species, *J. Biol. Chem.* **240**:997–1001.
Spiro, M. J., and Spiro, R. G., 1968, Glycoprotein biosynthesis: Studies on thyroglobulin, *J. Biol. Chem.* **243**:6520–6528.
Stratel, A., and Spooner, R. L., 1971, Isolation and properties of individual components of cattle transferrin: The role of sialic acid, *Biochem. Genetics* **5**:347–365.
Svensmark, O., 1961, Human-serum cholinesterase as a sialo-protein, *Acta Physiol. Scand.* **52**:267–275.
Svensmark, O., and Kristensen, P., 1962, Electrophoretic mobility of sialidase-treated human serum cholinesterase, *Danish Med. Bull.* **9**:16–17.
Svensmark, O., and Kristensen, P., 1963, Isoelectric point of native and sialidase-treated human-serum cholinesterase, *Biochim. Biophys. Acta* **67**:441–452.
Taoka, Y., and Fillios, L. C., 1971, Early effects of protein depletion on hepatic glycoprotein synthesis in the rat, *J. Nutr.* **101**:93–100.
Tuppy, H., Wiesbauer, U., and Wintersberger, E., 1963, Uber die Einwirkung von Neuraminidase auf die Serumoxytocinase, *Mh. Chem.* **94**:321–328.
van den Hamer, C. J. A., Morell, A. G., Scheinberg, I. H., Hickman, J., and Ashwell, G., 1970, Physical and chemical studies on ceruloplasmin. IX. The role of galactosyl residues in the clearance of ceruloplasmin from the circulation, *J. Biol. Chem.* **245**:4397–4402.
Vaitukaitis, J. L., and Ross, G. T., 1971, Altered biologic and immunologic activities of progressively desialylated human urinary FSH, *J. Clin. Endocrinol. Metab.* **33**:308–311.
Van Hall, E., Vaitukaitis, J. L., Ross, G. T., Hickman, J. W., and Ashwell, G., 1971, Immunological and biological activity of HCG following progressive desialylation, *Endocrinology* **88**:456–464.
Vermylen, J., Donati, M. B., de Gaetano, G., and Verstraete, M., 1973, Aggregation of human platelets by bovine or human factor. VIII: Role of carbohydrate side chains, *Nat. New Biol.* **244**:167–168.
Vermylen, J., de Gaetano, G., Donati, M. B., and Verstraete, M., 1974, Platelet-aggregating activity in neuraminidase-treated human cryoprecipitates: Its correlation with factor-VIII-related antigen. *Br. J. Haematol.* **26**:645–650.
Visser, A., and Emmelot, P., 1973, Studies on plasma membranes. XX. Sialidase in hepatic plasma membranes, *J. Membrane Biol.* **14**:73–84.
Warren, L., 1959, The thiobarbituric acid assay of sialic acids, *J. Biol. Chem.* **234**:1971–1975.
Whitten, W. K., 1948, Inactivation of gonadotropins. II. Inactivation of pituitary and chorionic gonadotropins by influenza virus and receptor destroying enzyme, *Austr. J. Sci. Res. B* **1**:388–390.

Winzler, R. J., 1955, Determination of Serum Glycoproteins. *Methods of Biochemical Analysis.* **2:**279–311.
Winzler, R., 1960, Glycoproteins, *in: The Plasma Proteins* (F. W. Putnam, ed.) pp. 309–347, Academic Press, New York.
Winzler, R., 1971, Glycoproteins in disease *in: Glycoproteins of Blood Cells and Plasma* (G. A. Jamieson and T. J. Greenwalt, eds.), pp. 204–213, Lippincott Co., Philadelphia.
Winzler, R., and Bocci, V., 1972, Turnover of plasma glycoproteins, *in: Glycoproteins, Part B* (A. Gottschalk, ed.), pp. 1228–1245, Elsevier Publishing Co., New York.
Woodman, R. J., 1974, Carbocyanine dye metachromasia of sialidase-sensitive polyanions in sera from normal and tumor-bearing mice, *Cancer Res.* **34:**2897–2905.
Woodruff, J. J., and Gesner, B. M., 1969, The effects of neuraminidase on the fate of transfused lymphocytes, *J. Exp. Med.* **129:**551–567.

Chapter 10
Sialidases

Abraham Rosenberg and Cara-Lynne Schengrund

I. BACKGROUND AND NOMENCLATURE

Sialidase (neuraminidase, N-acetylneuraminosyl glycohydrolase, EC 3.2.1.18) catalyzes the hydrolysis of sialic acid from sialic-acid-containing glycoproteins, glycolipids, and oligo- and polysaccharides. Sialidases recently have attracted very wide interest.

McCrea (1947) discovered that culture filtrates of *Clostridium welchii* (or *perfringens*) could cause erythrocytes to be nonagglutinable by influenza virus; so did Burnet and Stone (1947) with *Vibrio cholerae* filtrates. They called the agent in *Vibrio* filtrates the "receptor-destroying enzyme" because it destroyed viral receptors on erythrocytes, and it displayed enzyme-like characteristics. They pointed to its close similarity to an agent in viruses of the mumps–influenza group which had the ability likewise to destroy, progressively, the potentiality of erythrocytes to be agglutinated by other viruses of this group (Burnet *et al.*, 1946).

Influenza virus was found to liberate enzymatically from brain mucolipids a compound chemically characterized as sialic acid (Rosenberg *et al.*, 1956) whereupon in retrospect the low-molecular-weight compound found by Gottschalk and Lind (1949) to be released from ovomucin, a sialoglycoprotein, by influenza virus, also was recognized to have been sialic acid and the liberating enzyme, sialidase. This enzyme was thought to represent the eluting factor which Hirst (1942) had observed in influenza virus, during his studies on the viral agglutination of erythrocytes.

ABRAHAM ROSENBERG AND CARA-LYNNE SCHENGRUND • Department of Biological Chemistry, The Milton S. Hershey Medical Center, The Pennsylvania State University, Hershey, Pennsylvania 17033.

In 1956, Heimer and Meyer identified sialic acid as a compound released from bovine submaxillary mucoid by an extract from pneumococci. They suggested, at this point, that the name "sialidase" be employed generally for the responsible enzyme whenever the product of enzymatic action could be identified as sialic acid, and that the name "receptor-destroying enzyme" be applied when only biological effects were measured (Heimer and Meyer, 1956). In the following year, Gottschalk (1957) independently suggested the use of an alternate term, "neuraminidase," which has gained fairly wide acceptance.

We preferentially use the earlier, more correct term, "sialidase." Neuraminic acid, the unstable parent compound (cf. Chapter 1) now is known not to be the product of enzymatic action and has not been found *per se* in nature. The product generally is the free N- or N,O-substituted derivative of neuraminic acid, i.e., sialic acid (cf. Chapter 1).

In 1960, the occurrence of mammalian sialidase was reported by Warren and Spearing. Shortly thereafter, Ada and Lind (1961) found sialidase in the chorioallantoic membrane of the embryonic chicken, and Carubelli *et al.* (1962) found sialidase in several organs of the rat. Much work since has been done on microbial and on mammalian sialidases, and to a lesser degree, on avian sialidase, to determine intracellular and organellar locations of the enzyme, and its properties and biological functions.

Sections II–IV of this chapter cover microbial sialidases, and Section V, mammalian and avian sialidases.

II. BACTERIAL SIALIDASES

A. Occurrence of Microbial Sialidases

In the microbial world, occurrence of sialidase appears mostly to be restricted to ortho- and to paramyxoviruses and to members of two major taxonomic orders of bacteria, Eubacteriales and Pseudomonadales (for review, Drzeniek, 1972), but not entirely. Sialidase also has been uncovered in one of the Mycoplasmatales (gallisepticum, Sethi and Müller, 1972; Roberts, 1967), as well as in a structurally related L-phase variant of *Vibrio comma*, or *cholerae* (Madoff *et al.*, 1961); this latter belongs, however, to the Spirillaceae family of the aforementioned Pseudomonadales. Sialidase has been found also in a higher Protistan, *Trichomonas foetus* (Romanovska and Watkins, 1963) and in several strains of a fungus, *Streptomyces albus* (Myhill and Cook, 1972), of the order Actinomycetales. Future discoveries may further widen the pattern of distribution among eukaryotic as well as among prokaryotic microorganisms.

B. Organismic Characteristics and Induction of Bacterial Sialidases

The intracellular location of bacterial sialidase has not been firmly established. A striking feature of certain of the sialidase-producing bacteria is their ability to release large quantities of sialidase into growth media. The latter often are used for the initial isolation of bacterial sialidases for further purification and study. Moriyama and Barksdale (1967) found most of the sialidase of *Corynebacterium diphtheriae* to be located in the pellet obtained after centrifugation of disrupted washed cells, and they considered the enzyme to be bound to the cell membrane. However, Jamieson (1966) readily isolated the enzyme from a crude culture filtrate from the PW8 strain of this bacterium. Sialidase has been shown to be an extracellular enzyme for streptococci (Pinter *et al.*, 1968; Hayano and Tanaka, 1967*a*), and the initial discovery of sialidase was made using cell-free culture filtrates of *Vibrio cholerae* (Burnet and Stone, 1947) and *Clostridium perfringens* (McCrea, 1947). For pneumococci, it has been shown that, additionally, only a small fraction of the sialidase produced is cell-associated at any stage of the bacterial growth cycle. Dividing pneumococci appear actively to secrete the enzyme (Lee and Howe, 1966), and there is no correlation between the production of sialidase and the capsular type of these bacteria.

Although pneumococcal sialidase is secreted, its production may be stimulated (Kelly *et al.*, 1966) by addition to the culture medium of exogenous sialyllactose, sialic acid, or N-acetylmannosamine (a precursor for sialic acid biosynthesis, cf. Chapter 4). Production of sialidase has likewise been induced in *Klebsiella aerogenes* by addition of sialyllactose or a wide variety of sialoglycolipids and sialoglycoproteins to the culture medium (Pardoe, 1970). In this organism, the enzyme may be mostly cell-bound rather than secreted, and it also has been reported for *Pasteurella multocida* that sialidase for the greater part is not secreted, but remains bound to the cell (Drzeniek *et al.*, 1972). Nevertheless, such cells release free sialic acid from low concentrations of exogenous sialyllactose in the medium, which may indicate that the bound enzyme is on the cell surface. Interestingly, *Pasteurella multocida* secretes substantial quantities of N-acetylneuraminate-pyruvate lyase (cf. Chapter 4), which cleaves free sialic acid to pyruvate and N-acetylmannosamine. The latter compound has recently been proved to be the inducer for sialidase synthesis in *Pasteurella multocida* (Drzeniek *et al.*, 1972). Exogenous N-acetylmannosamine, as well as sialyl substrates, and free sialic acid, as mentioned previously, were shown to induce production of sialidase in pneumococci (Kelley et al., 1966), and these substances have been known for some time to stimulate secretion of sialidase by *Vibrio cholerae* (Ada and French, 1957, 1959a). It is possible that inducibility, secretion, and surface-binding of bacterial sialidases together represent a general and biologically important property of the sialidase-producing bacteria.

C. Purification of Bacterial Sialidases

Simultaneous early attempts at purification of the sialidase of *Vibrio cholerae*—after concentration of the enzyme from culture filtrates by adsorption and elution from red cells (Ada and French, 1959b; Schramm and Mohr, 1959), or directly from the culture filtrate by ammonium sulfate precipitation at various pH values (Rosenberg et al., 1960)—rapidly led to obtaining this enzyme in crystalline form. Further manipulation of the crystalline enzyme greatly increased its activity (Rosenberg et al., 1960). This phenomenon is not yet understood, nor has advantage yet been taken of the ready ability of *Vibrio* sialidase to crystallize, in order to perform X-ray studies of its structural properties. The crystalline *Vibrio* sialidase isolated early on by Schramm and Mohr (1959) and by Rosenberg et al. (1960) had remarkably low sedimentation constants, around 1 to 2 S, which suggested that they were very small proteins, but the sialidase crystallized by Ada and French (1959b) after induction by sialyllactose in strain 4Z of *Vibrio cholerae* (Ada et al., 1961) had a much higher S value, around 5. Interestingly, this larger enzyme appears not to be a polymer of that with the lower sedimentation constant (Laver et al., 1964). While there may be unusual genetic differences between the sialidase synthesized by *Vibrio cholerae* in the presence or in the absence of sialyllactose as an inducer, the possibility of proteolytic degradation as a reason for the occurrence of small but still highly catalytic sialidase also may be suggested. Aside from elution from erythrocytes (Schramm and Mohr, 1959; Ada and French, 1959b), and recent attempts at affinity chromatography (Cuatrecasas and Illiano, 1971a), which latter have not yet produced pure preparations, procedures for the isolation and purification of microbial sialidases generally have followed standard lines of approach for protein purification, both time-honored and recent.

It will be clear, from the descriptions arranged in chronological order that follow, that very little change in basic methodology for isolation of bacterial sialidases have come about over the years.

Schramm and Mohr (1959) achieved a high purification of sialidase activity from a crude culture filtrate of *Vibrio cholerae* by adsorption of the enzyme to erythrocytes, chromatography of the eluted enzyme on kieselgur, and fractional precipitation with ammonium sulfate. Ada and French (1959b) precipitated the enzyme from crude filtrates of *Vibrio cholerae* with methanol, adsorbed the enzyme onto erythrocytes, and after elution, precipitated it from buffered saline, finally subjecting it to chromatography on hydroxylapatite. Rosenberg *et al.* (1960) fractionally precipitated the enzyme directly from a culture filtrate of *Vibrio cholerae* with ammonium sulfate, first at pH 6.2, then at pH 6.6, and finally at pH 5.3, extracted the enzyme from the precipitate with water, purified it on DEAE-cellulose and lastly, reprecipitated the crystallizable enzyme with ammonium sulfate. Hughes and Jeanloz (1964) purified sialidase from filtrates of *Diplococcus pneumoniae* by alternating fractional precipitations of culture filtrate proteins with ammonium sulfate and chromatography on DEAE- and CM-cellulose. Cassidy *et al.* (1965) purified sialidase from cultures of *Clostridium perfringens* by ammonium sulfate precipitation and chromatography on Sephadex G-75 and DEAE-cellulose. Jamieson (1966) purified sialidase from *Corynebacterium diphtheriae* toxin by ammonium sulfate precipitation and chromatography on Sephadex G-200. Hayano and Tanaka (1967) precipitated sialidase from culture filtrates of group K streptococcus by ammonium sulfate precipitation and purified it by chromatography on DEAE-cellulose. White and Mellanby (1969) separated sialidase from the "edema-producing," "hemorrhage-producing," and lecithinase activities in filtrates of *Clostridium sordelli* CN3903 by ammonium sulfate precipitation and chromatography on Sephadex G-75. Jagielski (1969) purified sialidase from *Corynebacterium diphtheriae* by ammonium sulfate precipitation alone. Tannenbaum *et al.* (1969) separated multiple forms of pneumococcus sialidase by ammonium sulfate fractionation, ultrafiltration, heat treatment, and chromatography on DEAE-cellulose, CM-cellulose, and CM-Sephadex. Vertiev and Ezepchuk (1972) recently have isolated sialidase from the culture medium of a nontoxicogenic C_7 strain of *Corynebacterium diphtheriae* by precipitation with zinc chloride followed by chromatographic fractionation on bio-gel P-60. von Nicolai and Zilliken (1972) isolated sialidase from the soluble fraction of sonicated *Lactobacillus bifidus* by ammonium sulfate precipitation and chromatography on Sephadex G-75. Stahl and O'Toole (1972) purified sialidase from the culture filtrate of type I pneumococcus by ammonium sulfate precipitation and chromatography on DEAE-Sephadex and Sephadex G-100. In a manner analogous to those used for these bacterial sialidases, Kunimoto *et al.* (1974) have purified Streptomyces sialidase by ammonium sulfate precipitation and column chromatography on DEAE-cellulose, CM-cellulose, and Sephadex G-100.

Often these preparations of sialidase, although purified, are contaminated with other enzymatic and nonenzymatic proteins. For example, immunologically distinct sialidase and the toxin of *Corynebacterium diphtheriae* have been found by Moriyama and Barksdale (1967) to move together on CM-cellulose and presumably to have similar molecular weights. Commercial preparations of sialidase made according to methods similar to those described above have more recently been further purified by ion-exchange chromatography (Hatton and Regoeczi, 1973) to remove proteases, and an earlier study has described (Kraemer, 1968) the presence of cytotoxic, hemolytic, and phospholipase (White and Mellanby, 1964) activities still remaining in commercial preparations of sialidase from *Clostridium perfringens* prepared ostensibly by a method similar to that described by Cassidy *et al.* (1965). The wide use of commercial sialidases for the experimental alteration of biologically important sialic-acid-containing compounds and cell surfaces makes these considerations of purity quite important. A recent affinity chromatography procedure for the purification of microbial sialidases (Cuatrecasas and Illiano, 1971*a*) has proven, in our hands, to give preparations readily shown by acrylamide gel electrophoresis to be contaminated with numerous other proteins (Den *et al.*, 1975). The latter and isoelectric focusing have produced very pure large scale preparations (Nees *et al.*, 1975).

D. Size and Properties of Bacterial Sialidases

In view of the aforementioned heterogeneity of even highly purified preparations of bacterial sialidases, reliably precise values for their molecular weights generally are not yet available. Indications of (a) multiple forms with similar molecular weights, at least for pneumococcal sialidase (Tannenbaum, *et al.*, 1969; Tannenbaum and Sun, 1971), (b) the production of sialidases falling into two different molecular-weight classes by *Erysipelothrix insidiosa* (Müller, 1971*a*), and three by *Pasteurella multocida* (Müller, 1971*b*), lower-molecular-weight sialidase being found in older cultures, and (c) the great variations in molecular-weight parameters for *Vibrio cholerae* sialidase (Schramm and Mohr, 1959; Ada and French, 1959*b*, Rosenberg, *et al.*, 1960), possibly depending upon the bacterial strain or upon the organismic response to inducers, all serve to complicate the picture.

Estimates of molecular weight based upon behavior of purified preparations of *Diplococcus pneumoniae* sialidase during molecular sieving by gel filtration has produced a molecular-weight value of 70 × 10^3 daltons; and by acrylamide gel electrophoresis (Stahl and O'Toole, 1972), 69.8 × 10^3 daltons. Gel filtration had earlier produced a value of 65 × 10^3 daltons for the sialidase of *Corynebacterium diphtheriae* (Moriyama and Barksdale, 1967). For sialidase of *Vibrio cholerae*, sedimentation studies in the analytical ultracentrifuge (Ada *et al.*, 1961) and chain-size estimation on the basis of determination of *N*-terminal amino acids in a purified sample (Laver *et al.*, 1964) both have produced a value of 90 × 10^3 daltons, but rough estimates based on calculated sedimentation constants obtained in the analytical ultracentrifuge have led Schramm and Mohr (1959) and Rosenberg *et al.* (1960) to suggest values around 10 × 10^3 daltons for the sialidase produced by the cultures of *Vibrio cholerae* investigated by them. By gel filtration and analytical ultracentrifugation, Balke and Drzeniek (1969) have derived a value of 56 × 10^3 daltons for the sialidase of *Clostridium perfringens* purified to the stage described by Cassidy *et al.* (1965). However, Kraemer (1968) later gave evidence that such preparations run the danger of being still contaminated with other proteins. It is an interesting point of comparison that work done recently with *Streptomyces* sialidase has shown the purified sialidase of this microorganism to have a molecular weight of only 32 × 10^3 daltons, as compared with a weight of 57 × 10^3 daltons for the bacterium, *Clostridium perfringens*. The latter value corroborates that obtained by Balke and Drzeniek (1969). Nees *et al.*, (1975) have found an average of 64 × 10^3.

No evidence for subunit structure is yet available for bacterial sialidases, and where evidence has been obtained for multiple forms as described above, lower-molecular-weight forms do not appear to be subunits of higher-molecular-weight forms produced by the same type of organism. Bacterial sialidases, as single polypeptide chains, are smaller than viral sialidases with multiple polypeptide chains, falling generally in the range of 50,000 to 100,000 daltons, i.e., roughly one-half of the molecular-weight range reported for complete viral sialidases (see below).

Tannenbaum and Sun (1971) have reported that the isoenzymes of pneumococcal sialidase, which they studied, do not differ in molecular weight, but rather in charge. In contrast, sialidases of *Vibrio cholerae*, type K *Diplococcus pneumoniae*, and *Clostridium perfringens* were shown to be homogeneous upon isoelectric focusing, with isoelectric points at 4.80, 4.90, and 4.95, respectively. No generalizations are yet possible in this regard. There appears to be no serological cross-reaction for different bacterial sialidases (Madoff *et al.*, 1961) although, serologically, they may be type-specific (Lee and Howe, 1966) among a particular grouping of bacterial strains.

TABLE I. Kinetic Constants of Microbial Sialidases

Enzyme preparation[a]	V_{max}[b]	K_m[b,c]
VCS (purified)	25.0	5.6×10^{-5} M
VCS (commercial)	8.2	1.2×10^{-4} M
CPS (purified)	28.6	2.7×10^{-4} M
CPS (commercial)	3.2	2.2×10^{-5} M

[a] VCS, *V. cholerae* sialidase; CPS, *C. perfringens* sialidase.
[b] Assay mixtures were prepared in 0.01 M Tris-acetate, pH 6.5, with G_{M1}, G_{D1a}, G_{D1b}, G_{T1} as substrate. Substrate concentration ranged from 10^{-3} to 10^{-5} M. An optimum concentration of Ca^{2+} (4 mM) (Rosenberg *et al.*, 1960) was included in assay mixtures containing *V. cholerae* sialidase.
[c] The K_m values presented for *V. cholerae* sialidase refer to the aggregated, micellar form of the mixed substrate.

E. Mode of Action of Bacterial Sialidases

The simplest substrate in general use for measurement of the kinetic constants, K_m and V_{max}, for bacterial sialidase has been sialyllactose, that particular isomer in which sialic acid is in α, 2 → 3 linkage to galactose (cf. Chapter 1) being the most active. K_m values reported for sialidases with this substrate have varied over a tenfold range. Values in the neighborhood of 10^{-4} M appear to have been obtained most frequently for this and for most other substrates, and a generally acid pH optimum, from 4.5 to 7, is reported (for review, Drzeniek, 1972). However, it must be pointed out that such values may be valid only for the very specific conditions under which they have been obtained. We have found that, for *Vibrio cholerae* sialidase (Lipovac *et al*; 1973; Barton *et al.*, 1975) acting on gangliosides, V_{max} values may be manipulated at will by controlling the ionic strength of the reaction medium, and that for *Clostridium perfringens* sialidase, both the V_{max} and the pH optimum may be varied in a predictable fashion simply by controlling the ionic strength (Barton *et al.*, 1975). Both V_{max} and K_m also may vary widely with degree of purity of the enzyme preparation as well. For examples of these phenomena, see Table I and Figures 1 and 2. All bacterial sialidases examined so far hydrolyze sialic acid linked α-ketosidically (cf. Chapter 1) to an oxygen atom in the aglycone portion of the substrate (for review, Drzeniek, 1973). β-ketosides (Meindl and Tuppy, 1965) or α-N-ketosides and α-S-ketosides (cf. Chapter 1) apparently are not attacked (Khorlin *et al.*, 1970). Compounds in which N-

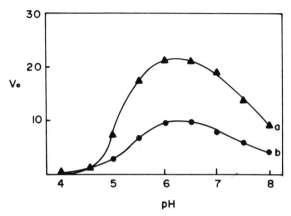

FIGURE 1. Initial velocity–pH relationship for *V. cholerae* sialidase. Assay mixtures containing 0.10 μg of enzyme protein, 3×10^{-4} M G_{M1}, G_{D1a}, G_{D1b}, G_{T1} and 4 mM Ca^{2+} were prepared in 0.01 M Tris, which had been adjusted to the desired pH values with HOAc in the presence and absence of added strong electrolyte (NaCl). Curve a, 0.01 M Tris-acetate alone; curve b, 0.01 M Tris-acetate containing 0.10 M NaCl.

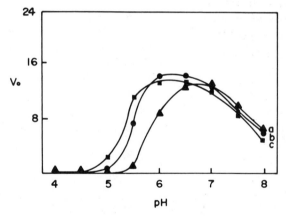

FIGURE 2. Initial velocity–pH relationship for *C. perfringens* sialidase. Assay mixtures containing 0.23 μg of enzyme protein and 2×10^{-4} M G_{M1}, G_{D1a}, G_{D1b}, G_{T1} were prepared in 0.01 M Tris, which had been adjusted to the desired pH values with HOAc in the presence and absence of added strong electrolyte (NaCl). Curve a, 0.01 M Tris-acetate alone; curve b, 0.01 M Tris-acetate containing 0.01 M NaCl; curve c, 0.01 M Tris-acetate containing 0.02 M NaCl.

acetylgalactosamine is linked glycosidically to position 4 of a galactose residue which bears sialic acid at position 3, such as G_{M2} ganglioside and its aceramido derivative (cf. Chapter 1), show great resistance toward hydrolytic removal of their sialyl residue by sialidases. This resistance has been ascribed to steric hindrance of the sialyl group by the substituent on the axial hydroxyl group at position 4 of the galactose residue (Ledeen and Salsman, 1965). However, we have found that molecular models do not indicate that availability of the somewhat hindered sialic acid residue is precluded, and we have proposed a competitive inhibitory mechanism (Lipovac and Rosenberg, 1967), reasoning from the identity of the constellation of protons, N-acetyl groups, and —C—O- functions in the sialyl and N-acetylgalactosaminyl residues, and the ability of free N-acetylgalactosamine to act as a competitive inhibitor. Wenger and Wardell (1973) now have shown that *Clostridium perfringens* sialidase in the presence of bile salts readily cleaves the "hindered" sialyl residue. The synthetic sialic acid analog, 2-deoxy-2,3-dehydrosialic acid, has proven to be a potent competitive inhibitor for the sialidase of *Vibrio cholerae* (Meindl and Tuppy, 1969).

The sialidase of *Vibrio cholerae* is activated by calcium ions at an optimum concentration of 4 mM and by cobaltous or manganous ions equally as well, and strongly inhibited by ferric or mercuric ions (Rosenberg *et al.*, 1960). Sialidase of group K streptococci (Hayano and Tanaka, 1967) appears to be activated by calcium and cobaltous ions, but inhibited by manganous ions, while sialidases from other groups of streptococci are seemingly not activated by divalent cations (Hayano and Tanaka, 1969). Sialidase of *Corynebacterium diphtheriae* (Moriyama and Barksdale, 1967) and *Diplococcus pneumoniae* (Hughes *et al.*, 1964) appears to require a residue of tightly bound calcium ion, or perhaps certain other divalent cations may be substituted as shown for *Vibrio* sialidase (Rosenberg *et al.*, 1960). Not surprisingly, EDTA, which can remove divalent cations, reversibly inhibits these enzymes (Boscham and Jacobs, 1965).

High ionic strength proportionally represses the activity of sialidase of *Clostridium perfringens* and *Vibrio cholerae* with certain substrates (see Figures 1 and 2). Other bacterial sialidases also may possibly be inhibited by high ionic strength, but careful study of this phenomenon as a generality for bacterial sialidases has not been made. As a point of comparison for these bacterial sialidases, the sialidase of Streptomyces has been found recently (Kunimoto *et al.*, 1974) to be inhibited by mercuric, ferrous, magnesium, calcium, zinc, cobaltous, manganous, nickelous, barium, and aluminum ions.

We have found that *Clostridium perfringens* sialidase appears to act on the micellar or "aggregate" form of sialoglycolipid substrates (Barton *et al.*, 1975), but in contrast, *Vibrio cholerae* sialidase clearly can attack sialyl residues in sialoglycolipids when they are not aggregrated into micelles (Lipovac *et al.*, 1973). For *Vibrio cholerae* sialidase, an underivatized (Brossmer and Holmquist, 1971) and unsubstituted carboxylate group in the sialic acid moiety of the substrate is necessary for enzymatic activity (cf. Chapter 1), and bulky N substituents in place of the N-acetyl group on the sialic acid inhibit activity very effectively (Faillard *et al.*, 1969), as does also an O-acetyl substituent on the hydroxyl group at position 4 of sialic acid. This latter substitution also inhibits the sialidase of *Clostridium perfringens* (for review, Drzeniek, 1973).

Thus, in some respects, it is possible to generalize concerning the mode of action of bacterial sialidases, but in numerous instances, the various bacterial enzymes show important differences, and inferences may not be made, as is done too often, concerning still uncharacterized bacterial sialidases based upon findings with *Clostridium* or *Vibrio* sialidases, or both. To date, the bulk of detailed study has been performed on these two latter examples. Certain other details and an extensive bibliography may be found in the comprehensive reviews by Drzeniek (1972; 1973) and Gottschalk and Drzeniek (1972).

F. Biological Roles for Bacterial Sialidases

Correlation between the sialidases elaborated by pathogenic bacteria and the life-styles of these organisms still is difficult. Clearly, sialidases are capable of altering the characteristics of important sialic-acid-containing macromolecules in mammalian cell surfaces and intercellular matrices, and they can attack circulatory sialic-acid-containing macroglobulins, enzymes, and hormones (see Table II). This ability may serve to provide more favorable environmental conditions for propagation of the bacteria, and the released sialic acid together with other glycoses which become available for attack by extracellular or surface-bound glycosidases after removal of sialic acid may serve *in situ* as a readily available energy source.

Solovev *et al.* (1972) have shown that noncholera vibrios have far lower sialidase activity than cholera vibrios and suggested that sialidase activity be used as an aid in *Vibrio* taxonomy. Thomas (1970) has proposed that sialidase might be the toxic factor elaborated by some mycoplasmas. Müller (1971*a,b*) has called attention to the similar

TABLE II. Experimental Removal of Sialic Acid from Biologically Important Sialic-Acid-Containing Substances by Microbial Sialidases

Source of sialidase	Substrate	Effect on substrate	Reference
Vibrio cholerae and A/Lee and swine S15 influenza viruses	Erythrocytes	Release of heterogenetic mononucleosis receptor	Springer and Rapaport, 1957
Vibrio cholerae	Erythropoietin	Inactivation	Lowy et al., 1960
Vibrio cholerae	FSH	Inactivation	Gottschalk et al., 1960
Vibrio cholerae	Erythrocytes	No correlation with hemolysis	Yachnin and Gardner, 1961
Asian influenza virus	Orosomucoid	Increased antigenicity	Athineos et al., 1962
Vibrio cholerae	Erythrocytes and ascites hepatoma cells	Decreased electrophoretic mobility	Fuhrmann et al., 1964
Vibrio cholerae	Lung tumor cells	Reduced metastasis	Gasic and Gasic, 1962
Vibrio cholerae	Transferrin	Reduced electrophoretic mobility	Parker and Bearn, 1967
Vibrio cholerae	Erythrocytes	Reduced electrophoretic mobility	Seaman and Uhlenbruck, 1962
Vibrio cholerae	Blood group antigens	Loss of M antigen	Yokoyama and Trams, 1962
Vibrio cholerae	Pseudocholinesterase	No change in enzyme activity	Ecobichon and Kalow, 1963
Vibrio cholerae	Transferrin	Reduced electrophoretic mobility	Kristjansson and Cipera, 1963
Vibrio cholerae	Erythrocytes and ascites tumor cells	Removal of surface sialic acid	Miller et al., 1963
Vibrio cholerae	Oxytocinase	No change in activity	Tuppy et al., 1963
Vibrio cholerae	Normal liver and ascites hepatoma cells	Release of sialic acid	Granzer et al., 1964
Vibrio cholerae	Erythrocytes and leukocytes	No effect on phagocytosis	Jensen and Moreno, 1964
Vibrio cholerae	Blood group antigens	No loss of N specificity	Hotta and Springer, 1964
Vibrio cholerae	Erythrocytes	Increased sensitivity to lysis	Perona et al., 1964
Vibrio cholerae	Serum alkaline phosphatase	Differential release of sialic acid from isoenzymes	Robinson and Pierce, 1964

Vibrio cholerae	Serum cholinesterase	Decreased electrophoretic mobility	Svensmark and Heilbronn, 1964
Vibrio cholerae	γ-Glutamyl transpeptidase	Increase in acid stability	Szewczuk and Connell, 1964
Clostridium perfringens	ATP-ase	Inactivation	Emmelot and Bos, 1965
Vibrio cholerae	Blood platelets	Spontaneous aggregation	Hovig, 1965
pneumococci	α_1-Antitrypsin	No effect	Laurell, 1965
Vibrio cholerae	Erythrocytes	Unmasking of ABH blood groups	Saber et al., 1965
Vibrio cholerae	Murine leukemia cells	Increased immunogenicity	Bagshawe and Currie, 1968
Vibrio cholerae	Kidney alkaline phosphatase	Changes in heterogeneity	Butterworth, 1966
Corynebacterium diphtheriae	Transferrin	Removal of sialic acid	Jamieson, 1966
Vibrio cholerae	Chinese hamster ovary cells	Removal of sialic acid	Kraemer, 1966
Clostridium perfringens	Alkaline phosphatases	Alteration in electrophoretic mobilities	Moss et al., 1966
Vibrio cholerae	Monocytes	Increased phagocytosis	Weiss et al., 1966
Vibrio cholerae	Cerebral tissue	No change in metabolic response to electric stimulation	Evans and McIlwain, 1967
Clostridium perfringens	Urinary glycoprotein	Increase in low P_K tyrosine group	Pape and Maxfield, 1967
Vibrio cholerae	Landschutz ascites tumor cells	Increase in immunogenicity	Currie and Bagshawe, 1968
Vibrio cholerae	Erythrocytes	Increased ingestion by macrophages	Lee, 1968
Vibrio cholerae	IgA and IgG	No acquisition of complement fixing properties	Vaerman and Heremans, 1968
Vibrio cholerae	Synaptic membranes	Decrease in electron microscopic density	Bondareff and Sjöstrand, 1969
Vibrio cholerae	Mouse sarcoma cells	Increased immunogenicity	Currie and Bagshawe, 1969
Vibrio cholerae	Lymphocytes	Decreased homing	Gessner et al., 1969

(continued)

TABLE II. (Continued)

Source of sialidase	Substrate	Effect on substrate	Reference
Vibrio cholerae	Tissue culture cells	Changes in electrophoretic mobility	Mayhew, 1969
Clostridium perfringens	Bone alkaline phosphatase	Changes in electrophoretic mobility	Smith et al., 1969
Vibrio cholerae	Transfused lymphocytes	Decrease in lymph nodes and spleen, increase in liver	Woodruff and Gessner, 1969
Vibrio cholerae	Py cells	Reduced invasiveness	Yarnell and Ambrose, 1969
Vibrio cholerae	Macrophages	Increased bacterial adhesion	Allen and Cook, 1970
Vibrio cholerae	Interferon	No effect on activity	Viti et al., 1970
Vibrio cholerae	Acid phosphatase	Increase in pH optimum and substrate affinity	Dziembor et al., 1970
Vibrio cholerae	Amylase	Changes in isoenzymes	Finnegan and Hope, 1970
Vibrio cholerae	Postcoital intravenous injection in mice	Suppression of pregnancy	Gasic and Gasic, 1970
Clostridium perfringens *Vibrio cholerae*	Ceruloplasmin	Loss of inhibition of viral hemagglutination	Johnson et al., 1970
Vibrio cholerae	Chick embryo muscle cells	Reduced aggregation	Kemp, 1970
Vibrio cholerae	Erythrocyte	Loss of protamine agglutination	Khan and Zinneman, 1970
Vibrio cholerae	Ascites hepatoma cells	Differences in charge	Kojima and Maekawa, 1970
Vibrio cholerae	Erythrocytes	Decreased electrophoretic mobility	Mehrishi, 1970
Clostridium perfringens	Uterine cervix carcinoma cells	Reduction in electrophoretic mobility	Vasudevan et al., 1970
Vibrio cholerae	Leukemia L1210 tumor cells	Increased immunogenicity	Bekesi et al., 1971
Clostridium perfringens	Luteinizing hormone	Loss of activity	Braunstein, 1971

Sialidase	Substrate/system	Effect	Reference
Vibrio cholerae or influenza virus $A_2/R_1/5$	Injection into rats	Thrombocytopenia	Choi et al., 1972
Vibrio cholerae	Human lymphocytes	Increased lysis susceptibility	Grothaus et al., 1971
Vibrio cholerae	Leukocytes	Increase in A and H, decrease in M and N antigens	Kassulke et al., 1971
Vibrio cholerae	Fibroblasts	No decrease in viability	Kemp, 1971
Vibrio cholerae	Lymphocytes	Increase in stimulatory capacity	Lundgren and Simmons, 1971
Vibrio cholerae	Mouse islets of Langerhans	Decreased insulin secretion	Maier and Pfeiffler, 1971
Clostridium perfringens	Circulating glycoproteins	Uptake by liver	Morell et al., 1971
Vibrio cholerae	Prothrombin	Reduced circulatory half-life	Nelsestuen and Suttie, 1971
Vibrio cholerae	Mouse mammary carcinoma cells	Increased immunogenicity	Oxley and Griffin, 1971
Clostridium perfringens	Liver plasma membranes glycoproteins	Reduced glycoprotein uptake Increased binding to liver cell membranes	Pricer and Ashwell, 1971
Vibrio cholerae	Lymphoid cells	Increased susceptibility to cytolysis by complement	Ray et al., 1971
Vibrio cholerae	Lymphoid cells	No unmasking of allogeneic antigens	Ray and Simmons, 1971
Clostridium perfringens	Brown fat cells	Insulin-like effect	Rosenthal and Fain, 1971
Vibrio cholerae	TA3 Mouse adenocarcinoma cells	Increased antigenecity	Sanford and Codington, 1971
Pasteurella multocida and hemolytica *Vibrio cholerae* *Diplococcus pneumoniae*	Transferrin	Changes in electrophoretic mobility	Scharmann et al., 1971
Vibrio cholerae	Erythropoietin	Loss of activity	Schooley and Garcia, 1971
Vibrio cholerae	Erythropoietin	Loss of activity *in vivo*	Schooley and Mahlmann, 1971
Vibrio cholerae	H-2 Alloantigens	No effect on reactivity	Shimada and Nathenson, 1971

(continued)

TABLE II. (Continued)

Source of sialidase	Substrate	Effect on substrate	Reference
Vibrio cholerae	Methylcholanthrene fibrosarcoma	Tumor regression	Simmons et al., 1971a
Vibrio cholerae	Trophoblast	No unmasking of histocompatibility antigens	Simmons et al., 1971b
Vibrio cholerae	Lymphoid cells	Increased Antigenicity	Simmons et al., 1971c
Clostridium perfringens	FSH	Loss of biologic activity	Vaitukaitis and Ross, 1971
Clostridium perfringens	HCG	Decrease in plasma half-life	Van Hall et al., 1971a
Clostridium perfringens	HCG	Loss of biological activity	Van Hall et al., 1971b
Vibrio cholerae	Brush border alkaline phosphatase	Loss of activity	Varute and Patil, 1971
Vibrio cholerae	Hepatoma cells	Activation of host lymphocytes	Watkins et al., 1971
Vibrio cholerae	Spleen cells	Increased ^{51}Cr release	Weiss and Cudney, 1971
Vibrio cholerae	Lymphoid cells	Blocked internalization of mitogens	Adler et al., 1972
Vibrio cholerae	Lymph node cells	Increased amyloid promotion	Ebbesen, 1972
Vibrio cholerae	Mouse ascites tumor cells	Decreased tumor forming ability	Hughes et al., 1972
Vibrio cholerae	"Protagon"	Loss of tetanus toxin binding	Kryzhanovskii and Sakharova, 1972
Vibrio cholerae	Polymorphonuclear leukocytes	No effect on phagocytosis	Noseworthy et al., 1972
Vibrio cholerae or Clostridium perfringens	Lymphoid cells	Increased susceptibility to cytolysis	Ray and Simmons, 1972
Vibrio cholerae	Leukemic myeloblasts	Agglutination	Sauter et al., 1972
Vibrio cholerae	Mouse methylcholanthrene fibrosarcoma	Immunospecific regression	Simmons and Rios, 1972

Sialidases

Vibrio cholerae or *Clostridium perfringens*	Chicken fibroblasts	Stimulation of division and sugar uptake	Vaheri *et al.*, 1972
Vibrio cholerae	Injected into hamsters	Decrease in rolling granulocytes	Atherton and Born, 1973
Clostridium perfringens	Rat peritoneal cells	Reduced antigen-induced release of histamine	Bach and Brashler, 1973
Vibrio cholerae	Intact mammalian cells	Removal of protein-bound sialic acid	Barton and Rosenberg, 1973
Vibrio cholerae	Human lymphocytes	Enhanced sheep erythrocyte binding	Bentwich *et al.*, 1973
Vibrio cholerae	Retinal cells	Decrease in electrophoretic mobility, loss of sialic acid	Connins *et al.*, 1973
Vibrio cholerae	Erythrocytes	Increased anti-A, -B, -H, -D and -C agglutinability	Greenwalt and Steane, 1973
Vibrio cholerae	Lymphocytes	Enhanced blastogenic response	Han, 1973
Vibrio cholerae	Mouse TA3 tumor cells	Enhanced cytotoxicity	Hughes *et al.*, 1973
Vibrio cholerae	Injection into mice	Delay of skin graft rejection	Madoff and McKenna, 1973
Vibrio cholerae	Trophoblastic cells	Pregnancy rejection	Nista *et al.*, 1973
Vibrio cholerae	Tumor and normal lymphoid cells	More sialic acid release from tumor cells	Ray and Simmons, 1973
Vibrio cholerae	Methylcholanthrene fibrosarcoma cells	Immunospecific regression	Rios and Simmons, 1973
Clostridium perfringens	Human spermatozoa	Reduced electroconductivity	Rosado *et al.*, 1973
Vibrio cholerae	Mouse TA3-Ha tumor cells	Increased antigenicity	Sanford, 1973
Vibrio cholerae	Ribosome-free membranes	Loss of ability to reattach ribosomes	Scott-Burden and Hawtrey, 1973
Vibrio cholerae or *Clostridium perfringens*	Mouse leukemia 1210	Increased immunogenicity	Sethi and Brandis, 1973
Vibrio cholerae	Mouse mammary carcinoma cells	Increased immunogenicity	Simmons and Rios, 1973

(*continued*)

TABLE II. (Continued)

Source of sialidase	Substrate	Effect on substrate	Reference
Vibrio cholerae	Normal and transformed tissue culture cells	Removal of bovine enterovirus receptors	Stoner *et al.*, 1973
Vibrio cholerae or *Clostridium perfringens*	Human T lymphocytes	Enhanced binding of sheep erythrocytes	Weiner *et al.*, 1973
Vibrio cholerae	Sheep, horse, and chick erythrocytes	Enhanced immunogenicity of sheep and horse, but not chick erythrocytes	Barth and Singla, 1974
Clostridium perfringens	Synaptic structure	Abolishes synaptic transmission	Tauc and Hinzen, 1974
Vibrio cholerae	Human lymphocytes	Broader cytotoxicity with alloantisera	Reisner *et al.*, 1974
Clostridium perfringens	Pancreatic islets	Altered response to insulin secretagogues	Hahn *et al.*, 1974
Vibrio cholerae	X-ray killed B-16 cells	Enhanced growth of B16 tumors	Froese *et al.*, 1974
Vibrio cholerae	Granulation tissues, embryonic fibroblasts	Decreased incorporation of proline into collagen	Aalto *et al.*, 1974
Vibrio cholerae	Walker tumor cells	No change in tumor distribution	Weiss *et al.*, 1974
Vibrio cholerae	Adrenal cells	Increased response to cholera enterotoxin	Haksar *et al.*, 1974
Vibrio cholerae	Tetanotoxin fixed to brain	*In vitro* release	Kryzhanovskii *et al.*, 1974
Clostridium perfringens	Nervous system cells in culture	Activation of ectoacetyl and butyryl cholinesterases	Rosenberg *et al.*, 1975
Clostridium perfringens	Nervous system cells in culture	Activation of ectopyrophosphatase	Stefanovic *et al.*, 1975

clinical picture brought about in infection by organisms such as *Erysipelothrix insidiosa, Pasteurella multocida*, and certain streptococci, and has suggested that these may relate to the sialidase which is elaborated by these organisms. He has found (Müller, 1971b) immunoelectrophoretic changes in the sialic-acid-containing macromolecules, acid α-glycoprotein, haptoglobin, transferrin, hemopexin, and IgA in a human infected with *Pasteurella,* and has measured the effects of the sialidase elaborated by this organism upon the serum sialoglycoproteins of an injected guinea pig. O'Toole *et al.* (1971) found that high cerebrospinal fluid levels of free sialic acid in bacterial meningitis correlate with an adverse prognosis, and suggested that this may be due to the activity of sialidase elaborated by the pneumococci. Pardoe (1970), working with *Klebsiella aerogenes*, has suggested that sialidase elaborated by such bacteria, in response to mammalian sialic-acid-containing compounds as inducers, may represent a means by which the bacteria gain access in the infected organism. In the same year, Kelly and Grieff (1970) identified pneumococcal sialidase as the toxic factor for mice elaborated by type I pneumococci, noting that pneumococci that are isolated directly from clinical infections always possess sialidase activity (Kelly *et al.*, 1967). This activity may not be demonstrable upon further culture in artificial growth media. As early as 1959, Laurell reported that cultures of pneumococci, streptococci, and *Pasteurella tuberculosis* split off sialic acid and altered the electrophoretic mobility of serum α- and β-globulins. Thonard *et al.* (1965) have suggested that sialidase elaborated by human oral bacteria may serve to help break down gingival epithelium and account for the histopathological aspects of gingival diseases. These findings, still in their infancy, together are suggestive of a role for bacterial sialidase in the aggressive survival of bacterial pathogens.

III. VIRAL SIALIDASES

A. Morphology and Genetics of Viral Sialidases

Viral sialidases occur in orthomyxoviruses, including avian, porcine, equine, and human influenza viruses, and in paramyxoviruses, such as Sendai, Simian 5, mumps, and Newcastle disease viruses and, possibly, also in a newly emerging third subgroup, the metamyxoviruses. These latter have helical diameters falling between 9 and the 18 nm in size, respectively, for the helices of the *ortho* and *para* subgroups (Melnick, 1972). Putative examples in the *meta* subgroup are murine pneumonia and respiratory syncytial viruses.

Sialidase is currently recognized to be one of the surface antigens of myxoviruses. Designation of the antigenic relatedness of the sialidase of a particular strain of virus now is conventionally indicated in the viral nomenclature (Kendal and Madelay, 1969; Madelay *et al.*, 1971; Wildy, 1971).

Sialidase in undisrupted viruses generally can attack exogenous substrates. The enzyme is described, along with the other major viral surface antigen, hemagglutinin, as being associated with spikelike projections on the outer surface of the virion (Laver and Valentine, 1969), based on work done with, e.g., the BEL strain of influenza virus A. The spikes are thought to be imbedded in the lipophilic portion of the surface membranes (Kendal *et al.*, 1968), as determined with a variant of influenza virus strain A_2/Sing/I/57. Treatment of the hemagglutinating virus of Japan with trypsin has been found to remove the surface projections from this virus with concomitant loss of sialidase, and also, hemagglutinin activities (Maeno *et al.*, 1970); likewise, a protease from *Streptomyces griseus* has been shown to remove the sialidase- (and hemagglutin-) containing spikes from the paramyxovirus, SV5 (Chen *et al.*, 1971). Although easy, morphological generalizations are risky. It would appear that hydrophilic projections from the surface of the enveloping membranes of myxoviruses have incorporated into their structures sialidase and hemagglutinating proteins.

As a major surface antigen in myxoviruses, sialidase may undergo antigenic drift (Meier-Ewert *et al.*, 1970; Fridman *et al.*, 1969), that is, gradual mutational change in structure conferring enhanced growth potential upon a virus in the presence of preexisting antibodies, and sialidase also may figure in major epidemic-causing antigenic shifts resulting in the appearance of ostensibly new viruses, which may be due to the formation of viral recombinants between human and avian, or animal, viruses during natural transmission. Sialidases of some avian influenza viruses and equine influenza viruses, on the one hand, and avian and human influenza viruses, on the other, are related antigenically. The relatedness of the sialidase of the latter viruses and, also, swine influenza viruses, have been taken as indications of viral recombination (for review, Webster, 1972). Interrelatedness of sialidase of various human strains of influenza virus without a counterpart in animals also has been observed (Baars *et al.*, 1971). Antibodies to sialidase of a certain equine influenza have been found in man (Fedson *et al.*, 1972), and human influenza-type sialidases have been found in viruses causing disease in ducks (Higgins and Schild, 1972). Antibodies to the sialidase of strains of human influenza virus have been found in pelagic birds (Laver and Webster, 1972). Antigenic interrelationships have been shown for the sialidases of human influenza B viruses isolated over a long (28 year) period (Chakraverty, 1972); likewise for a number of strains of human influenza A virus (Schulman and Kilbourne, 1969). Not always, but often, there has been genetic conservatism with regard to viral sialidases. The serological similarity between the sialidases and distinctness of the hemagglutinins of Rostock strain fowl plague virus and swine influenza virus, and the ability of the latter virus to immunize chickens against the former suggests that sialidase is involved in the immune response of the host organism (Rott et al., 1974). As shown early on by Rafelson et al. (1962; for review, Webster, 1972) and by Scholtissek and Rott (1964) and Laver and Kilbourne (1966), the information for the viral sialidase is encoded in the viral genome, although there has been recent indication that a small portion of the viral sialidase may be encoded in the host genome as well (Zhumatov et al., 1972).

Lipkind and Tsvetkova (1967) found an "eclipse" of sialidase and hemagglutinin activities during cellular adsorption of Newcastle disease virus propagated in chick embryo cell cultures. They concluded from this finding that synthesis of sialidase is controlled by the viral genome, and that the viral sialidase is not acquired from a cellular supply of the host. Seto (1964) found that incomplete Japan 305 virus had decreased sialidase and hemagglutinin activities, and they considered the sialidase to be inserted between spikes of hemagglutinin during assembly of the viral envelope. Early kinetic evidence has been obtained which indicates that sialidase synthesis corresponds with self-replication of viral templates during influenza virus infection of chorioallantoic membranes (Noll et al., 1961). Brydak et al. (1971) have more recently found that freshly isolated influenza virus can undergo increase and a final decrease in sialidase activity upon serial passage in chicken embryos. Working with influenza A_2 virus, Maeno and Kilbourne (1970) observed that sialidase and hemagglutinin appeared coincidentally 4 hr after infection in the cytoplasm of an aneuploid clone, 1-5C-4, of cultured human cells. Antibody directed against the hemagglutinin blocked the staining of the sialidase in the cells. They concluded that the two cytoplasmic viral antigens, i.e., the sialidase and hemagglutinin, were closely related physically and may be initially associated as a larger protein in the host cytoplasm. Scheid and Choppin (1974) recently have suggested that the behavior of sialidase and hemagglutinating activity as properties of a large single protein retained and eluted from a sialo-protein affinity column indicates identity of the active site for these two activities. Interestingly, Lipkind and Tsvetkova (1971) found that in Newcastle disease virus propagated in chick embryo cell cultures inhibition of protein synthesis by actidione or puromycin was attended by a diminution in both sialidase and hemagglutinin activities. They suggested that this functional lability of sialidase and hemagglutinin may result from their incorporation into larger structures in the newly forming viral membrane. Very early, Howe *et al.* (1961) found that heat-treated, "indicator" influenza virus of a variety of types showed evidence of independent lability of sialidase and hemagglutinin activities, and later, Tsvetkova and Lipkind (1970) showed differences in the heat lability of the hemagglutinin and the sialidase in two variants of influenza A_2 virus. More recently, Frösner and Gerth (1972) have shown differences in heat lability of these two viral activities in some 16 strains of influenza A_2 virus. However, Pierce and Haywood (1973) have made the interesting finding that the rates of heat inactivation of hemagglutinin and sialidase of Newcastle disease virus are coupled, and they have concluded that a factor in the viral envelope can simultaneously influence the heat stability of both activities. Although, once again, generalizations are risky, a heavy weight of circumstantial evidence bears on the thesis that there is close physical proximity between the sialidase and hemagglutinin of myxoviruses in general, and that the viral genome codes independently for the viral sialidase, probably a glycoprotein, which appears in the host cell's cytoplasm and then is incorporated as a polymer, along with the hemagglutinin, into projections on the surface of the viral envelope.

B. Purification of Viral Sialidases

The review by Drzeniek (1972) carries a useful description of approaches toward purification of viral sialidases. Recent (Oxford, 1973) attempts at purification of enzymes associated with influenza B virus have involved purification of the viruses grown in 10-day-old embryonated hen eggs, disruption of the viruses by SDS and β-mercaptoethanol, and polyacrylamide gel electrophoresis under reducing conditions. This procedure gives sialidase migrating as a single band. Influenza A viruses dating from 1957, 1960, and 1969 were similarly grown and purified and treated with proteolytic enzymes. The sialidase was purified some 16- to 32-fold by density gradient centrifugation and gave single bands on SDS-acrylamide gel electrophoresis under reducing conditions (Kendal and Kiley, 1973). A similar approach was used for influenza B/Lee virus (Wrigley *et al.*, 1973). The paramyxovirus, Sendai, was grown similarly in eggs or in calf kidney cell culture, purified by density gradient centrifugation, disrupted with Triton X-100, purified on Sepharose 6-B, and by equilibrium centrifugation, to give a single fraction (Tozawa *et al.*, 1973). The A_0/WSN strain of influenza virus grown in cultures of chick embryo fibroblasts yielded, when disrupted with the detergent nonidet P-40, an active sialidase which could be purified by gradient centrifugation in glycerol, or by column chromatography on DEAcellulose, or Sephadex G-200. In contrast with that of other viruses, the sialidase activity of this virus was destroyed when the virus was disrupted with the anionic detergent, SDS, or by treatment with trypsin (Gregoriades, 1972). High purification of the sialidase from an influenza A_2 virus strain has been achieved by disruption of the egg-grown, density-gradient-purified virus preparation with Tween-80 and ether followed by pronase digestion and chromatography on DEAE-cellulose (Bachmayer, 1972). Single bands of sialidase in SDS-acrylamide electrophoresis under reducing conditions were also obtained from a recombinant, similarly grown and purified, influenza virus simply by proteolytic digestion of the virus with nagarse followed by sucrose density gradient centrifugation of the products (Kendal and Eckert, 1972). Highly purified sialidase was obtained from the B/Lee strain of influenza virus grown in eggs or calf kidney cell cultures and concentrated by density gradient centrifugation. These viruses were disrupted by SDS, or by trypsin, which was thought to produce lower-molecular-weight artifacts of sialidase, and the enzyme isolated by SDS-acrylamide gel electrophoresis under reducing conditions (Lazdins *et al.*, 1972). Methods similar to those outlined above have been employed for the isolation and purification of sialidase from numerous types and strains of myxoviruses (Iinuma *et al.*, 1971; Skehel and Schield, 1971; Thacore and Youngner, 1971; Maeno *et al.*, 1970; for review, Drzeniek, 1972). Isolation and purification of sialidase of the paramyxoviruses, Sendai and Newcastle disease virus, has been achieved by nonionic detergent treatment and chromatography on agarose gel beads (Brostrom *et al.*, 1971). Influenza A_2 virus sialidase has been separated by preparative electrophoresis after disruption of the purified virus preparation with ether (Hjerteń *et al.*, 1970). Recently, the large glycoprotein of simian virus 5 possessing both hemagglutinating and sialidase activities has been purified by affinity chromatography on the sialoprotein, fetuin, linked to Sepharose (Scheid and Choppin, 1974). For a survey of other variations in methodology in the older literature, see Drzeniek, 1972.

These methods are not always as straightforward as they seem. Conflicting results on the selective degradation of either the sialidase (Brand and Skehel, 1972) or the hemagglutinin (Bachmayer and Schmidt, 1972) by treatment of influenza (A_2) virus by bromelain have been presented. Sialidases of the A_1 and DSP strains of influenza virus A were found to be destroyed by trypsin, those of the PR8 and swine strains were partially resistant to trypsin, and the A_2 and Lee strain of influenza virus B were completely resistant to both trypsin and pronase (Hoyle and Almeda, 1971).

Clearly, sialidases of myxoviruses are readily removed from the outer surface of the virion by treatment with lipophilic solvents or with various kinds of detergent, anionic or neutral, or by proteolytic digestion. Column chromatography or gel filtration of the released enzyme have been the general routes for subsequent purification of the enzyme. As indicated in the examples cited above, the sialidases of different myxoviruses can show an extremely wide variability in their degree of stability during such procedures, especially after treatment of the whole virus with a proteolytic enzyme or with an anionic detergent. For this reason, published estimates of molecular parameters, physical and kinetic properties, and levels of activity of purified viral sialidases must be viewed with caution.

C. Size of Viral Sialidases

Molecular weights differ somewhat for the various myxovirus sialidases. Preparations subjected to proteolytic digestion in the course of isolation and purification may or may not have suffered unsuspected degradations. Molecular complexes having sialidase activity have been isolated and characterized. Often, however, dissociation to enzymatically active monomeric units has been the standard prelude to determination of molecular weight. Sizes ranging from 1 to 2×10^5 daltons have been estimated in the past for the high-molecular-weight complexes, while 5 to 6×10^4 covers the range generally reported for the monomeric enzymes (Drzeniek, 1972). Recent values for sialidase from influenza B/Lee virus obtained by polyacrylamide gel electrophoresis are 6.3×10^4 daltons for sialidase from SDS-disrupted virus (Oxford, 1973) and 6×10^4 for another preparation not exposed to proteolytic digestion, but 4.8×10^4 after treatment with trypsin (Wrigley et al., 1973). Interestingly, Lazdins et al. (1972) have found that B/Lee sialidase may be a polymer composed of four glycoprotein molecules weighing 6.3×10^4 daltons, and that the monomers are linked into pairs by disulfide bridges. They obtained evidence that trypsin treatment releases a carbohydrate-rich fragment, reducing the monomer weight to 5.6×10^4 daltons, and destroys ability of the monomer to aggregate. Evidence has been obtained for the occurrence, in preparations visualized in the electron microscope, of an aggregate structure consisting of four coplanar (40 Å)³ glycoprotein units weighing an estimated total of 2.4×10^5 daltons for the sialidase of B/Lee undigested by proteolytic enzymes (Wrigley et al., 1973). For Sendai virus sialidase, a glycoprotein having a molecular weight of 6.7×10^4 daltons was obtained by polyacrylamide gel electrophoresis (Tozawa et al., 1973). For late N_2 influenza A viruses, glycoprotein subunits weighing 5.1 to 5.2×10^4 daltons were obtained for sialidase dissociated in urea and dithiothreitol. The sulfhydryl groups on the proteins were blocked by reaction with labeled iodoacetamide (Kendal and Kiley, 1973). This monomer would give a tetramer weighing somewhat over 2×10^5 daltons. The enzyme was found to degrade to smaller monomeric units upon storage. For A_0/WSN sialidase, which is destroyed by trypsin, a rough estimate of 1.8×10^5 daltons was obtained on Sephadex G-200 in the presence of the detergent, nonidet (Gregoriades, 1972). A molecular weight of 5.4×10^4 daltons was estimated in SDS–acrylamide gel electrophoresis for subunits obtained after dissociation and carboxyamidomethylation of influenza A_2/1957 virus sialidase isolated from a nagarse-treated recombinant virus, X-7F1 (Kendal and Eckert, 1972). The molecule contained about 21 cysteine residues, apparently linked by disulfide bridges. An estimated value for a tetramer therefore is again, somewhat over 2×10^5 daltons. For the recombinant X-7F1, dissociated by SDS, urea, and β-mercaptoethanol, without protease treatment, Skehel and Schild (1971) have found the sialidase to be made up of two polypeptides weighing 7 and 8×10^4 daltons. Earlier, Haslam et al. (1969) found a lower, 3.8×10^4 dalton, molecular weight for sialidase of the paramyxovirus, Newcastle disease virus. For a useful, less recent, discussion of the molecular weights of viral sialidases, see the review by Drzeniek (1972).

D. Properties of Viral Sialidases

For a recent comprehensive general coverage of the kinetic properties and substrate specificities of viral sialidases, see the review by Drzeniek (1973). Optimum pH for viral sialidases often varies with the substrate and like the bacterial sialidases, is generally acidic: Values as low as 3.5 have been obtained for sialidase from paramyxovirus Yucaiba acting upon fetuin (Brostrom et al., 1971). Generally, K_m values around 10^{-4}M are obtained with the small molecule, sialyl (α, 2→3) lactose (cf. Chapter 1), and higher, about 10^{-3}M, for sialyl (α, 2→6) lactose as substrates. Most viral sialidases appear to hydrolyze the sialyl (α, 2→6) Gal linkage less readily than the (α, 2→3) linkage (Drzeniek, 1972). Neither the sialidase of Newcastle disease virus nor that of fowl plague virus appear to be capable of hydrolyzing the sialyl (α, 2→6) GlcNac linkage found in milk oligosaccharides (Huang and Orlick, 1972), and these compounds act as competitive inhibitors. It is not known whether the (α, 2→6) GlcNac linkage is resistant for all viral sialidases. K_m values with glycoprotein substrates are generally more favorable than with sialyllactose. K_m with fetuin or orosomucoid ranges around 10^{-6}M for the sialidases of mumps, Sendai, Newcastle disease, and Yucaiba paramyxoviruses (Brostrom et al., 1971). Bovine submaxillary mucin, which contains 7- and 8-O-acetylsialic acids (cf. Chapter 1) is generally more resistant to hydrolysis. K_m values may increase sharply when the enzyme is released by disruption of the virus (Brostrom et al., 1971), which is not surprising in view of the possible effects of the disrupting agents upon the enzyme. V_{max} values (Drzeniek, 1972, 1973; Brostrom et al., 1971) also may be much higher with glycoprotein as compared with sialyllactose substrate. Reduction of the chain length of enzymatically susceptible sialyl residues, by selective periodate oxidation, greatly reduces their enzymatic cleavage rate (Suttajit and Winzler, 1971).

Calcium ion (often tightly bound) is necessary for the activity of many viral sialidases (for review, Drzeniek, 1972, 1973). Factors such as temperature optima may vary widely, even among variants of a particular strain (Golubev et al., 1973).

Many potent inhibitors of viral sialidase are known. These include free sialic acid and its analogs such as 2-deoxy-2,3-dehydrosialic acid, substrate analogs with bulky N-substituents on the sialic acid, or lactonized sialic acid derivatives, oxamic acids, certain heavy metals (Hg^{2+}), and polyanionic compounds (for review, Drzeniek, 1972). 2-Deoxy-2,3-dehydro-N-trifluoroacetylneuraminic acid can also act as a potent inhibitor of myxovirus growth in tissue culture (Palese et al., 1974). Isoquinoline derivatives apparently do not, as once thought, inhibit, but rather interfere with the analysis for free sialic acid (Shinkai and Nishimura, 1972).

E. Possible Biological Roles for Viral Sialidases

While many roles have been proposed quite early for viral sialidase (viral penetration into the host cell, provision of lower-molecular-weight metabolites for viral propagation, release of newly formed virus from the host cell, destruction of substances which protect the cell surface from virus binding, viral binding to the host cell surface by enzyme–substrate interaction), conflicting lines of evidence for each hypothesis have prevented the certain adoption of any of them. For details in an old and a new review, see those by Kelly (1963) and Drzeniek (1972). More recently Tsvetkova and Lipkind (1973) have suggested that the viral sialidase may function in elution of preabsorbed virus. Such eluted virus could attack fresh cells and increase the efficiency of viral propagation by circumventing localized "overkill." It has been known for some time that sialic-acid-containing compounds inhibit viral hemagglutination. Experiments with erythrocytes coated with several different serum sialoglycoproteins have shown that sialyl groups probably are necessary components of the cellular binding sites for myxoviruses, but that purified viral sialidase does not adsorb to the cell surface (Huang et al., 1973) when it is coated with sialoglycoprotein.

While it is a far cry from "purified" isolated viral sialidase to the native sialidase in the viral surface, the current evidence is strong that sialidase does not mediate viral binding to the cell surface. Interestingly, the sialidase of Sendai virions (Tozawa et al., 1973) has been found to be identical with the viral protein having blocking activity for hemagglutination inhibition antibody, which in a sense, serves to support the latter findings. Yet, there is mounting information which shows correlation between immunity to viral infection and the presence of antisialidase activity (Morein et al., 1973; Murphy et al., 1972; Monto and Kendal, 1973) in man and animals. However, "antigenically identical" influenza virus recombinants with greatly differing amounts of sialidase per virion have recently been isolated by Palese and Schulman (1974). Interestingly, two temperature-sensitive mutants of a similar recombination type have been shown to have low infectivity and to lack sialidase and hemagglutinating activity when grown at nonpermissive temperature (Palese et al., 1974). Very recent work (Dowdle et al., 1974), corroborating an hypothesis made a little earlier (Recht et al., 1971) has presented evidence that antisialidase antibodies function to prevent viral release by binding newly formed viruses to each other and to the surface membrane of the cell. Interferon recently has been found to suppress the production of sialidase in a single cycle of growth of influenza A_2/HK/I/ 68 virus and in the recombinant influenza virus, X7F1, in chicken embryo cell cultures (Sedmak and Grossberg, 1973).

The picture still is far from clear. The uniqueness of myxoviruses in having a surface enzyme such as sialidase lead one to presuppose that this enzyme helps to carry out a special myxoviral function. We propose a function different from those tested and found wanting. The newly synthesized sialidase molecules, for which the viral genome codes, moves to and is incorporated into the surface membrane of the host cell, and occupies a position similar to the native mammalian plasma membrane sialidase (see Section V). By removing contiguous sialic acid groups from the surface of the host cell membrane, the enzyme may serve to alter the physicochemical properties of the plasma membrane so as to permit budding and envelopment of the formed and exiting virus. This hypothesis is open to experimental verification.

IV. EXPERIMENTAL USE OF MICROBIAL SIALIDASES

A great wave of current research concerns itself with the enzymatic removal of sialic acid from mammalian cell surfaces and sialic-acid-containing hormones, enzymes, and inter- and intracellular macromolecules. The biological effects of the removal of sialic acid from such substances by sialidase often are profound. Except for the information given in Table II, which lists some examples of the experimental removal of sialic acid from biologically important substances and cell structures by microbial sialidases, consideration of this currently proliferative area of research will lie beyond the purview of this chapter. We should like to point out, in passing, that interpretation of the results of treatment of whole cells with microbial sialidases must take into account the often overlooked findings of Nordling and Mayhew (1966) on the uptake of active *Vibrio* sialidase into mammalian cells and the recent observation of McQuiddy and Lilien (1973) that at least the *Vibrio* enzyme binds to cells and to tissues when added to their medium.

V. MAMMALIAN SIALIDASES

Over the past ten years there have been many papers published about mammalian sialidase and a few about avian sialidase. Organellar and subcellular distribution studies have been reported upon, as have substrate specificity, purification, physical properties and possible effects of enzyme action. Much of the work is suggestive but not complete, i.e., characterization of the physical properties such as the kinetic parameters, K_m and V_{max}, has only been done on crude enzyme preparations. Composition, molecular weight, and mechanism of action are not known.

Several workers have tried to ascertain the effects of mammalian sialidase action on endogenous substrates. Studies of the effects of soluble microbial sialidases on cell-surface properties and on the activities of sialocompounds have been performed in order to try to determine, by analogy, the role of mammalian sialidase. This part of the chapter will summarize what is known about mammalian, and where possible, avian sialidase; and will suggest biological roles for the enzyme.

A. Organ Distribution of Mammalian Sialidases

Sialidase has been found in many avian and mammalian organs. Warren and Spearing (1960) showed the existence of a mammalian sialidase in commercial preparations of human and bovine glycoprotein. They observed that the sialidase acted on the sialyl-containing proteins in the preparation and that the amount of sialic acid released increased only slightly in the presence of added sialoglycoprotein substrates. Ada and Lind (1961), in studies of avian tissue, found sialidase activity in the chorioallantoic membrane of chicken embryos. Subsequently, Cook and Ada (1963) found that tissues of the chicken embryo that were derived from the ectodermal layer had sialidase activity, while other tissues had very low levels. Carubelli *et al.* (1962) determined sialidase activity in different organs of the rat. The highest levels of sialidase activity were found in lactating mammary gland, followed by brain and liver, with reduced amounts of activity in the kidney and small intestine. Only traces of sialidase activity were observed in the spleen and testes. The substrate used was sialyllactose, and the enzyme preparation was the supernatant fraction of the tissue homogenate obtained after centrifugation at $105,000g$ for 1 hr. These early papers, describing the existence of sialidase in mammalian cells served as the basis for following work.

Several papers have been published describing the presence of sialidase in brain. Morgan and Laurell (1963) measured sialidase activity with endogenous substrate in guinea pig, bovine, and human brains and found that sialic acid was released. Svennerholm (1967) observed sialidase activity in a homogenate from a human child's brain. Kelly and Grieff (1965) found sialidase in brains of mice, while Zvetkova (1965) found activity in kidneys, liver, and brain, with lesser amounts in the spleen, lung, and testes. More recently, research on the distribution of sialidase in different regions of the brain has shown sialidase activity to be concentrated in the gray matter. Gielen and Harprecht (1969) found the highest sialidase activity in the gray matter and the cerebellum of bovine brains. In 1970, Roukema and Heijlman found that both endogenous and exogenous sialidase activity were much greater in gray than in white matter of bovine brains. Öhman (1971a) studied the distribution of sialidase active toward human brain ganglioside substrate and found that it was almost entirely located in the gray matter, in agreement with the earlier results with bovine brain.

Other organs which have been reported to have sialidase activity are rat liver (Mahadevan *et al.*, 1967), rat and rabbit kidneys (Mahadevan *et al.*, 1967; Kirschbaum and Bosmann, 1973; Kuratowska and Kubicka, 1967), ram testis (Roston *et al.*, 1966), human intestinal mucosa (Ghosh *et al.*, 1968), pig corpus luteum (Unbehaun, 1970), sheep urinary tract transitional epithelium (Candiotti *et al.*, 1972), rat heart muscle (Tallman and Brady, 1973), and rat eyes (Tulsiani *et al.*, 1973). Sialidase activity has been found in saliva (Perlitsh and Glickman, 1966; Menguy *et al.*, 1970; Fukui *et al.*, 1973). The question of whether the sialidase activity found in saliva is due to the presence of oral bacteria or is a true salivary enzyme was raised by Menguy *et al.* (1970). Fukui *et al.* (1973) found a salivary sialidase with properties different from the bacterial enzyme. Erythrocytes, platelets (Gielen *et al.*, 1969; Bosmann, 1972), and leukocytes (Gielen *et al.*, 1970; Yeh *et al.*, 1971) also have sialidase activity.

The preceding survey of mammalian organs which have sialidase activity indicates that it is broadly distributed and may be an integral component of all organs containing sialyl compounds.

Sialidase has been reported to be a soluble enzyme, as well as to be associated with particulate cell fractions. The weight of evidence shows it mainly to be a cell-surface component. It is associated with its endogenous substrate in the plasma membrane of the cell with, nevertheless, some sialidase activity also present in lysosomes. The "soluble" sialidase may represent small membrane fragments.

B. Subcellular Distribution of Mammalian Sialidases

Morgan and Laurell (1963) found that brain sialidase had an acid pH optimum, which suggested that it might be a lysosomal constituent. In an early study Jibril and McCay (1965) measured the activities of several acid hydrolases in vitamin-E-deficient rabbits. They observed that sialidase active toward sialyllactose was enriched in the liver "lysosomal" fraction obtained by the procedure of DeDuve *et al.* (1955). Sandhoff and Jatzkewitz (1967) found sialidase in a crude "mitochondrial" fraction of rat liver, but the fractionation procedure was gross, and conclusions about its subcellular location are dubious. Taha and Carubelli (1967), Horvat and Touster (1968), and Bernacki and Bosmann (1973), studying rat livers, and Tulsiani and Carubelli (1972), studying chicken livers, also found sialidase in what they called the lysosomal fractions. Tulsiani and Carubelli (1970) found both a "lysosomal" and a "soluble" sialidase in rat liver. None of these workers have characterized their subfractions thoroughly, nor did they separate the plasma membranes from the other crude fractions in order to determine whether the enzyme was truly lysosomal, whether it was bimodally distributed in the lysosomes and plasma membranes, or whether it was only in the plasma membrane fraction. Schengrund et al. (1972) found that sialidase active toward added ganglioside substrate was enriched in the plasma membrane of rat liver. They isolated the plasma membranes using the procedure developed by Touster et al. (1970), and in a second preparation, isolated lysosomes using the procedure of Leighton et al. (1968). Their results indicated that sialidase active toward gangliosides is enriched in the plasma membrane. Visser and Emmelot (1973) isolated plasma membranes and lysosomes from rat liver. Based on differences in K_m values for sialyllactose and activity at different pH's, they concluded that plasma membranes and lysosomes contain different sialidases.

The intracellular distribution of sialidase in brain has also been studied in detail. In 1968, Preti et al. found sialidase activity in a myelin-containing fraction and in a fraction containing nerve terminals. Schengrund and Rosenberg (1970) found a sialidase active toward ganglioside substrate concentrated in the synaptosomal fraction obtained from bovine brain. Within the synaptosomal fraction, the sialidase activity was localized in the synaptosomal membrane. Shortly thereafter, Öhman (1971b) found that sialidase active toward ganglioside substrate was either bound to the limiting structure of the nerve endings, localized in a "low-density lysosome fraction," or in lysosomes trapped within the synaptosomes isolated from human brains. In 1972, Tettamanti et al., studying rat brains, corroborated our finding (Schengrund and Rosenberg, 1970) that sialidase active toward ganglioside substrate is localized in the synaptosomal plasma membrane. This finding is reiterated by Tettamanti et al. (1973) in studies with rabbit cortex.

In studies with other systems, Glick et al. (1971) found sialidase, active toward fetuin, to be enriched in the lysosomal fraction of L cells; Tulsiani and Carubelli (1971) found sialidase, active toward sialyllactose, not only enriched in "lysomes" isolated from rat mammary glands, but also in the soluble fraction; Gielen et al. (1973) found a membrane-bound sialidase in leukocytes that was very active toward glycoproteins in the leukocyte homogenate; and Bosmann (1974) found sialidase activity associated with the plasma membranes of human erythrocytes. Recently, Kishore et al. (1975) have reported the presence of a sialidase in Golgi isolated from rat liver.

Whether there is one sialidase active toward the various substrates or several sialidases specific for certain substrates and localized in certain organelles has not been shown. It is not inconceivable that the same sialidase may be found in the plasma membrane and in lysosomes since part of the plasma membrane is taken up into lysosomes in the natural sequence of events during turnover.

C. Purification of Mammalian Sialidases

To date, little about the purification of sialidase from mammalian cells has been published. The membrane-bound sialidase must be "solubilized" before any of the usual purification procedures can be applied. The "solubilized" sialidase that has been purified probably is still associated with a small membrane fragment. The "soluble" enzyme can be readily concentrated by standard procedures. To determine the mechanism of action of the enzyme and its physical characteristics, it may eventually be necessary somehow to obtain sialidase in truly purified form.

Kuratowska and Kubicka (1967) partially purified sialidase from rabbit kidney. Their starting material was the 105,000g supernatant fraction. The sialidase precipitated at 0.5–0.75 ammonium sulfate saturation, and after dialysis and passage through a Sephadex G-75 column, the specific activity of the enzyme had increased about 4000-fold. Leibovitz and Gatt (1968) extracted calves' brains with cold acetone and found sialidase, active toward ganglioside substrate, in the acetone powder. They extracted the powder twice with sodium cholate and once with Triton X-100 to obtain in the latter extraction a supernatant fraction in which sialidase specific activity had increased fivefold. Tettamanti and Zambotti (1968) purified sialidase, active toward ganglioside substrates, from pig brain. After homogenization of the pig brains in isotonic KCl, the sialidase did not sediment when centrifuged at 15,000g for 30 min. Addition of ammonium sulfate caused precipitation of the sialidase activity at 35–55% saturation. Further purification steps included: dialysis of the ammonium sulfate precipitate to remove the ammonium sulfate; addition of solid $CaCl_2$ to a concentration of 0.5%, and heating at 52°C for 10 min followed by cooling and centrifugation; chromatography of the supernatant on hydroxylapatite; and finally, chromatography on Sephadex G-200 of the fractions containing sialidase activity. The sialidase thus obtained was purified approximately 600-fold. Tuppy and Palese (1968) used pig kidney as a source of sialidase. They started with a particulate fraction sedimented between 1500 and 22,000g for 1 hr. About 60% of the sialidase activity was solubilized from the particulate fraction by treatment with n-butanol. The solubilized sialidase was further purified by ammonium sulfate fractionation. The 18–40% ammonium sulfate precipitate was dialyzed and then passed through a bio-gel P-300 column. The fractions with sialidase activity were chromatographed on sulfoethyl cellulose. The sialidase recovered after the last chromatographic procedure was more than 1000 times as pure as the sialidase in the initial particulate fraction. In more recent work, Tallman and Brady (1973) purified, from rat heart muscle, sialidase active toward ganglioside substrates including the normally resistant substrate, G_{M2}. Briefly, (1) the muscle was homogenized, followed by low-speed centrifugation (1000g for 10 min) to remove a heavy particulate fraction; (2) the supernatant was sonicated and centrifuged at 34,000g for 40 min; (3) the supernatant solution, containing most of the sialidase activity, was adjusted to pH 3.7 and stirred in the cold (4°C) for 1 hr and then centrifuged at 10,000g for 10 min; (4) the pellet was taken up in phosphate buffer at pH 6.3 and applied to a Sephadex G-150 column; (5) the active fractions were combined and put through a carboxymethyl Sephadex G-50 column; (6) there was isoelectric focusing in ampholine over the pH range 7–10, after which the enzyme activity was found to be concentrated in the gradient between pH 7.0 and 7.3. These fractions were freed of ampholine and substrate by passage through a Sephadex G-150 column. This procedure allowed recovery of about 15% of the sialidase activity toward disialoganglioside, G_{D1a}, substrate. The enzyme had been purified about 3400-fold. About 3% of the sialidase activity toward Tay–Sachs ganglioside was recovered, purified about 660-fold. This seems to be the most purified mammalian, "solubilized," sialidase preparation to date.

D. Assay of Mammalian Sialidases

The reliability of values reported for sialidase activity often is quite seriously undermined by failure to recognize chromogenic artifacts in the assay. The assay procedure described below is the one routinely used by us to determine sialidase activity. This procedure minimizes errors due to (1) nonenzymatic release of sialic acid, (2) release of endogenous sialic acid when measuring exogenous sialidase activity, and (3) chromogens which interfere in the thiobarbituric acid assay for free sialic acid. Source of the enzyme and substrate used may necessitate a change in pH, ionic strength, or reaction time.

To determine endogenous activity, samples containing 1 to 2 mg of protein are incubated at 37°C in 1 ml of buffer (for nerve-ending membrane preparations, we use 0.01 M sodium acetate buffer, pH 3.9) for a time period which has previously been found to give an initial reaction rate. A control sample is kept at 4°C to account for nonenzymatically released sialic acid. The reaction is terminated by cooling the samples and adjusting the pH to 7. The samples are passed through a column of Dowex 1-X10 ion exchange resin (Horvat and Touster, 1968) in order to remove substances which might give interfering chromogens in the Warren (1959) thiobarbituric acid assay for free sialic acid.

To measure sialidase activity toward added substrate, enzyme which has been incubated at 37°C to allow it to deplete available endogenous substrate is assayed. From 1 to 2 mg of the preincubated protein is incubated at 37°C with added substrate in 1 ml of buffer. Two controls are run to account for any nonenzymatic release of sialic acid. One control consists of enzyme in buffer, the other of substrate and buffer. The time is chosen to give an initial reaction rate. Again, the reaction is terminated by cooling and adjustment of the pH to 7, followed by passage through the ion exchange column before determination of free sialic acid by the thiobarbituric acid procedure. If substrate specifically labeled (cf. Chapter 1) in the sialic acid residue is available, the enzymatically released labeled sialic acid can be determined by liquid scintillation spectrometry (van Lenten and Ashwell, 1971; Tallman and Brady, 1973). With appropriate controls, this eliminates interference from other chromogenic substances or ions which might interfere with the Warren reaction.

E. Physical Properties of Mammalian Sialidases

Most of the studies on the properties of sialidase have been done with particulate cellular fractions which have sialidase activity. The enzyme environment is influenced by the membrane components and the results obtained for the particulate enzyme may be quite different from those obtained if the enzyme could be studied free from membrane components.

1. pH. Many workers have studied the pH optimum of mammalian sialidases located in different cell fractions and acting upon different substrates. Table III is a compilation of the data obtained. In all the results obtained, the pH optimum for sialidase appears to be acidic, usually between pH 4 and 5. There are a few soluble preparations which have reported pH optima greater than 5, which may possibly indicate contaminating microbial sialidase. At higher ionic strengths, the pH optimum for microbial sialidases can become more acidic (Barton *et al.*, 1975). This possibility should be carefully investigated for soluble preparations.

2. Substrate Specificity. The particulate (membrane-bound) sialidase has available to it endogenous sialic-acid-containing substrates upon which it acts before it acts on added substrate. Morgan and Laurell (1963) characterized the product they obtained after allowing sialidase present in brain homogenates to act on endogenous substrates and identified it as N-acetylneuraminic acid. After the early work of Morgan and Laurell (1963), several measurements of sialidase activity toward added sialic-acid-containing compounds were made. Care should be exercised in selecting an exogenous substrate. Sialyllactose is frequently the substrate tested. The activities obtained with this substrate may be misleading, i.e., brain sialidase exhibits reduced activity toward sialyllactose, while gangliosides (G_{D1a}, G_{T1}) are a good substrate for the brain enzyme. Leibovitz and Gatt (1968) found that sialidase present in the acetone powder of gray matter was active toward added di- and trisialogangliosides and hematoside but not toward G_{M1}, G_{M2}, sialyllactose, or sialoglycoproteins. Several workers have determined an apparent K_m value of 10^{-3} M for membrane-bound sialidase with sialyllactose as the substrate (Tuppy and Palese, 1968; Unbehaun, 1970; Schengrund and Rosenberg, 1970; Tulsiani and Carubelli, 1970; Visser and Emmelot, 1973). Tulsiani and Carubelli (1970) found a K_m with sialyllactose of 6×10^{-4} M for the "lysosomal" sialidase obtained from rat mammary glands. Visser and Emmelot (1973) found a similar K_m for plasma membrane-bound sialidase from rat liver. All of these values are in the range of 10^{-3} M. For different ganglioside substrates, however, the K_m values obtained for sialidase from human brain in an acetone powder were $1-3 \times 10^{-5}$ M for G_{T1}, G_{D1a}, G_{M3-NAN}, and G_{M3-NGN}, and 1×10^{-4} for G_{D1b}, which was a poorer substrate than these other gangliosides for this enzyme (Öhman et al., 1970). The sialidase appeared to act on gangliosides below the critical micelle concentration. The latter workers also observed that the sialidase was active toward sialyllactose, human transferrin, and ovine submaxillary mucin. For the general class structures for these substrates, see Figures 8 and 9 of Chapter 4. For an acetone powder preparation of bovine brain sialidase, Schengrund and Rosenberg (1970) obtained $K_m = 8 \times 10^{-5}$ M with mixed di- and trisialogangliosides as the substrate. Yeh et al. (1971), in studies with sialidase in leukocytes, found that sialyllactose sulfate and sialoglycopeptides (from a pronase digest of ovine submaxillary glycoprotein) were the best substrates, with sialyllactose or intact ovine submaxillary glycoprotein being acted upon at about one-third and one-sixth the rate, respectively, of sialyllactose sulfate. Bovine brain gangliosides were poor substrates, and the enzyme was inactive toward fetuin. Tallman and Brady (1973) determined K_m values for sialidase purified from heart, which not only was active toward G_{D1a} and G_{D1b} but also toward Tay–Sachs ganglioside, G_{M2}. For G_{D1a} and G_{D1b}, they obtained values of $K_m = 10^{-4}$ M; for G_{T1}, $K_m = 9 \times 10^{-6}$ M; and for G_{M2}, $K_m = 4.3 \times 10^{-5}$ M. They observed that the purified sialidase was active toward added fetuin but showed no activity with sialyllactose as the substrate. The ability of sialidase to act upon a specific added substrate may depend upon several factors such as (1) the location and linkage of the

TABLE III. pH Optimum and Sources for Some Mammalian Sialidases

Source	Fraction	Substrate	pH optimum	Reference
Brain				
Guinea pig, bovine and human brains	Whole brain homogenate	Endogenous sialyl-containing compounds	3.5–4.0	Morgan and Laurell (1963)
Calf brain	Gray matter acetone powder	Di- and trisialo-gangliosides and hematoside	4.4	Leibovitz and Gatt (1968)
Pig brain	Purified (~600-fold)	Gangliosides	4.9	Tettamanti and Zambotti (1968)
Pig brain		Sialyllactose	4.7	Tettamanti and Zambotti (1968)
Pig brain		Ovine submaxillary glycoprotein	4.4	Tettamanti and Zambotti (1968)
Rabbit brain	105,000g for 1 hr particulate	Disialoganglioside, G_{D1a}	4.0	Tettamanti et al., (1970)
Bovine brain	Lysosomal-mitochondrial	Bovine brain gangliosides	4.3	Gielen and Harprecht (1969)
Human brain	Gray matter acetone powder	Endogenous and exogenous gangliosides	4.4	Öhman et al. (1970)
Bovine brain	Synaptic membranes	Endogenous and exogenous gangliosides	4.3	Schengrund and Rosenberg (1970)
Liver				
Rat Liver	Lysosomes	Sialyllactose	4.4	Mahadevan et al. (1967)
		Glycoproteins	4.0	Mahadevan et al. (1967)
Rat liver	Lysosomes	Sialyllactose	4.4	Tulsiani and Carubelli (1970)
	Soluble	Sialyllactose	5.8	
Chicken liver	Lysosomal	Sialyllactose	3.8	Tulsiani and Carubelli (1972)
	Soluble	Sialyllactose	4.4	

Source	Fraction	Substrate	pH	Reference
Rat liver	Plasma membranes	Gangliosides	4.2	Schengrund et al. (1972)
Rat liver	Plasma membranes	Bovine brain gangliosides	4.2	Visser and Emmelot (1973)
Rat liver	Golgi	Sialyllactose	3.9	Visser and Emmelot (1973)
Rat liver	Lysosomes	Sialyltrisaccharides	4.2	Kishore et al. (1975)
Kidney		Bovine brain gangliosides	4.2	Visser and Emmelot (1973)
Rabbit kidney	Soluble	Seromucoid or sialyllactose	5.5	Kuratowska and Kubicka (1967)
Rat kidney	Lysosomes	Sialyllactose	4.4	Mahadevan et al. (1967)
		Glycoproteins	4.0	
Pig kidney	Lysosomal-mitochondrial fraction	Sialyllactose	4.4	Unbehaun (1970)
Other				
Rat mammary glands	Soluble	Sialyllactose	5.8	Tulsiani and Carubelli (1971)
Leukocytes	Lysosomal	Sialyllactose	4.4	
	Particulate	Sialyllactose	4.0	Yeh et al. (1971)
Transitional epithelium of sheep urinary tract	7710g supernatant	Sialyllactose	4.4	Candiotti et al. (1972)
Rat eyes	Soluble	Sialyllactose	5.8	Tulsiani et al. (1973)
	Particulate lens	Sialyllactose	4.2	Tulsiani et al. (1973)
Rat heart muscle	Purified 3400-fold	Ganglioside, G_{D1a}	5.0	Tallman and Brady (1973)
Rat heart muscle	Purified 600-fold	Ganglioside, G_{M2}	5.0	Tallman and Brady (1973)
Human erythrocyte	Plasma membrane; purified 470-fold	Fetuin and 5-acetamido 3,5-dideoxy-L-arabino-2-hep-tulosonic acid	4.2	Bosmann (1974)

sialic acid within the substrate [G_{D1a} is a better substrate then G_{D1b} for human brain sialidase (Öhman et al., 1970)], (2) in particulate enzyme preparations, the ability of the substrate to fit into the membrane and become available to the enzyme may be important, and (3) the degree of purification of the enzyme may affect the availability of substrate to enzyme. In order to approach *in vivo* conditions, the added substrate presumably should be similar to the sialyl compounds normally present in that tissue, i.e., sialidase from brain might best be tested on brain ganglioside substrates, while that from liver, on hematoside (G_{M3}). See Chapter 1 for chemical structures.

Recently, investigators have studied the endogenous sialidase activity of mammalian sialidase preparations. Gangliosides (G_{M3}, G_{D1a}, G_{T1}) appear to be readily accessible endogenous substrates, and after a longer reaction time, sialoglycoproteins are acted upon. In 1970, Öhman et al. reported the release of sialic acid when they allowed sialidase present in their acetone powder preparation of gray matter from human brain to act on endogenous substrate. They found that the sialic acid was released from intrinsic ganglioside substrate. Schengrund and Rosenberg (1970) also reported release of endogenous sialic acid from subcellular fractions obtained from bovine brain gray matter that had sialidase activity. After allowing the enzyme to act on endogenous substrate, they found that monosialoganglioside (G_{M1}) and a trace of disialoganglioside remained. The sialidase did not act on added substrate until it had first depleted some of the endogenous substrate. Preti et al. (1970) reported sialic acid release from endogenous substrate of a particulate fraction from rabbit brain. Lombardo et al. (1970) showed that the endogenous sialic acid released from a particulate fraction isolated from rabbit brain was from gangliosides. In 1972, Schengrund et al. found that sialidase present in a plasma membrane fraction obtained from rat liver was active toward endogenous hematoside (G_{M3}) substrate. Heijlman and Roukema (1972) reported release of sialic acid from endogenous brain gangliosides, sialoglycoproteins, and sialoglycopeptides. They observed that the release of sialic acid from gangliosides was much faster initially then sialic acid release from sialoglycopeptides. After lengthy incubation (48 hr), 40% of the total sialic acid was released from both substrates (pH 4.0 and 37°C). Irwin et al. (1973) reported enzymatic release of sialic acid from endogenous ganglioside and sialoglycoprotein substrates present in a particulate fraction from rat brains. Preti et al. (1973) have also been able to show sialidase activity toward endogenous sialoglycoproteins in a particulate fraction obtained from calf brain.

The findings on endogenous sialidase activity show that the membrane-bound enzyme is closely associated with endogenous sialic-acid-containing substrate which it must partially deplete (Schengrund and Rosenberg, 1970) before it can act on added exogenous substrates. Therefore, when making kinetic measurements of sialidase activity toward added exogenous substrate, the enzyme should be preincubated to allow for release of endogenous substrate before the exogenous substrate is added. In this manner, more meaningful initial rates for the exogenously directed reactions can be obtained.

3. Inhibitors and Activators of Mammalian Sialidase Activity. Several reports about the effects of ions on sialidase activity have appeared in the literature. The observed effects vary with the source and purity of the enzyme and with the substrate used, which may in part reflect different environments for the sialidase. Kuratowska and Kubicka (1967) observed that Ca^{2+} over the concentration range of 0.45–450 μM had no effect on rabbit kidney sialidase, and Gielen and Harprecht (1969) found no effect of Ca^{2+} on bovine brain sialidase activity. Tulsiani and Carubelli (1970), examining the effects of ions on sialidase activity from rat liver, found that the divalent cations Cu^{2+}, Hg^{2+}, and Zn^{2+} inhibited the soluble enzyme while the monovalent cations Li^+, Na^+, and K^+ inhibited the lysosomal enzyme with sialyllactose as the substrate. They found similar ion effects on sialidase from rat mammary glands (1970). Öhman et al. (1970) found 0.05 M Cd^{2+} and Zn^{2+} inhibited sialidase from human brain that was active toward ganglioside substrate. Of the several anions that they tested (SO_4^{2-}, Cl^-, Br^-, I^-, PO_4^{3-}, NO_3^-, and acetate) only SO_4^{2-} was inhibitory. Yeh et al. (1971) found that Mg^{2+}, Zn^{2+}, Cu^{2+}, and Hg^{2+} over the concentration range of 10^{-6}–10^{-3} M had no effect on the activity of sialidase from leukocytes toward sialyllactose. Hg^{2+} at 10^{-3} M gave 25% inhibition and Cl^- at 10^{-1} M inhibited the sialidase activity 50%. Tulsiani and Carubelli (1972) found that 1 mM Cu^{2+} or Hg^{2+} inhibited the soluble sialidase obtained from chicken livers, while the same ions had negligible effect on the activity of the particulate enzyme toward sialyllactose. Mg^{2+} stimulated the particulate sialidase, and Ca^{2+} and Zn^{2+} had no effect on the activity of either the soluble or particulate enzyme. Visser and Emmelot (1973) found that 1 mM Cu^{2+} inhibited the sialidase activity of liver plasma membrane toward ganglioside substrate. The lysosomal sialidase activity was differently inhibited. Cu^{2+} inhibited the release of endogenous sialic acid from the plasma membrane at least 50%. A comparatively high Li^+ concentration (0.2 M) was found to inhibit moderately the membrane sialidase at pH 4.4. In work with purified sialidase from rat heart muscle acting on labeled substrates, Tallman and Brady (1973) found that Cu^{2+} and Fe^{3+} also inhibited this enzyme. Preti et al. (1974) found that Na^+ and Li^+ had no effect on sialidase activity in a crude preparation of membrane-bound sialidase obtained from calf brains. NH_4^+ started inhibiting at 10^{-2} M, Hg^{2+} from 10^{-6} M, Cu^{2+} from 10^{-5} M, and Ca^{2+} from 10^{-3} M. G_{D1a} was used as substrate.

In our studies (Schengrund and Rosenberg, 1973; Schengrund and Nelson, 1975) on the effect of ions on the sialidase activity in the synaptic membrane from bovine brains, we have observed that Na^+, K^+, Ca^{2+}, Zn^{2+}, Mg^{2+}, and Mn^{2+} had no effect on endogenous sialidase activity over the concentration range of 10^{-5}–10^{-3} M. The aforementioned divalent cations were inhibitory for sialidase activity toward added ganglioside substrate at 10^{-3} M, but at 10^{-4} M, optimum activity toward added ganglioside substrate was observed. 10^{-5} M K^+ was also suboptimum for exogenous activity. This may indicate that a certain ionic strength is necessary for exogenously directed nerve ending membrane sialidase activity. The higher (10^{-3} M) concentrations of divalent cations were found to cause increased aggregation of the ganglioside substrate, and this may in part account for the observed inhibitory effect. Increased monovalent cation concentration did not inhibit the exogenously directed sialidase activity nor did it have any observed effect on the aggregation of the ganglioside substrate.

Several detergents have been tested for their effect on sialidase activity. Generally, detergents at certain concentrations have been found to activate the particulate enzyme, perhaps by building into the particle and giving rise to a geometry for the enzyme within the membrane which is best for interaction of the sialidase with exogenous substrate. Horvat and Touster (1968) found a slight increase in rat liver sialidase activity toward sialyllactose in the presence of Triton X-100. Tettamanti *et al.* (1970) found that 0.3% w/v, Triton X-100 activated rabbit brain particulate sialidase, to give a 300% increase in specific activity with gangliosides as the substrate. Öhman *et al.* (1970) found that Triton X-100 over a narrow concentration range stimulated human brain sialidase activity toward ganglioside substrate and that Triton CF-54, at a concentration of 0.5 mg/200 ml of assay mixture, activated the system optimally. Tulsiani and Carubelli (1970) found that Triton X-100 caused mild stimulation of the soluble sialidase from rat mammary glands and strongly inhibited the lysosomal sialidase activity toward sialyllactose. In contrast Yeh *et al.* (1971) found that low concentration (0.05%, v/v) of Triton X-100 inhibited the particulate sialidase of leukocytes and that sodium deoxycholate was somewhat inhibitory. Tallman and Brady (1973) found that Triton X-100, sodium taurocholate, and Cutscum were inhibitory for purified heart muscle sialidase activity toward ganglioside substrate. Tettamanti *et al.* (1974) found that sodium dodecylsulfate, cholate, deoxycholate, and Triton X-100 acted as non-competitive inhibitors of sialidase prepared from calf brain. The Triton X-100 at low concentrations was found to activate the enzyme. The detergents Triton QS-31 and Lubrol WX had different effects on sialidase activity. The Triton QS-31 altered the interaction between the enzyme and ganglioside substrate, allowing the enzyme to act on the micellar and submicellar forms of the substrate. The Lubrol WX acted still differently, increasing the amount of substrate needed before enzyme activity could be measured.

Other compounds tested for their ability to inhibit or activate sialidase include iodoacetamide (10^{-3} M). N-ethylmaleimide (10^{-2} M) and Mersalyl (10^{-4} M). On partially purified sialidase from pig kidney, these had no effect (Tuppy and Palese, 1968); α-chymotrypsin with sialidase from bovine brain caused an increase in activity (Gielen and Harprecht, 1969); EDTA, p-hydroxymercuribenzoate (formerly was believed to be p-chloromercuribenzoate), tested on sialidase in rat liver and kidney, had no effect (Mahadevan et al., 1967); the bacterial inhibitors 2-deoxy-2,3-dehydroneuraminic acid and p-nitrophenyloxamic acid tested on purified sialidase from rat heart muscle had no effect (Tallman and Brady, 1973) while 1-(4-methoxyphenoxymethyl)-3,4-dehydroisoquinoline and p-hydroxymercuribenzoate were inhibitory with ganglioside G_{D1a} substrate. The effect of specific inhibitors on purified sialidase may give useful information about the active site, an unknown entity at this time.

4. Stability. Mammalian sialidase preparations have been found to be quite stable. Leibovitz and Gatt (1968) found that sialidase in an acetone powder preparation from calf brain was fully active for at least 2 months when stored at $-20°$C, and after 9 months the preparation retained approximately 85% of the initial sialidase activity. At $0°$C, the preparation lost 15–20% of its activity within 6 hr. Öhman et al. (1970) found that Triton X-100 extracts of human brain could be stored for over 1 year at $-29°$C with retention of more than 90% of its initial sialidase activity. We (Schengrund and Rosenberg, unpublished results) have found that sialidase present in an acetone powder preparation of steer brain which had been preincubated at $37°$C for 1 hr to hydrolyze endogenous substrate is stable to moderate heating ($55°$C for 10 min) with retention of 70% of its initial sialidase activity toward ganglioside substrate. After heating at $65°$C for 10 min, however, essentially all exogenous activity was lost. Yeh et al. (1971) found that sialidase in leukocytes could be refrigerated or frozen for 24 hr with small loss of activity. Preincubation of the leukocytes at $37°$C and pH 4 for 1 hr resulted in an apparent 40% decrease in sialidase activity (perhaps due to loss of endogenous substrate). Repeated freezing and thawing caused a large decrease. Tulsiani and Carubelli (1972) found that soluble sialidase from chicken livers lost 15–20% of its activity toward sialyllactose when stored for 3 days at 0°C. Lysosomal sialidase from the same tissue lost 20–25% of its initial activity upon storage at 0° for 24 hr. They also found that preincubation of the sialidase sample at 37°C, to allow it to act on endogenous substrate, gave a loss of activity for the soluble sialidase while activity in the particulate fraction increased. Visser and Emmelot (1973) found that incubation of sialidase from rat liver organelles at 37°C inhibited lysosomal sialidase activity toward sialyllactose but had no effect on the activity of the plasma membrane-bound sialidase activity toward sialyllactose. However, the sialidase activity of the plasma membrane toward ganglioside substrate was reduced by 16%. The differences in stability are sometimes cited as evidence for two different enzymes, but this may as well reflect the environment of the sialidase in these crude preparations. At this point, it cannot be told whether there are different sialidases specific for given substrates.

F. Developmental Studies of Mammalian Sialidases

Sialidase in chickens has been found to be fully active at birth, whereas in mammalian organs, sialidase activity is found to increase over a period of time after birth until adult levels are reached. This probably reflects the relative stages of development at birth for different species.

The first studies of the appearance of sialidase during development were done by Cook and Ada (1963), who compared 11- and 18-day-old chicken embryos and also studied 7-day-old chickens. They observed that the nonneural embryonic tissues derived from the ectodermal layer showed a ten- to twentyfold increase in the specific activity of sialidase towards sialyllactose, while neural tissue showed a two- to threefold increase. These differences may reflect a difference in exogenous substrate specificity of the sialidase from different tissues. However, in these studies, there was no preincubation to allow for hydrolysis of endogenous substrate. Sialidase activity decreased after hatching. Low activity was found in other, nonectodermal, tissue at all stages of development. In more recent studies with acetone powder preparations of chicken brains and eyes, we (Schengrund and Rosenberg, 1971) have found endogenous sialidase activity in brain at all stages of development from the 5-day-old embryo to the adult chicken. However, activity toward added, exogenous, ganglioside substrate was first detected in the brains of 13-day-old embryos. Specific activity increased until hatching and then, by 2 days after birth, had decreased about 15%. There was little change in sialidase activity from 2-day-old to adult chickens. We found no sialidase activity, at any stage of development, in acetone preparations of chicken eyes. Tulsiani and Carubelli (1972) followed appearance of sialidase activity toward added sialyllactose in chicken livers. They found both a soluble and a particulate sialidase. The soluble sialidase was active during the last prenatal week, and the specific activity gradually increased during the first week after hatching. A gradual decrease in specific activity occurred from the second to the sixth week of life. For the particulate sialidase essentially no change in specific activity was observed from a week before hatching through six weeks after hatching.

Several studies have reported on sialidase activity in developing rat brains. Carubelli (1968) found that sialidase, active toward sialyllactose and present in the 100,000g supernatant, increased in a linear fashion from the low level found during the first week of life and reached the specific activity found in the mature animal by the end of the third week of life. These observations were confirmed by Quarles and Brady (1970) and by Roukema *et al.* (1970), who also showed that the sialidase activity remained essentially constant for 150 days. Carubelli and Tulsiani (1971) found that soluble sialidase activity in rat liver is low in fetal liver, undergoes a sharp increase during 3 to 4 days prior to birth, and is essentially fully active at birth. In contrast, the particulate enzyme is fully active during the last week of prenatal development. In this work, they also studied the particulate sialidase in brain. In contrast to Carubelli's (1968) earlier finding for the soluble sialidase (which they repeated), they found that the particulate sialidase had full activity a week before birth.

Öhman and Svennerholm (1971) measured the appearance of sialidase activity toward ganglioside substrate in developing human brain. Activity was first detected in the fetal brain at 15 to 20 weeks and by term had increased to half of the adult level. During the first year of life sialidase activity may decrease slightly, and then it increases until it approaches the adult level at 5 years. The appearance of sialidase activity in the later stages of development may reflect the need for the enzyme to alter, or to help in degradation of, the more complex sialyl compounds synthesized, and these latter in turn may serve to induce sialidase activity.

G. Possible Biological Roles of Mammalian Sialidases

A broad range of possibilities are currently being explored to try to determine the biological effects of sialidase activity upon endogenous substrates. Some areas in which sialidase activity may play an important role are reproduction, alterations of the half-life of circulating sialo compounds (cf. Chapter 9), blood clot formation, interaction of hormones with their target cells, neurotransmission, cell–cell interactions, and cellular transformation (cf. Chapters 7 and 8).

1. Reproduction. Sialic acid has been found to play a role in the activity of hormones involved in reproduction. Sialic acid is found in human chorionic gonadotropin (HCG) and in follicle-stimulating hormone (FSH). Removal of sialic acid residues causes a loss of FSH and HCG activity (Brossmer and Walter, 1958; Sairam and Moudgal, 1971). Inhibitors of each hormone have been found, and each contains sialidase activity. The FSH inhibitor has been found in the urine of the bonnet monkey, *Macaca radiata* (Sairam and Moudgal, 1971) and HCG inhibitor, in bovine pituitary fractions (Reichert *et al.*, 1971). Yaginuma, 1972, has shown that sialidase treatment of HCG allows for its rapid removal from the circulation by the liver. This results in greatly reduced uptake by the ovary, which results in a very low biological activity for the treated HCG. These results parallel those found by van den Hamer *et al.* (1970) in studies with ceruloplasmin. They measured the clearance of ceruloplasmin from the circulation after treatment with sialidase and found that it was rapidly removed by the liver. This may also explain the inactivation of plasma erythropoietin by sialidase as reported by Lowy *et al.* in 1960. More information about circulating sialylcompounds is presented in Chapter 9.

The activity of several lytic enzymes, including sialidase, is necessary for fertilization. While sperm are in the seminal plasma, inhibitors of these enzymes are present in the decapacitation factor (DF). Sialidase was found to be a component of sperm acrosomes by Srivastava *et al.* (1970). They found sialidase activity and a sialidase-like factor which rendered bound sialic acid reactive in the thiobarbituric acid assay (Warren, 1959) but did not release it from the molecule and, upon purification, appeared to be converted to sialidase. An inhibitor of the sialidase-like factor was found in a partially purified DF preparation. The DF is lost during capacitation, which can be broadly defined to include the changes sperm undergo to enable them to penetrate the vitelline membrane of the egg. Gould *et al.* (1971) have proposed that the role of sialidase in fertilization is to alter the tertiary and quaternary structure of sialoglycoprotein present in the *zona pellucida* to cause the *zona pellucida* to be less penetrable by spermatozoa and serve to reduce or block polyspermy.

2. Blood Clot Formation. In 1962, Chandrasekhar *et al.* found that fibrinogen lost about 18% of its sialyl residues during clot formation and that treatment of fibrinogen with microbial sialidase increased the ability of fibrinogen to clot. The sialic acid released during clotting was not free, as it could be assayed by the thiobarbituric acid assay only after acid hydrolysis. Laki and Chandrasekhar (1963) reported that clots formed by sialidase-treated fibrinogen were more readily solubilized than those formed from normal fibrinogen. Recently, Vermylen et al. (1974) have shown that fibrinogen is not involved in the aggregation of platelets, that the carbohydrate-containing compound factor VIII is. Sialidase treatment of factor VIII gave rise to a compound with strong platelet-aggregating ability.

3. Interaction of Hormones with Target Cells. Changes in sialic acid content in the plasma membrane of the cell have been found in some cases to block the ability of the cell to respond to some hormones. DeMoor et al. (1970) reported that only two factors showed a definite relationship to age of discovery of diabetes mellitus. Blood glucose levels were higher in men with early onset of diabetes while serum sialic acid levels were lower (which might indicate the presence of an active sialidase). In 1971b, Cuatrecasas and Illiano found that brief treatment of fat cells with microbial sialidase caused an increase in glucose transport, mimicking the physiological effects of insulin, while treatment with a greater dosage of microbial sialidase caused a reduction in the rate of glucose transport and abolished the effect of insulin on glucose transport and lipolysis. The sialidase treatment had no effect on the interaction of membrane receptors with insulin, but only on the transmission of effects of insulin binding on glucose transport across the membrane and on lipolysis.

Haksar et al. (1973) treated rat adrenal cells with sialidase and observed a decrease in response to adrenocorticotropic hormone (ACTH) stimulation as measured by corticosterone production. No effect was observed on the cell's response to cyclic AMP or dibutyryl cyclic AMP, indicating that the observed effect of sialidase treatment must be due to impairment of some step prior to cyclic AMP formation. They proposed two possibilities: (1) the sialic acid residues on the cell membrane provide affinity for the ACTH, thereby aiding in activation of the receptor, or (2) the sialic acid facilitates the "transmission" of the signal from the ACTH–receptor interaction to adenylcyclase. Similar studies with other hormones and isolated cell systems might indicate involvement of cell surface sialyl residues in transduction of other hormonal messages. Sialyltransferase and sialidase present in the cell membrane may serve to provide the environment and controls for transduction of the hormonal message.

4. Neurotransmission. Increasing evidence indicates that sialic acid is involved in neuronal transmission. Several workers (Heilbronn, 1962; Svensmark and Kristensen, 1962, 1963; Augustinsson and Ekedahl, 1962; Heilbronn and Cedergren, 1970; Brodbeck et al., 1973) have shown that cholinesterase contains sialic acid. Heilbronn and Cedergren (1970) studied the uptake of acetylcholine by brain slices. They observed that treatment of the brain slice with microbial sialidase reduced but did not abolish acetylcholine uptake. These results are incomplete, as the sialidase may have acted on many different sialic-acid-containing compounds present in the brain slice. We have shown (Schengrund and Rosenberg, 1970) that sialidase is present within the synapse and can act on endogenous lipid structure. Over a longer time period it can act on endogenous sialoglycoproteins (Heijlman and Roukema, 1972). In 1973, Brodbeck et al. observed that removal of sialyl residues from partially purified cholinesterase increased its specificity for acetylcholine. This may be a function of synaptic membrane-bound sialidase.

Based on our observations of the effect of cations and pH upon the activity of sialidase in the neuronal synaptic membrane, we have suggested the following series of events involving the action of sialidase upon synaptic transmission: (1) a local decrease in pH activates the synaptic membrane sialidases, which results in a partial depletion of presynaptic membrane ganglioside sialic acid. (2) The depletion of endogenous membrane substrate allows the enzyme to act on extramembrane substrates. (3) The influx of sodium ions followed by other cations serves to activate the sialidase. (4) Release of sialic acid residues reduces the screen of fixed negative charge in the synapse and permits diffusion of positively charged neurotransmitter molecules into the postsynaptic membrane. (5) Influx of sufficient calcium ions inactivates the extramembrane sialidase activity. Resialylation can reestablish the negative charge completing the cycle (Schengrund and Nelson, 1975).

Lipid-bound sialic acid has been implicated as part of the 5-hydroxytryptamine receptor complex in smooth muscle (Woolley and Gommi, 1966; Wesemann and Zilliken, 1967; Vaccari and Vertua, 1970; Vaccari, et al., 1971). Vaccari et al., proposed the following scheme for the interaction of serotonin with its receptor to give rise to contraction: (1) The drug interacts with the receptor, causing electrostatic changes in the membrane that allow cations to interact with gangliosides. (2) The gangliosides serve to carry Ca ions across the membrane into the cytoplasm, where the Ca ions are released, and contraction is activated. Their reasoning was based on the observation that sialidase treatment did not block the binding of serotonin, but did block muscular contraction. Treatment with lysergic acid diethylamide (LSD) blocked uptake of serotonin and contraction, but it did not block eledoisin receptors and, in the presence of eledoisin, contraction proceeded normally. This hypothesis for the function of sialyl groups is analogous to that proposed for transduction of hormone effects.

This brief overview of some of the investigations of changes in biological activity that occur upon removal of sialyl residues points to a need for further investigations on the action of membrane-bound sialidases on various endogenous substrates in order to determine whether the foregoing are valid control mechanisms.

H. Sialidase Activity in Cells in Tissue Culture

Sialic acid located on the cell surface has been implicated in cell–cell recognition (Hakomori, 1970; and Kemp, 1968) cf. Chapter 7, and sialic-acid-containing compounds have been shown to change in quantity upon cell transformation (Warren et al., 1973; Yogeeswaran et al., 1972; Brady et al., 1973) cf. Chapter 8. Glick et al. (1971) reported the presence of sialidase activity in L cells. These reports, plus our finding of sialidase in the plasma membrane of liver cells (Schengrund et al., 1972), led to our investigation of sialidase activity in normal and transformed fibroblasts in tissue culture. In initial studies, we (Schengrund et al., 1973) found sialidase activity toward endogenous substrate in both normal and transformed hamster embryo fibroblasts. However, sialidase activity toward added, exogenous, ganglioside substrate was observed only in the transformed fibroblasts. The exogenous sialidase activity was associated with the particulate fraction of the cell and showed a pH optimum of 4. Degree of confluency of the cultures at harvest, freezing and thawing of the cells, or harvesting the cells with trypsin had no effect on the sialidase activity. In subsequent work, we (Schengrund et al., 1974) observed that exogenously directed sialidase activity appeared to parallel the degree of oncogenicity of a series of herpes simplex virus-transformed hamster embryo fibroblasts. These results contribute to the growing speculation about the role of sialic acid in cell transformation, and they point to a possible function for sialidase in helping to maintain integrity of the cell surface.

Acknowledgment

This work was supported by National Institutes of Health, Grants NS08258 and CA14319.

VI. REFERENCES

Aalto, M., Rönnemaa, T., and Kulonen, E., 1974, Effect of neuraminidase on the synthesis of collagen and other proteins, *Biochim. Biophys. Acta* **342**:247–253.

Ada, G. L., and French, E. L., 1957, Stimulation of the production of the receptor destroying enzyme (RDE) of *V. cholerae* by neuraminic acid derivatives, *Austr. J. Sci.* **19**:227–228.

Ada, G. L., and French, E. L., 1959a, Stimulation of production of neuraminidase in *Vibrio cholerae* cultures by *N*-acetyl mannosamine, *J. Gen. Microbiol.* **21**:561.

Ada, G. L., and French, E. L., 1959b, Purification of bacterial neuraminidase (receptor-destroying enzyme), *Nature* **183**:1740–1741.

Ada, G. L., French, E. L., and Lind, P. E., 1961, Purification and properties of neuraminidase from *V. cholerae*, *J. Gen. Microbiol.* **24**:409–421.

Ada, G. L., and Lind, P. E., 1961, Neuraminidase in the chorioallantois of the chick embryo, *Nature* **190**:1171.

Adler, W. H., Osunkoya, B. O., Takiguchi, T., and Smith, R. T., 1972, The interactions of mitogens with lymphoid cells and the effect of neuraminidase on the cells' responsiveness of stimulation, *Cell. Immunol.* **3**:590–605.

Allen, J. M., and Cook, G. M. W., 1970, A study of the attachment phase of phagocytosis by murine macrophages, *Exp. Cell Res.* **59**:105–116.

Atherton, A., and Born, G. V. R., 1973, Effects of Neuraminidase and *N*-acetyl neuraminic acid on the adhesion of circulating granulocytes and platelets in venules, *J. Physiology* **234**:66P–67P.

Athineos, E., Thornton, M., and Winzler, R. J., 1962, Comparative antigenicity of native and "desialized" orosomucoid in rabbits, *Proc. Soc. Exp. Biol. Med.* **111**:353–356.

Augustinsson, K-B., and Ekedahl, G., 1962, The properties of neuraminidase-treated serum cholinesterase, *Biochim. Biophys. Acta* **56**:392–393.

Baars, A. J., Frankena, H., and Masurel, N., 1971, Antigenic relationship of influenza-virus neuraminidases from Asian, Hong Kong, and equi-2 strains, *Antonie van Leeuwenhoek* **37**:209–218.

Bach, M. K., and Brashler, J. R., 1973, On the nature of the presumed receptor for IgE on mast cells. I. The effect of sialidase and phospholipase C treatment on the capacity of rat peritoneal cells to participate in IgE-mediated, antigen-induced histamine release *in vitro*, *J. Immunol.* **110**:1599–1608.

Bachmayer, H., 1972, Effect of tryptophan modification on the activity of bacterial and viral neuraminidase, *FEBS Lett.* **23**:217–219.

Bachmayer, H., and Schmidt, G., 1972, Selective removal of neuraminidase from influenza A_2 viruses, *Med. Microbiol. Immunol.* **158**:91–94.

Bagshawe, K. D., and Currie, G. A., 1968, Immunogenicity of L 1210 murine leukemia cells after treatment with neuraminidase, *Nature* **218**:1254–1255.

Balke, E., and Drzeniek, R., 1969, Untersuchungen über die *Clostridium perfringens*—Neuraminidase, *Z. Naturforsch.* **24**:599–603.

Barth, R. F., and Singla, O., 1974, Alterations in the immunogenicity and antigenicity of mammalian erythrocytes following treatment with neuraminidase, *Proc. Soc. Exp. Biol. Med.* **145**:168–172.

Barton, N. W., and Rosenberg, A., 1973, Action of *Vibrio cholerae* neuraminidase (sialidase) upon the surface of intact cells and their isolated sialolipid components, *J. Biol. Chem.* **248**:7353–7358.

Barton, N. W., Lipovac, V., and Rosenberg, A., 1975, Effects of strong electrolyte upon the activity of *Clostridium perfringens* sialidase toward siallactose and sialoglycolipids, *J. Biol. Chem.* **250**:8462–8466.

Becht, H., Hammerling, U., and Rott, R., 1971, Undisturbed release of influenza virus in the presence of univalent antineuraminidase antibodies, *Virology* **46**:337–343.

Bekesi, J. G., St-Arneault, G., and Holland, J. F., 1971, Increase of leukemia L 1210 immunogenicity by *Vibrio cholerae* neuraminidase treatment, *Cancer Res.* **31**:2130–2132.

Bentwich, Z., Douglas, S. D., Skutelsky, E., and Kunkel, H. G., 1973, Sheep red cell binding to human lymphocytes treated with neuraminidase; enhancement of T cell binding and identification of a subpopulation of B Cells, *J. Exp. Med.* **137**:1532–1537.

Bernacki, R. J., and Bosmann, H. B., 1973, Rat-liver-sialidase activity utilizing a tritium-labeled sialic-acid derivative of glycoprotein substrates, *Eur. J. Biochem.* **34**:425–433.

Bondareff, W., and Sjostrand, J., 1969, Cytochemistry of synaptosomes, *Exp. Neurol.* **24**:450–458.

Boschman, T. A. C., and Jacobs, J., 1965, The influence of ethylene-diaminetetraacetate on various neuraminidase. *Biochem. Z.* **342**:532–541.

Bosmann, H. B., 1972, Platelet adhesiveness and aggregation. II. Surface sialic acid, glycoprotein: N-Acetylneuraminic acid transferase, and neuraminidase of human blood platelets, *Biochim. Biophys. Acta* **279**:456–474.

Brady, R. O., Fishman, P. H., and Mora, P. T., 1973, Membrane components and enzymes in virally transformed cells. *Fed. Proc.* **32**:102–108.

Brand, C. M., and Skehel, J. I., 1972, Crystalline antigen from the influenza virus envelope, *Nat. New Biol.* **238**:145–147.

Braunstein, G. D., Reichert, L. E., Jr., Van Hall, E. V., Vaitukaitis, J. L., and Ross, G. T., 1971, The effect of desialylation of the biologic and immunologic activity of human pituitary luteinizing hormone, *Biochem. Biophys. Res. Commun.* **42**:962–967.

Brodbeck, U., Gentinetta, R., and Lundin, S. J., 1973, Multiple forms of a cholinesterase from body muscle and possible role of sialic acid in cholinesterase reaction specificity, *Acta Chem. Scand.* **27**:561–572.

Brossmer, R., and Walter, K., 1958, Enzymatische abspaltung von lactaminšaure and inactivierung von choriongonadotropinpräparaten, *Klin. Wschi.* **36**:925.

Brossmer, R., and Holmquist, L., 1971, On the specificity of neuraminidase, *Hoppe-Seyl. Z.* **352**:1715–1719.

Brostrom, M., Bruening, G., and Bankowski, R. A., 1971, Comparison of neuraminidases of paramyxoviruses with immunologically dissimilar hemagglutinins, *Virology* **46**:856–865.

Brydak, L., Semkow, R., and Zgorzelska, K., 1971, Studies on the adaptation of freshly isolated strains of influenza virus, *Med. Dosw. Mikrobiol.* **23**:33–38.

Burnet, F. M., McCrea, J. F., and Stone, J. D., 1946, Modification of human red cells by virus action. I. The receptor gradient for virus action in human red cells, *Br. J. Exp. Path.* **27**:228–236.

Burnet, F. M., and Stone, J. D., 1947, The receptor-destroying enzyme of *V. cholerae*, *Austr. J. Exp. Biol. Med. Sci.* **25**:227–233.

Butterworth, P. J., 1966, Action of neuraminidase on human kidney alkaline phosphatase, *Nature* **209**:805–806.

Candiotti, A., Ibanez, N., and Monis, B., 1972, Sialidase of transitional epithelium of sheep urinary tract, *Experientia* **28**:541–542.

Carubelli, R., 1968, Changes in rat brain neuraminidase during development, *Nature* **219**:955–956.

Carubelli, R., Trucco, R. E., and Caputto, R., 1962, Neuraminidase activity in mammalian organs, *Biochim. Biophys. Acta* **60**:196–197.

Carubelli, R., and Tulsiani, D. R. P., 1971, Neuraminidase activity in brain and liver of rats during development, *Biochim. Biophys. Acta* **237**:78–87.

Cassidy, J. T., Jourdian, G. W., and Roseman, S., 1965, The sialic acids. VI. Purification and properties of sialidase from *Clostridium perfringens, J. Biol. Chem.* **240:**3501–3506.

Chakraverty, P., 1972, Antigenic relationships between the neuraminidases of influenza B virus, *Bull. World Health Org.* **46:**473–476.

Chandrasekhar, N., Warren, L., Osbahr, A. J., and Laki, K., 1962, Role of sialic acid in fibrinogen, *Biochim. Biophys. Acta* **63:**337–339.

Chen, C., Compans, R. W., and Choppin, P. W., 1971, Parainfluenza virus surface projections: Glycoproteins with haemagglutinin and neuraminidase activities, *J. Gen. Virol.* **11:**53–58.

Choi, S. I., Simone, J. V., and Journey, L. J., 1972, Neuraminidase-induced thrombocytopenia in rats, *Br. J. Haematol.* **22:**93–101.

Collins, M. F., Holland, K. D., and Sanchez, R., 1973, Regeneration of sialic acid-containing components of embryonic cell surfaces, *J. Exp. Zool.* **183:**217–224.

Colobert, L., and Fontanges, R., 1963, Activité neuraminidasique des myxovirus influenzae A et B et de myxovirus parainfluenzae I (virus Sendai) sur le mucopolysaccharide de la glande sous-maxillaire de boeuf, *Ann. Inst. Pasteur* **1963:**734–745.

Cook, B., and Ada, G. L., 1963, Neuraminidase in tissues of the chick embryo and chick, *Biochim. Biophys. Acta* **73:**454–461.

Cuatrecasas, P., and Illiano, G., 1971a, Purification of neuraminidases from *Vibrio cholerae, Clostridium perfringens,* and influenza virus by affinity chromatography, *Biochem. Biophys. Res. Commun.* **44:**178–184.

Cuatrecasas, P., and Illiano, G., 1971b, Membrane sialic acid and the mechanism of insulin action in adipose tissue cells, *J. Biol. Chem.* **246:**4938–4946.

Currie, G. A., and Bagshawe, K. D., 1968, The effect of neuraminidase on the immunogenicity of the landschütz ascites tumour: Site and mode of action, *Br. J. Cancer* **22:**588–594.

Currie, G. A., and Bagshawe, K. D., 1969, Tumour specific immunogenicity of methylcholanthrene-induced sarcoma cells after incubation in neuraminidase, *Br. J. Cancer* **23:**141–149.

Darrell, R. W., and Howe, C., 1964, The neuraminidase of parainfluenza virus (type 2), *Proc. Soc. Exp. Biol. Med.* **116:**1091–1094.

Dawson, P. S., and Patterson, D. S. P., 1967, Neuraminidase activity of a bovine strain of parainfluenza 3 virus, *Nature* **213:**185–186.

DeDuve, C., Pressman, B. C., Gianetto, R., Wattiaux, R., and Appelmans, F., 1955, Tissue fractionation studies. 6. Intracellular distribution pattern of enzymes in rat-liver tissue, *Biochem. J.* **60:**604–617.

DeMoor, P., Bouillon, R., and Van Mieghem, W., 1970, Transcortin activity as related to the age at discovery of diabetes mellitus, *Clin. Chim. Acta* **30:**627–633.

Den, H., Malinzak, D., and Rosenberg, A., 1975, Cytotoxic contaminants in commercial *Clostridiuum perfringens* neuraminidase preparations purified by affinity chromatography, *J. Chromatogr.* **111:**217–222.

Dowdle, W. R., Downie, J. C., and Laver, W. G., 1974, Inhibition of virus release by antibodies to surface antigens of influenza viruses, *J. Virol.* **13:**269–275.

Drzeniek, R., 1972, Viral and bacterial neuraminidases, *Curr. Top. Microbiol. Immunol.* **59:**35–74.

Drzeniek, R., 1973, Substrate specificity of neuraminidases, *Histochem. J.* **5:**271–290.

Drzeniek, R., Scharmann, W., and Balke, E., 1972, Neuraminidase and N-acetylneuraminate pyruvate-lyase of *Pasteurella multocida, J. Gen. Microbiol.* **72:**357–368.

Ebbesen, P., 1972, The influence of neuraminidase on casein-induced amyloidosis in C_3H mice, *Acta Path. Microbiol. Scand. A* **80**:854–856.
Ecobichon, D. J., and Kalow, W., 1963, The effects of sialidase on pseudocholinesterase types, *Can. J. Biochem. Physiol.* **41**:969–974.
Emmelot, P., and Bos, C. J., 1965, Differential effect of neuraminidase on the Mg^{2+}-ATPase, Na^+-K^+-Mg^{2+}-ATPase, and 5′-nucleotidase of isolated plasma membranes, *Biochim. Biophys. Acta* **99**:578–580.
Evans, W. H., and McIlwain, H., 1967, Excitability and ion content of cerebral tissues treated with alkylating agents, tetanus toxin, or a neuraminidase, *J. Neurochem.* **14**:35–44.
Faillard, H., DuAmaral, C. F., and Blohm, M., 1969, Untersuchen zur enzymatischen spezifität der neuraminidase und N-acyl-neuraminatlose in bezug auf die N-substitution, *Z. Physiol. Chem.* **350**:792–802.
Fedson, D. S., Huber, M. A., Kasel, I. A., and Webster, R. G., 1972, Presence of A 1 equi-2 hemagglutinin and neuraminidase antibodies in man, *Proc. Soc. Exp. Biol. Med.* **139**:825–826.
Finnegan, D. J., and Hope, R. M., 1970, The role of sialic acid in the serum amylase isoenzyme pattern of the marsupial mouse *Sminthopsis crassicaudata, Austr. J. Exp. Biol. Med. Sci.* **48**:237–240.
Fridman, E. A., Vishnevsky, V. G., Sinitsky, A. A., 1969, Some aspects of influenza A_2 virus sensitivity to non-specific α-inhibitors of Francis. Heat resistance of the sialidase of different variants of influenza virus group 02, *Vop. Virus* **14**:691–695.
Froese, G., Berczi, I., and Sehon, A. H., 1974, Neuraminidase-induced enhancement of tumor growth in mice, *J. Natl. Cancer Inst.* **52**:1905–1908.
Frosner, G. G., and Gerth, H.-J., 1972, Heat sensitivity of haemagglutinin and neuraminidase of influenza A strains, *Arch. Gesamt. Virusforsch.* **37**:167–175.
Fuhrmann, G. F., Granzer, E., Bey, E., and Ruhenstroth-Bauer, G., 1964, Zur Anderung der Oberflachenladung von Zellmembranen durch chemische und physikalische Einwirkungen, *Z. Naturforsch.* **19**:613–621.
Fukui, Y., Fukui, K., and Moriyama, T., 1973, Source of neuraminidase in human whole saliva, *Infect. Immun.* **8**:329–334.
Gasic, G., and Gasic, T., 1962, Removal of sialic acid from the cell coat in tumor cells and vascular endothelium and its effects on metastasis, *Proc. Natl. Acad. Sci. U.S.* **48**:1172–1177.
Gasic, G. J., and Gasic, T. B., 1970, Total suppression of pregnancy in mice by postcoital administration of neuraminidase, *Proc. Natl. Acad. Sci. U.S.* **67**:793–798.
Gesner, B., Woodruff, J. J., and McCluskey, R. T., 1969, An autoradiographic study of the effect of neuraminidase or trypsin on transfused lymphocytes, *Am. J. Path.* **57**:215–230.
Ghosh, N. K., Kotowitz, L., and Fishman, W. H., 1968, Neuraminidase in human intestinal mucosa, *Biochim. Biophys. Acta* **167**:201–204.
Gielen, W., and Harprecht, V., 1969, Die neuraminidase-activität in einigen Regionen des Rindergehirns, *Hoppe-Seyl. Z.* **350**:201–206.
Gielen, W., Etzrodt, H., and Uhlenbruck, G., 1969, Uber eine Neuraminidase in den Thrombozyten und Erythrozyten des Rindes, *Thromb. Diath. Haemorrh.* **22**:203–207.
Gielen, W., Schaper, R., and Pink, H., 1970, Neuraminidase und cytidinmonophosphat-N-acetylneuraminat Synthetase in Rinderleukozyten, *Hoppe-Seyl. Z.* **351**:768–770.
Gielen, W., Schaper, R., and Uhlenbruck, G., 1973, Uber die spezifität der leukozyten-neuraminidase, *Blut* **26**:54–60.

Glick, M. C., Comstock, C. A., Cohen, M. A., and Warren, L., 1971, Membranes of animal cells. VIII. Distribution of sialic acid, hexosamines and sialidase in the L-cell, *Biochim. Biophys. Acta* **233**:247–257.

Golubev, D. B., Aptekareva, M. N., Ivanova, N. A., Bykov, S. E., and Ivannikov, Yu, G., 1973, Temperature optima of neuraminidase activity of influenza viruses, *Acta Virol.* **17**:281–286.

Gottschalk, A., 1957, Neuraminidase: The specific enzyme of influenza virus and *Vibrio cholerae, Biochim. Biophys. Acta* **23**:645–646.

Gottschalk, A., and Lind, P. E., 1949, Product of interaction between influenza virus enzyme and ovomucin, *Nature (Lond.)* **164**:232–233.

Gottschalk, A., Whitten, W. K., Graham, E. R. B., 1960, Inactivation of follicle-stimulating hormone by enzymic release of sialic acid, *Biochim. Biophys. Acta* **38**:183–184.

Gottschalk, A., and Drzeniek, R., 1972, Neuraminidase as a tool in structural analysis, *in: Glycoproteins,* Vol. 5, Part A (A. Gottschalk, ed.), 2nd ed., pp. 381–402, Elsevier Publishing Co., New York.

Gould, K. G., Srivastava, P. N., Cline, E. M., and Williams, W. L., 1971, Inhibition of *in vitro* fertilization of rabbit ova with naturally occurring antifertility agents, *Contraception* **3**:261–267.

Granzer, E., Fuhrmann, G.-F., and Ruhenstroth-Bauer, G., 1964, Untersuchungen uber die mit Neuraminidase abspaltbaren Neuraminsauerderivate aus Oberflachenmembranen normaler Leberzellen und Asciteshepatomzellon von Ratten, *Neuraminsaurederivate Zellmembranen* **337**:52–56.

Greenwalt, T. J., and Steane, E. A., 1973, Quantitative haemagglutination. IV. Effect of neuraminidase treatment on agglutination by blood group antibodies, *Br. J. Haematol.* **25**:207–215.

Gregoriades, A., 1972, Isolation of neuraminidase from the WSN strain of influenza virus, *Virology* **49**:333–336.

Grothaus, M., Flye, W., Yunis, E., and Amos, D. B., 1971, Human lymphocyte antigen reactivity modified by neuraminidase, *Science* **173**:542–544.

Hahn, H.-J., Hellman, B., Lernmark, Å., Sehlin, J., and Taljedahl, I.-B., 1974, The pancreatic β-cell recognition of insulin secretagogues, *J. Biol. Chem.* **249**:5275–5284.

Hakomori, S., 1970, Cell density-dependent changes of glycolipid concentrations in fibroblasts, and loss of this response in virus-transformed cells, *Proc. Natl. Acad. Sci. U.S.* **67**:1741–1747.

Haksar, A., Baniukiewicz, S., and Peron, F. G., 1973, Inhibition of ACTH-stimulated steroidogenesis in isolated rat adrenal cells treated with neuraminidase, *Biochem. Biophys. Res. Commun.* **52**:956–966.

Haksar, A., Maudsley, D., and Peron, F. G., 1974, Neuraminidase treatment of adrenal cells increases their response to cholera enterotoxin, *Nature* **251**:514–515.

Han, T., 1973, Enhancement of *in vitro* lymphocyte response by neuraminidase, *Clin. Exp. Immunol.* **13**:165–170.

Haslam, E. A., Cheyne, I. M., and White, D. O., 1969, The structural proteins of Newcastle disease virus, *Virology* **39**:118–129.

Hatton, M. W. C., and Regoeczi, E., 1973, A simple method for the purification of commercial neuraminidase preparations free from proteases, *Biochim. Biophys. Acta* **327**:114–120.

Hayano, S., and Tanaka, A., 1967*a*, Sialidase-like enzymes produced by group A, B, C, G and streptocci and by *Streptococcus sanguis, J. Bacteriol.* **97**:1328–1333.

Hayano, S., and Tanaka, A., 1967b, Streptococcal sialidase. 1. Isolation and properties of sialidase produced by group K streptococcus, *J. Bacteriol.* **93**:1753–1757.
Heijlman, J., and Roukema, P. A., 1972, Action of calf brain sialidase on gangliosides, sialoglycoproteins and sialoglycopeptides, *J. Neurochem.* **19**:2567–2575.
Heilbronn, E., 1962, Treatment of horse serum cholinesterase with sialidase, *Acta Chem. Scand.* **16**:516.
Heilbronn, E., and Cedergren, E., 1970, Chemically induced changes in the acetylcholine uptake and storage capacity of brain tissue, *in:* Conference on the Effect of Cholinergic Mechanisms in the *CNS* (E. Heilbronn and A. Winter, eds.), pp. 245–269 Skoklostu, Sweden.
Heimer, R., and Meyer, K., 1956, Studies on sialic acid of submaxillary mucoid, *Proc. Natl. Acad. Sci. U.S.* **42**:728–734.
Higgins, P. A., and Schild, G. C., 1972, Characterization of the haemagglutinin and neuraminidase antigens of some recent asian type A influenza virus isolates from Hong Kong, *Bull. World Health Org.* **47**:531–534.
Hirst, G. K., 1942, Absorption of influenza haemagglutinins and virus by red blood cells, *J. Exp. Med.* **76**:195–209.
Hjertén, S., Höglund, G., and Rulttsay-Nedecky, G., 1970, Purification, characterization, and structural studies of influenza $+_2$ virus components, *Acta Virol.* **14**:89–101.
Horvat, A., and Touster, O., 1968, On the lysosomal occurrence and the properties of the neuraminidase of rat liver and Ehrlich ascites tumor cells, *J. Biol. Chem.* **243**:4380–4390.
Hotta, K., and Springer, G. F., 1964, Blood group *N* specificity and sialic acid, *Sangre* **9**:183–187.
Hovig, T., 1965, The effect of various enzymes on the ultrastructure, aggregation, and clot retraction ability of rabbit blood platelets, *Haemorrhag.* **13**:84–113.
Howe, C., Lee, J. T., and Rose, H. M., 1961, Collocalia mucoid: A substrate for myxovirus neuraminidase, *Arch. Biochem. Biophys.* **95**:512–520.
Hoyle, L., and Almeida, J. D., 1971, The chemical reactions of the haemagglutinins and neuraminidases of different strains of influenza viruses. III. Effects of proteolytic enzymes, *J. Hyg. Comb.* **69**:461–469.
Huang, R. T. C., and Orlich, M., 1972, Substrate specificities of the neuraminidases of Newcastle disease and fowl plague viruses, 2., *Physiol. Chem.* **358**:318–322.
Huang, R. T. C., Rott, R., and Klenk, H.-D., 1973, On the receptor of influenza viruses. 1. Artificial receptor for influenza virus, *Z. Naturforsch.* **28**:342–345.
Hughes, R. C., and Jeanloz, R. W., 1964, The extracellular glycosidases of *Diplococcus pneumoniae*. I. Purification and properties of a neuraminidase and a β-galactosidase. Action on the α_1-acid glycoprotein of human plasma, *Biochemistry* **3**:1535–1543.
Hughes, R. C., Sanford, B., and Jeanloz, R. W., 1972, Regeneration of the surface glycoproteins of a transplantable mouse tumor cell after treatment with neuraminidase, *Proc. Natl. Acad. Sci. U.S.* **69**:942–945.
Hughes, R. C., Palmer, P. D., and Sanford, B. H., 1973, Factors Involved in the cytotoxicity of normal guinea pig serum for cells of murine tumor TA3 sublines treated with neuraminidase, *J. Immunol.* **111**:1071–1080.
Iinuma, M., Yoslinda, T., Nagai, Y., Maeno, K., Matsumoto, T., and Hoshino, M., 1971, Subunits of NDV. Haemagglutinin and neuraminidase subunits of Newcastle disease virus, *Virology* **46**:663–677.
Irwin, L. N., Mancini, J., and Hills, D., 1973, Sialidase activity against endogenous substrate in rat brain, *Brain Res.* **53**:88–491.

Jagielski, M., 1969, *Corynebacterium diptheriae*. 1. Purification and physical properties, *Med. Dosiv. Mikrobiol.* **21**:137–150.

Jamieson, G. A., 1966, Reaction of diptheria toxin neuraminidase with human transferrin, *Biochim. Biophys. Acta* **121**:326–337.

Jensen, W. N., and Moreno, G., 1964, Etude de l'erythrophagocyteose après traitement des cellules par la neuraminidase, *Nouv. Rev. Fr. Hematol.* **4**:183–486.

Jibril, A. O., and McCay, P. B., 1965, Lysosomal enzymes in experimental encephalomalacia, *Nature* **205**:1214–1215.

Johnson, C. A., Lovstad, R. A., Walaas, E., and Walaas, O., 1970, The Properties of neuraminidase-treated crystalline ceruloplasmin, *Experientia* **26**:134–135.

Kassulke, J. T., Stutman, O., and Yunis, E. J., 1971, Blood-group isoantigens in leukemic cells: Reversibility of isoantigenic changes by neuraminidase, *J. Natl. Can. Inst.* **46**:1201–1208.

Kelly, R., 1963, Influenza virus neuraminidase, *Marquette Med. Rev.* **29**:86–89.

Kelly, R. T., and Greiff, D., 1965, Neuraminidase and neuraminidase-labile substrates in experimental influenza virus encephalitis, *Biochim. Biophys. Acta* **110**:548–553.

Kelly, R. T., and Greiff, D., 1970, Toxicity of pneumococcal neuraminidase, *Infect. Immun.* **2**:115–117.

Kelly, R. T., Greiff, D., and Farmer, S., 1966, Neuraminidase activity in *Diplococcus pneumoniae*, *J. Bact.* **91**:601–603.

Kelly, R. T., Farmer, S., and Greiff, D., 1967, Neuraminidase activities of clinical isolates of *Diplococcus pneumoniae*, *J. Bact.* **94**:272–273.

Kemp, R. B., 1968, Effect of the removal of cell surface sialic acids on cell aggregation *in vitro*, *Nature* **218**:1255–1256.

Kemp, R. B., 1970, The effect of neuraminidase (3.2.1.18) on the aggregation of cells dissociated from embryonic chick muscle tissue, *J. Cell Sci.* **6**:751–766.

Kemp, R. B., 1971, Studies on the role of cell surface sialic acids in intercellular adhesion, *Fol. Histochem. Cytochem.* **9**:25–30.

Kendal, A. P., Biddle, F., and Belyavin, G., 1968, Influenza virus neuraminidase and the viral surface, *Biochim. Biophys. Acta* **165**:419–431.

Kendal, A. P., and Madeley, C. R., 1969, A comparative study of influenza virus neuraminidases, using automated techniques, *Biochim. Biophys. Acta* **185**:163–177.

Kendal, A. P., and Eckert, E. A., 1972, The separation and properties of ^{14}C carboxamide-methylated subunits from A_2/1957 influenza neuraminidase, *Biochim. Biophys. Acta* **258**:484–495.

Kendal, A. P., and Kiley, M. P., 1973, Characterization of influenza virus neuraminidases: Peptide changes associated with antigenic divergence between early and late N2 neuraminidases, *J. Virol.* **12**:1482–1490.

Khan, M. Y., and Zinneman, H. H., 1970, The role of sialic acid in hemagglutination, *Am. J. Clin. Path.* **54**:715–719.

Khorlin, A. Ya., Privalova, I. M., Zakstelskaya, L. Ya., Molibog, E. V., and Evstigneeva, N. A., 1970, Synthetic inhibitors of *Vibrio cholerae* neuraminidase and neuraminidases of some influenza virus strains, *FEBS Lett.* **8**:17–19.

Kirschbaum, B. B., and Bosmann, H. B., 1973, Renal membrane biosynthesis and degradation. II. Localization and characterization of neuraminidase activity in rat kidney, *Nephron* **11**:26–39.

Kishore, G. S., Tulsiani, D. R. P., Bhavanandan, V. P., and Carubelli, R., 1975, Membrane-bound neuraminidases of rat liver, neuraminidase activity in Golgi apparatus, *J. Biol. Chem.* **250**:2655–2659.

Kojima, K. and Maekawa, A., 1970, Difference in electrokinetic charge of cells between

two cell types of ascites hepatoma after removal of sialic acid, *Can. Res.* **30:**2858–2862.
Kraemer, P. M., 1966, Regeneration of sialic acid on the surface of Chinese hamster cells in culture. I. General characteristics of the replacement process, *J. Cell. Physiol.* **68:**85–90.
Kraemer, P. M., 1968, Cytotoxic, hemolytic, and phospholipase contaminants of commercial neuraminidases, *Biochim. Biophys. Acta* **167:**205.
Kristjansson, F. K., and Cipera, J. D., 1963, The effect of sialidase on pig transferrins, *Can. J. Biochem. Physiol.* **41:**2523–2527.
Kryzhanovskii, G. N., and Sakharova, O. P., 1972, Effect of neuraminidase on the protagon-tetanus toxin complex, *Byull. Eksp. Biol. Med.* **73:**36–38.
Kryzhanovskii, G. N., Rozanov, Ya. A., and Bondarchuk, G. N., 1974, In vitro release by neuraminidase of tetanotoxin fixed on brain structures, *Byull. Eksp. Biol. Med.* **76:**26–29.
Kunimoto, S., Aoyagi, T., Takeuchi, T., and Umezaiwa, H., 1974, Purification and characterization of streptomyces sialidases, *J. Bacteriol.* **119:**394–400.
Kuratowska, Z., and Kubicka, T., 1967, Purification and some properties of the neuraminidase from rabbit kidney, *Acta Biochim. Polonica* **14:**255–259.
Laki, K., and Chandrasekhar, N., 1963, Sialic acid in fibrinogen and "vulcanization" of the fibrin clot, *Nature* **197:**1267–1268.
Laurell, A.-B., 1959, Neuraminidase-like factors in cultures of pneumococci, alpha-streptococci and *Pasteurella tuberculosis*, *Acta Pathol. Microbiol. Scand.* **47:**182–190.
Laurell, A.-B., 1965, Effect of neuraminidase, acetone and chloroform on α_1-antitrypsins, *Scand. J. Clin. Lab. Invest.* **17:**297–298.
Laver, W. G., Pye, J., and Ada, G. L., 1964, The molecular size of neuraminidase from *Vibrio cholerae* (strain 4Z), *Biochim. Biophys. Acta* **81:**177–180.
Laver, W. G., and Kilbourne, E. D., 1966, Identification in a recombinent, influenza virus of structural proteins derived from both parents, *Virology* **30:**493–501.
Laver, W. G., and Valentine, B. C., 1969, Morphology of the isolated hemagglutinin and neuraminidase sub-units of influenza virus, *Virology* **38:**105–119.
Laver, W. G., and Webster, R. G., 1972, Antibodies to human influenza virus neuraminidase (the A-Asian-57 H2N2 strain) in sera from Australian pelagic birds, *Bull. World Health Org.* **47:**535–541.
Lazdins, I., Haslam, E. A., and White, D. O., 1972, The polypeptides of influenza virus. VI. composition of the neuraminidase, *Virology* **49:**758–765.
Ledeen, R., and Salsman, K., 1965, Structure of Tay–Sachs ganglioside, *Biochemistry* **4:**2225–2232.
Lee, A., 1968, Effect of Neuraminidase on the phagocytosis of heterologous red cells by mouse peritoneal macrophages, *Proc. Soc. Exp. Biol. Med.* **128:**891–894.
Lee, L. T., and Howe, C., 1966, Pneumococcal neuraminidase, *J. Bact.* **91:**1418–1426.
Leibovitz, Z., and Gatt, S., 1968, Enzymatic hydrolysis of sphingolipids. VII. Hydrolysis of gangliosides by a neuraminidase from calf brain. *Biochim. Biophys. Acta* **152:**136–143.
Leighton, F., Poole, B., Beaufay, H., Baudhuin, P., Coffey, J. W., Fowler, S., and DeDuve, C., 1968, The large-scale separation of peroxisomes, mitochondria and lysosomes from the livers of rats injected with Triton WR1339, *J. Cell Biol.* **37:**482–513.
Lipkind, M., and Tsvetkova, I. V., 1967, Intracellular synthesis of myxovirus neuraminidase in chick embryo cell monolayer culture, *J. Virol.* **1:**327–333.

Lipkind, M., and Tsvetkova, I. V., 1971, Disappearance of neuraminidase and hemagglutinin activities in NDV-infected chick embryo cell monolayer culture with inhibitors of protein synthesis, *Arch. Ges. Virusforch.* **35**:303–307.

Lipovac, V., and Rosenberg, A., 1967, The mechanism by which sialic acid in gangliosides is rendered immune to sialidase, *Proceedings of the First International Meeting of the International Society for Neurochemistry,* Strasbourg, p. 138.

Lipovac, V., Barton, N., and Rosenberg, A., 1973, Control of the action of *Vibrio cholerae* sialidase on mammalian brain gangliosides by ionic strength, *Biochemistry* **12**:1858–1861.

Lombardo, A., Preti, A., and Tettamanti, G., 1970, Assay of brain particulate neuraminidase. II. Identification of the endogenous substrates of the enzyme, *Ital. J. Biochem.* **19**:386–396.

Lowy, P. H., Keighley, G., and Borsook, H., 1960, Inactivation of erythropoietin by neuraminidase and by mild substitution reactions, *Nature* **185**:102–103.

Lundgren, G., and Simmons, R. L., 1971, Effect of neuraminidase on the stimulatory capacity of cells in human mixed lymphocyte cultures, *Clin. Exp. Immunol.* **9**:915–926.

Madeley, C. R., Allan, W. H., and Kendal, A. P., 1971, Studies with avian influenza A viruses: Serological relations of the haemagglutinin and neuraminidase antigens of ten virus isolates, *J. Gen. Virol.* **12**:69–78.

Madoff, M. A., Eylar, E. H., and Weinstein, L., 1960, Serological studies of the neuraminidases of *Vibrio cholerae, Diplococcus pneumoniae,* and influenza virus, *J. Immunol.* **85**:603–613.

Madoff, M. A., Annenberg, S. M., and Weinstein, L., 1961, Production of neuraminidase by L forms of *Vibrio cholerae, Proc. Soc. Exp. Biol. Med.* **107**:776–777.

Madoff, A., McKenna, J. J., and Monaco, A. V., 1973, Neuraminidase induced delay of mouse skin graft rejection, *Transplantation* **16**:157–160.

Maeno, K., and Kilbourne, E. D., 1970, Developmental sequence and intracellular sites of synthesis of three structural protein antigens of influenza A_2 virus, *J. Virol.* **5**:153–164.

Maeno, K., Yoshida, T., Iinusra, M., Nagai, Y., Matsumoto, T., and Osai, J., 1970, Isolation of haemagglutinin and neuraminidase subunits of haemagglutinating virus of Japan, *J. Virol.* **6**:492–499.

Mahadevan, S., Nduaguba, J. C., and Tappel, A. L., 1967, Sialidase of rat liver and kidney, *J. Biol. Chem.* **242**:4409–4413.

Maier, V., and Peiffer, E. F., 1971, Influence of α-neuraminidase treatment on biosynthesis and secretion of insulin from isolated mouse islets of langerhans, *Hoppe-Seyl. Z.* **352**:1733–1734.

Mayhew, E., 1969, Effect of ribonuclease and neuraminidase on the electrophoretic mobility of tissue culture cells in parasynchronous growth, *J. Cell. Physiol.* **69**:311–320.

McCrea, J. F., 1947, Modification of red-cell agglutinability by *Cl. welchii* toxins, *Austr. J. Exp. Biol. Med. Sci.* **25**:127–136.

McQuiddy, P., and Lilien, J. E., 1973, The binding of exogenously added neuraminidase to cells and tissues in culture, *Biochim. Biophys. Acta* **291**:774–779.

Mehrishi, J. N., 1970, Action of pronase and neuraminidase on the electrophoretic mobility of erythrocytes from normal cats and those with spontaneous tumours, *Vox Sang.* **18**:27–33.

Meier-Ewert, H., Gibbs, A. J., and Dimmock, N. J., 1970, Studies on antigenic variation of the haemagglutinin and neuraminidase of swine influenza virus isolates, *J. Gen. Virol.* **6**:409–419.

Meindl, P., and Tuppy, H., 1965, Uber synthetische Ketoside der *N*-acetyl-D-neuraminsaure, *Mk. Chem.* **96**:816–827.
Meindl, P., and Tuppy, H., 1969, Uber 2-desoxy, 2,3-dehydrosialinsauren. II. Konpetetive hemmung der *Vibrio cholerae*-neuraminidase durch 2-desoxy, 2,3-dehydro-*N*-acylneuraminsauren, *Z. Physiol. Chem.* **350**:1088–1092.
Melnick, J. L., 1972, Classification and nomenclature of viruses, *Prog. Med. Virol.* **14**:321–332.
Menguy, R., Masters, Y. F., and Desbaillets, L., 1970, Human salivary glycosidases, *Proc. Soc. Exp. Bio. Med.* **134**:1020–1025.
Miller, A., Sullivan, J. F., and Katz, J. H., 1963, Sialic acid content of the erythrocyte and of an ascites tumor cell of the mouse, *Cancer Res.* **23**:485–490.
Monto, A. S., and Kendal, A. P., 1973, Effect of neuraminidase antibody on Hong Kong influenza, *Lancet* **1973**:623–625.
Morein, B., Hoglund, S., and Bergman, R., 1973, Immunity against parainfluenza-3 virus in cattle: Anti-neuraminidase activity in serum and nasal secretion, *Infect. Immun.* **8**:650–656.
Morell, A. G., Gregoriadis, G., Scheinberg, I. H., Hickman, J., and Ashwell, G., 1971, The role of sialic acid in determining the survival of glycoproteins in the circulation, *J. Biol. Chem.* **246**:1461–1467.
Morgan, E. H., and Laurell, C.-B., 1963, Neuraminidase in mammalian brain, *Nature* **197**:921–922.
Moriyama, T., and Barksdale, L., 1967, Neuraminidase of *Corynebacterium diphtheriae*, *J. Bacteriol.* **94**:1565–1581.
Moss, D. W., Eaton, R. H., Smith, J. K., and Whitby, L. G., 1966, Alteration in the electrophoretic mobility of alkaline phosphatases after treatment with neuraminidase, *Biochem. J.* **98**:32c–33c.
Muller, H. E., 1971*a*, Uber das Vorkommen von Neuraminidase bei *Erysipelothrix insidiosa*, *Path. Microbiol.* **27**:241–248.
Muller, H. E., 1971*b*, Untersuchen *in vitro* uber die Neuraminidase der *Pasteurella multocida*, *Zeitblatt Bakt. Hyg.* **217**:326–344.
Murphy, B. R., Kasel, J. A., and Chanock, R. M., 1972, Association of serum anti-neuraminidase antibody with resistance to influenza in man, *New Engl. J. Med.* **286**:1329–1332.
Myhill, M., and Cook, T. M., 1972, Extracellular neuraminidase of sterptomyces albus, *Can. J. Microbiol.* **18**:1007–1014.
Nelsestuen, G. L., and Suttie, J. W., 1971, Properties of asialo- and aglycoprothrombin, *Biochem. Biophys. Res. Commun.* **45**:198–203.
Nees, S., Veh, R. W., and Schauer, R., 1975, Purification and characterization of neuraminidase from *Clostridium perfringens*, *Z. Physiol. Chem.* **356**:1027–1042.
Nista, A., Sezzi, M. L., and Bellelli, L., 1973, Pregnancy rejection induced by neuraminidase-treated placental cells, *Oncology* **28**:402–410.
Noll, H., Aoyagi, T., and Orlando, J., 1971, Intracellular synthesis of neuraminidase following infection of chorioallantoic membranes with influenza virus, *Virology* **1**:141–143.
Nordling, S., and Mayhew, E., 1966, On the intracellular uptake of neuraminidase, *Exp. Cell Res.* **44**:552–562.
Noseworthy, J., Jr., Korchak, H., and Karnovsky, M. L., 1972, Phagocytosis and the sialic acid of the surface of polymorphonuclear leukocytes, *J. Cell. Physiol.* **79**:91–96.
Öhman, R., Rosenberg, A., and Svennerholm, L., 1970, Human brain sialidase, *Biochemistry* **9**:3774–3782.

Öhman, R., 1971a, The activity of ganglioside sialidase in different regions of human brain, *J. Neurochem.* **18**:531–532.

Öhman, R., 1971b, Subcellular fractionation of ganglioside sialidase from human brain, *J. Neurochem.* **18**:89–95.

Öhman, R., and Svennerholm, L., 1971, The activity of ganglioside sialidase in the developing human brain, *J. Neurochem.* **18**:79–87.

Orlova, T. G., Orlova, N. G., and Eremkina, E. I., 1969, Sialidase activity of chick and mouse embryo tissue culture cells infected with myxoviruses and the effect of actinomycin D on this activity, *Acta Virol.* **13**:363–370.

O'Toole, R. D., Goode, L., and Howe, C., 1971, Neuraminidase activity in bacterial meningitis, *J. Clin. Invest.* **50**:979–985.

Oxford, J. S., 1973, Polypeptide composition of influenza B viruses and enzymes associated with the purified virus particles, *J. Virol.* **12**:827–835.

Oxley, S. B., and Griffen, W., Jr., 1971, Immunogenic enhancement by neuraminidase, *Surg. Forum* **22**:113–114.

Palese, P., and Schulman, J., 1974, Isolation and characterization of influenza virus recombinants with high and low neuraminidase activity, *Virology* **57**:227–237.

Palese, P., Schulman, J. L., Bodo, G., and Meindl, P., 1974, Inhibition of influenza and parainfluenza virus replication in tissue culture by 2-deoxy-2,3-dehydro-N-trifluoroacetyl neuraminic acid (FANA) *Virology* **59**:490–498.

Palese, P., Tobita, K., Ueda, M., and Compans, R. W., 1974, Characterization of temperature sensitive influenza virus mutants defective in neuraminidase, *Virology* **61**:397–410.

Pape, L., and Maxfield, M., 1967, Neuraminidase action on the human urinary glycoprotein of Tamm and Horsfall, *Biochim. Biophys. Acta* **133**:574–581.

Pardoe, G. I., 1970, The inducible neuraminidase CN-acyl-neuraminyl hydrolase (Ec 3.2.18) of *Klebsiella aerogenes* NCID9479. *Path. Microbiol.* **35**:361–376.

Parker, W. C., and Bearn, A. G., 1962, Studies on the transferrins of adult serum, cord serum, and cerebrospinal fluid; the effect of neuraminidase, *Exp. Med.* **115**:83–105.

Perlitsh, M. J., and Glickman, I., 1966, Salivary neuraminidase. I. The presence of neuraminidase in human saliva, *J. Periodont.* **37**:368–373.

Perona, G., Cortesi, S., Xodo, P., Scandellari, C., Ghiotto, G., and DeSandre, G., 1964, Variations of *in vivo* survival, acetylcholinesterase activity, and sensitivity to acid lysis in human erythrocytes treated with proteolytic enzymes and neuraminidase, *Acta Isotop.* **4**:287–295.

Pierce, J. S., and Haywood, A. M., 1973, Thermal inactivation of Newcastle disease virus. I. Coupled inactivation rates of hemagglutinating and neuraminidase activities, *J. Virol.* **11**:168–176.

Pinter, J. K., Hayashi, J. A., Bahn, A. N., 1968, Extracellular Streptococcal neuraminidase, *J. Bacteriol.* **95**:1491–1492.

Preti, A., Tettamanti, G., and DiDonato, S., 1968, Presenza nel cervello di ratto di due neuraminidase aventi differente localizzazione subcellulare, *Boll. Soc. Ital. Biol. Sper.* **44**:1143–1147.

Preti, A., Lombardo, A., and Tettamanti, G., 1970, Assay of brain particulate neuraminidase. I. Determination of N-acetylneuraminic acid released by the enzyme from endogenous substrates, *Ital. J. Biochem.* **19**:371–385.

Preti, A., Lombardo, A., Tettamanti, G., and Zambotti, V., 1973, Removal of N-acetylneuraminic acid from particulate sialoglycoproteins by endogenous membrane-bound neuraminidase in calf brain, *J. Neurochem.* **21**:1559–1562.

Preti, A., Lombardo, A., Cestaro, B., Zambotti, S., and Tettamanti, G., 1974, Studies on brain membrane-bound neuraminidase. 1. General properties of the enzyme prepared from calf brain, *Biochim. Biophys. Acta* **350**:406–414.

Pricer, W. E., Jr., Ashwell, G., 1971, The binding of desialylated glycoproteins by plasma membranes of rat liver, *J. Biol. Chem.* **246**:4825–4833.

Quarles, R. H., and Brady, R. O., 1970, Sialoglycoproteins and several glycosidases in developing rat brain, *J. Neurochem.* **17**:801–807.

Rafelson, M., Wilson, and Schneir, M., 1962, The neuraminidases of influenza virus. *Presbyterian St. Luke Hosp. Med. Bull.* **1**:34–39.

Rafelson, M. E., Gold, S., and Priede, I., 1966, Neuraminidase (sialidase) from influenza virus, *Methods Enzymol.* **8**:677–680.

Ray, P. K., Gewurz, H., and Simmons, R. L., 1971, Complement sensitivity of neuraminidase-treated lymphoid cells, *Transplantation* **12**:327–328.

Ray, P. K., and Simmons, R. L., 1971, Failure of neuraminidase to unmask allogeneic antigens on cell surfaces. *Proc. Soc. Exp. Bio. Med.* **138**:600–604.

Ray, P. K., and Simmons, R. L., 1972, Comparative effect of viral and bacterial neuraminidase on the complement sensitivity of lymphoid cells, *Clin. Exp. Immunol.* **10**:139–150.

Ray, P. K., and Simmons, R. L., 1973, Differential release of sialic acid from normal and malignant cells by *Vibrio cholerae* neuraminidase or influenza virus neuraminidase, *Cancer Res.* **33**:936–939.

Reichert, L. E., Jr., Gavin, J. R., III, and Neill, J. D., 1971, Studies on a neuraminidase containing gonadotropin inhibitor substance in bovine putuitary extracts, *Endocrinology* **88**:1497–1502.

Reisner, E. G., Flye, M. W., Su Chung, K. S., and Amos, D. B., 1974, The cytotoxic reactivity and sialic acid content of human lymphoid cells, *Tissue Antigens* **4**:7–20.

Rios, A., and Simmons, R. L., 1973, Immunospecific regression of various syngeneic mouse tumors in response to neuraminidase-treated tumor cells, *J. Natl. Cancer Inst.* **51**:637–644.

Roberts, D. H., 1967, Neuraminidase-like enzyme present in *Mycoplasma Gallisepticum*, *Nature* **213**:87–88.

Robinson, J. C., Pierce, J. E., 1964, Differential action of neuraminidase on human serum alkaline phosphatases, *Nature* **204**:472–473.

Romanovska, E., and Watkins, W. M., 1963, Fractionation of neuraminidases in extracts of *Trichimonas foetus*, *Biochem. J.* **87**:37–38.

Rosado, A., Velazquez, A., and Lara-Ricalde, R., 1973, Cell polarography. II. Effect of neuraminidase and follicular fluid upon the surface characteristics of human spermatozoa. *Fert. Ster.* **24**:349–354.

Rosenberg, A., Howe, C., and Chargaff, E., 1956, Inhibition of influenza virus haemagglutination by a brain lipid fraction, *Nature* **177**:234–235.

Rosenberg, A., Binnie, B., and Chargaff, E., 1960, Properties of a purified sialidase and its action on brain mucolipid, *J. Am. Chem. Soc.* **82**:4113–4114.

Rosenberg, A., Stefanovic, V., and Mandel, P., 1975, Acetyl and butyryl cholinesterases of intact cultured neuroblastoma and glial cells and activation upon cellular desialylation, *Fed. Proc.* **34**:244.

Rosenthal, J. W., and Fain, J. N., 1971, Insulin-like effect of clostridial phospholipase C, neuraminidase, and other bacterial factors on Brown fat cells, *J. Biol. Chem.* **246**:5888–5895.

Roston, C. P. J., Caygill, J. C., and Jevons, F. R., 1966, Degradation of mucoprotein carbohydrate by ram testis enzymes, *Life Sci.* **5**:535–540.

Rott, R., Becht, H., and Orlich, M., 1974, The Significance of influenza virus neuraminidase in immunity, *J. Gen. Virol.* **22**:35–41.
Roukema, P. A., and Heijlman, J., 1970, The regional distribution of sialoglycoproteins, gangliosides and sialidase in bovine brain, *J. Neurochem.* **17**:773–780.
Roukema, P. A., van den Eijnden, D. H., Heijlman, J. and van der Berg, G., 1970, Sialoglycoproteins, gangliosides and related enzymes in developing rat brain, *FEBS Lett.* **9**:267–270.
Saber, M. S., Drzeniek, R., and Krupe, M., 1965, I. Mitt.: Freilegung von ABH-Blutgruppen determinierenden Kohlenhydraten an Erythrozyten durch Neuraminidase-Einwirkung, *Z. Naturforsch.* **20**:965–973.
Sandhoff, K., and Jatzkewitz, H., 1967, A particle-bound sialyl lactosidoceramide splitting mammalian sialidase, *Biochim. Biophys. Acta* **141**:442–444.
Sanford, B. H., 1973, Effect of neuraminidase on tumor development and growth, *J. Natl. Cancer Inst.* **51**:1393–1394.
Sanford, B., and Codington, J. F., 1971, Further studies on the effect of neuraminidase on tumor cell transplantability, *Tissue Antigens* **I**:153–161.
Sauter, Chr., Lindenmann, J., and Gerber, A., 1972, Agglutination of leukemic myeloblasts by neuraminidase, *Eur. J. Cancer* **8**:451–453.
Sairam, M. R., and Moudgal, N. R., 1971, On the mechanism of action of the monkey urinary follicle stimulating hormone inhibitor—Its sialidase activity, *Ind. J. Biochem. Biophys.* **8**:141–146.
Scharmann, W., Bruckler, J., and Blobel, H., 1971, Wirkung bakterieller Neuraminidasen auf Transferrin vom Menschen, Rind Und Kaninchen, *Biochim. Biophys. Acta* **229**:136–142.
Scheid, A., and Choppin, P. W., 1974, The hemagglutinating and neuraminidase protein of a paramyxovirus: Interaction with neuraminic acid in affinity chromatography, *Virology* **62**:125–133.
Schengrund, C.-L., and Rosenberg, A., 1970, Intracellular location and properties of bovine brain sialidase, *J. Biol. Chem.* **245**:6196–6200.
Schengrund, C.-L., and Rosenberg, A., 1971, Gangliosides, glycosidases, and sialidase in the brain and eyes of developing chickens, *Biochemistry* **10**:2424–2428.
Schengrund, C.-L., Jensen, D. S., and Rosenberg, A., 1972, Localization of sialidase in the plasma membrane of rat liver cells. *J. Biol. Chem.* **247**:2742–2746.
Schengrund, C.-L., and Rosenberg, A., 1973, Effect of cations on the sialidase activity of nerve ending membranes, *Trans. Am. Soc. Neurochem.* **4**:90.
Schengrund, C.-L., Lausch, R. N., and Rosenberg, A., 1973, Sialidase activity in transformed cells, *J. Biol. Chem.* **248**:4424–4428.
Schengrund, C.-L., Duff, R., and Rosenberg, A., 1974, Sialidase activity of oncogenic cells transformed by herpes simplex virus, *Virology* **53**:595–599.
Schengrund, C.-L., and Nelson, J. T., 1975, Influence of cation concentration on the sialidase activity of neuronal synaptic membranes, *Biochem. Biophys. Res. Commun.* **63**:217–223.
Schooley, J. C., and Garcia, J. F., 1971, The destruction by neuraminidase of the biological activity of erythropoietin when complexed with antierythropoietin, *Proc. Soc. Exp. Bio. Med.* **138**:66–68.
Schooley, J. C., and Mahlmann, L. J., 1971, Inhibition of the biologic activity of erythropoietin by neuraminidase *in vivo*, *J. Lab. Clin. Med.* **78**:765–770.
Schottissek, C., and Rott, R., 1964, Behavior of virus-specific activities in tissue cultures infected with myxoviruses after chemical changes of ribonucleic acid, *Virology* **22**:169–176.

Schramm, G., and Mohr, E., 1959, Purification of neuraminidase from *Vibrio cholerae*, *Nature* **183**:1677–1678.

Schulman, J. L., and Kilbourne, E. D., 1969, The antigenic relationship of the neuraminidase of Hong Kong to that of other human strains of influenza A virus, *Bull. World Health Org.* **41**:425–428.

Scott-Burden, T., and Hawtrey, A. O., 1973, The effect of neuraminidase treatment of ribosome-free membranes on their ribosomal reattachment ability, *Biochem. Biophys. Res. Commun.* **54**:1288–1295.

Seaman, G. V. F., and Uhlenbruck, G., 1962, Die elektrophoretische Beweglichket von Erythrocyten nach Behandlung mit Verschiedenen Enzymen und Antiseren, *Kurz. Wiss. Mit.* **40**:699–701.

Sedmak, J. J., and Grossberg, S. E., 1973, Comparative enzyme kinetics of influenza neuraminidases with the synthetic substrate methoxyphenylneuraminic acid, *Virology* **56**:658–661.

Sedmak, J. J., and Grossberg, S. E., 1973, Interferon bioassay: Reduction in yield of myxovirus neuraminidases, *J. Gen. Virol.* **21**:1–7.

Sethj, K. K., and Muller, H. E., 1972, Neuraminidase activity in *Mycoplasma Gallisepticum*, *Infect. Immun.* **5**:260–262.

Sethi, K. K., and Brandis, H., 1973, Neuraminidase induced loss in the transplantability of murine leukaemia L 1210, induction of immunoprotection and the transfer of induced immunity to normal DBA/2 mice by serum and peritoneal cells, *Br. J. Cancer* **27**:106–113.

Seto, J. T., 1964, Sialidase (neuraminidase) activity of standard and incomplete virus, *Biochim. Biophys. Acta* **90**:420–427.

Shimada, A., and Nathenson, S. G., 1971, Removal of neuraminic acid from H-2 alloantigens without effect on antigenic reactivity, *J. Immunol.* **107**:1197–1199.

Shinkai, K., and Nishimura, T., 1972, Inability of isoquinoline derivatives to inhibit virus neuraminidase activity, *J. Gen. Virol.* **16**:227–229.

Simmons, R. L., and Rios, A., 1971, Combined use of BCG and neuraminidase in experimental tumor immunotherapy, *Surg. Forum* **22**:113–114.

Simmons, R. L., and Rios, A., 1972, Immunospecific regression of methylcholanthrene fibrosarcoma using neuraminidase. III. Synergistic effect of BCG and neuraminidase treated tumor cells, *Ann. Surg.* **176**:188–194.

Simmons, R. L., and Rios, A., 1973, Differential effect of neuraminidase on the immunogenicity of viral associated and private antigens of mammary carcinomas, *J. Immunol.* **111**:1820–1825.

Simmons, R. L., and Rios, A., Ray, P. K., and Lundgren, G., 1971*a*, Effect of neuraminidase on growth of a 3-methylcholanthrene-induced fibrosarcoma in normal and immunosuppressed syngeneic mice. *J. Natl. Cancer Inst.* **47**:1087–1094.

Simmons, R. L., Lipschultz, M. L., Rios, A., and Ray, P. K., 1971*b*, Failure of neuraminidase to unmask histocompatability antigens on trophoblast, *Nat. New Biol.* **231**:111–112.

Simmons, R. L., and Rios, A., and Ray, P. K., 1971*c*, Immunogenicity and Antigenicity of lymphoid cells treated with neuraminidase, *Nat. New Biol.* **231**:179–181.

Skehel, J. J., and Schild, G. C., 1971, The polypeptide composition of influenza A viruses, *Virology* **44**:396–408.

Smith, I., Perry, J. D., and Lightstone, P. J., 1969, Disc electrophoresis of alkaline phosphatases: Mobility changes caused by neuraminidase. *Clin. Chim. Acta* **25**:17–19.

Solovev, V. D., Domaradskii, I. V., Shimanyuk, N. Ya., Bichul, K. G., and Kurennaya,

I. I., 1972, Function of neuraminidase in *Vibrio* taxonomy. *Byull. Eksp. Biol. Med.* **73**:61–64.

Springer, G. F., and Rapaport, M. J., 1957, Specific release of heterogenetic "mononucleosis receptor" by influenza viruses, receptor destroying enzyme and plant proteases, *Proc. Soc. Exp. Biol. Med.* **96**:103–107.

Srivastava, P. N., Zaneveld, L. J. D., and Williams, W. L., 1970, Mammalian sperm acrosomal neuraminidases, *Biochem. Biophys. Res. Commun.* **39**:575–582.

Stahl, W. L., and O'Toole, R. D., 1972, Pneumococcal neuraminidase: Purification and properties, *Biochim. Biophys. Acta* **268**:480–487.

Stefanovic, V., Mandel, P., and Rosenberg, A., 1975, Ecto-pyrophosphatase activity of nervous system cells in tissue culture and its enhancement by removal of cell surface sialic acid, *Trans. Am. Soc. Neurochem.* **6**:102.

Stoner, G. D., Williams, B., Kniazeff, A., and Shimkin, M. B., 1973, Effect of neuraminidase pretreatment on the susceptibility of normal and transformed mammalian cells to bovine enterovirus 261, *Nature* **245**:319–320.

Suttajit, M., and Winzler, R., 1971, Effect of modification of N-acetylneuraminic acid on the binding of glycoproteins to influenza virus and on susceptibility to cleavage by neuraminidase, *J. Biol. Chem.* **246**:3398–3404.

Svennerholm, L., 1967, The metabolism of gangliosides in cerebral lipidoses, in: *Inborn Disorders of Sphingolipid Metabolism* (S. M. Aronson and B. W. Volk, eds.), pp. 169–186, Pergamon Press, New York.

Svensmark, O., and Kristensen, P., 1962, Electrophoretic mobility of sialidase-treated human serum cholinesterase, *Dan. Med. Bull.* **9**:16–17.

Svensmark, O., and Kristensen, P., 1963, Isoelectric point of native and sialidase-treated human serum cholinesterase, *Biochim. Biophys. Acta* **67**:441–452.

Svensmark, O., and Heilbronn, E., 1964, Electrophoretic mobility of native and neuraminidase-treated horse-serum cholinesterase, *Biochim. Biophys. Acta* **92**:400–402.

Szewczuk, A., and Connell, G. E., 1964, The effect of neuraminidase on the properties of γ-glutamyl transpeptidase, *Biochim. Biophys. Acta* **83**:218–223.

Taha, B. H., and Carubelli, R., 1967, Mammalian neuraminidase: Intracellular distribution and changes of enzyme activity during lactation, *Arch. Biochem. Biophys.* **119**:55–61.

Tallman, J. F., and Brady, R. O., 1973, The purification and properties of a mammalian neuraminidase (sialidase), *Biochim. Biophys. Acta* **293**:434–443.

Tannenbaum, S. W., and Gulbinsky, J., Katz, M., and Sun, S.-C., 1970, Separation purification and some properties of pneumoccal neuraminidase isoenzymes, *Biochim. Biophys. Acta* **198**:242–254.

Tannenbaum, S. W., and Sun, S.-C., 1971, Some molecular properties of pneumococcal neuraminidase isoenzymes, *Biochim. Biophys. Acta* **229**:824–828.

Tauc, L., and Hinzen, D. H., 1974, Neuraminidase: Its effect on synaptic transmission, *Brain Res.* **80**:340–344.

Tettamanti, G., and Zambotti, V., 1968, Purification of neuraminidase from pig brain and its action on different gangliosides, *Enzymologia* **35**:61–74.

Tettamanti, G., Lombardo, A., Preti, A., and Zambotti, V., 1970, Effect of temperature and Triton X-100 on the activity of particulate neuraminidase from rabbit brain, *Enzymologia* **39**:65–71.

Tettamanti, G., Morgan, I. G., Gombos, G., Vincendon, G., and Mandel, P. 1972, Subsynaptosomal localization of brain particulate neuraminidase, *Brain Res.* **47**:515–518.

Tettamanti, G., Preti, A., Lombardo, A., Bonali, F., and Zambotti, V., 1973, Parallelism of subcellular location of major particulate neuraminidase and gangliosides in rabbit brain cortex, *Biochim. Biophys. Acta* **306**:466–477.

Tettamanti, G., Cestaro, B., Lombardo, A., Preti, A., Venerando, B., and Zambotti, V., 1974, Studies on brain membrane-bound neuraminidase. II. Effect of detergents on the kinetics of the enzyme prepared from calf brain, *Biochim. Biophys. Acta* **350**:415–424.

Thacore, H. R., and Youngner, J. S., 1971, Cells persistently infected with Newcastle disease virus. III. Chemical stability of hemagglutinin and neuraminidase of a mutant isolated from persistently infected L cells, *J. Virol.* **1**:53–58.

Thomas, L., 1970, The toxic properties of *M. neurolyticum* and *M. gallisepticum*, in: *The Role of Mycoplasmas and L Forms of Bacteria in Disease* (J. T. Sharp, ed.), Charles Thomas, Springfield, Mass.

Thonard, J. C., Hefflin, C. M., and Steinberg, A. I., 1965, Neuraminidase activity in mixed culture supernatant fluids of human oral bacteria, *J. Bact.* **89**:924–925.

Touster, O., Aronson, N. N., Jr., Dulaney, J. T. and Hendrickson, M., 1970, Isolation of rat liver plasma membranes, *J. Cell Biol.* **47**:604–618.

Tozawa, H., Watanabe, M. and Ishida, N., 1973, Structural components of sendai virus. Serological and physicochemical characterization of hemagglutinin subunit associated with neuraminidase activity, *Virology* **55**:242–253.

Tsvetkova, I. V., and Lipkind, M. A., 1970, The difference in thermostability of haemagglutinin and neuraminidase of two variants of influenza A_2 virus, *Acta Virol.* **14**:86.

Tsvetkova, I. V., and Lipkind, M. A., 1973, Studies on the role of myxovirus neuraminidase in virus-cell receptor interaction by means of direct determination of sialic acid split from cells, *Archiv. Virusforschung.* **42**:125–138.

Tulsiani, D. R. P., and Carubelli, R., 1970, Studies on the soluble and lysosomal neuraminidases of rat liver, *J. Biol. Chem.* **245**:1821–1827.

Tulsiani, D. R. P., and Carubelli, R., 1971, Studies on the soluble and lysosomal neuraminidases of rat mammary glands, *Biochim. Biophys. Acta* **227**:139–153.

Tulsiani, D. R. P., and Carubelli, R., 1972, Soluble and lysosomal neuraminidases in the liver of developing chicks, *Biochim. Biophys. Acta* **284**:257–267.

Tulsiani, D. R. P., Nordquist, R. E., and Carubelli, R., 1973, The neuraminidase of rat eyes, *Exp. Eye Res.* **15**:93–103.

Tuppy, H., Wiesbauer, U., and Wintersberger, E., 1963, Uber die Einwirkung von Neuraminidase auf die Serumoxytocinase, *Mh. Chem.* **94**:321–328.

Tuppy, H., and Palese, P., 1968, Neuraminidase aus Schweinenieren. *Hoppe Seyl. Z.* **349**:1169–1178.

Unbehaun, V., 1970, Activität von Neuraminidase im Corpus luteum vom Schwein, *Hoppe-Seyl. Z.* **351**:705–710.

Vaccari, A., and Vertua, R., 1970, ^{14}C-5-Hydroxytryptamine and ^{3}H-D-amphetamine: Uptake and contraction by the rat stomach fundus *in vitro*, *Biochem. Pharmacol.* **19**:2105–2115.

Vaccari, A., Vertua, R., and Furlani, A., 1971, Decreased calcium uptake by rat fundal strips after pretreatment with neuraminidase or LSD *in vitro*, *Biochem. Pharmacol.* **20**:2603–2612.

Vaerman, J.-P., and Heremans, J. F., 1968, Effect of neuraminidase and acidification on complement-fixing properties of human IgA and IgG, *Int. Arch. Allergy* **34**:49–52.

Vaheri, A., Ruoslahti, E., and Nordling, S., 1972, Neuraminidase stimulates division and sugar uptake in density-inhibited cell cultures, *Nat. New Biol.* **238**:211–212.

Vaitukaitis, J., and Ross, G. T., 1971, Altered biologic and immunologic activities of progressively desialylated human urinary FSH, *J. Clin. Endocrinol.* **33**:308–311.

van den Hamer, C. J. A., Morell, A. G., Scheinberg, I. H., Hickman, J., and Ashwell, G., 1970, Physical and chemical studies on ceruloplasmin. IX. The role of galactosyl residues in the clearance of ceruloplasmin from the circulation, *J. Biol. Chem.* **245**:4397–4402.

Van Hall, E. V., Vaitukaitis, J. L., Ross, G. T., Hickman, J. W., and Ashwell, G., 1971a, Effects of progressive desialylation on the rate of disappearance of immunoreactive HCG from plasma in rats, *Endocrinology* **89**:11–15.

Van Hall, E. V., Vaitukaitis, J. L., Ross, G. T., Hickman, J. W., and Ashwell, G., 1971b, Immunological and biological activity of HCG following progressive desialylation, *Endocrinology* **88**:456–464.

van Lenten, L., and Ashwell, G., 1971, Studies on the chemical and enzymatic modification of glycoproteins, *J. Biol. Chem.* **246**:1889–1894.

Varute, A. T., and Patil, V. A., 1971, Histochemical analysis of molluscan stomach and intestinal alkaline phosphatase: A sialoglycoprotein, *Histochemie* **25**:77–90.

Vasudevan, D. M., Balakrishnan, K., and Talwar, G. P., 1970, Effect of neuraminidase on electrophoretic mobility and immune cytolysis of human uterine cervix carcinoma cells, *Int. J. Cancer* **6**:506–516.

Vermylen, J., de Gaetano, G., Donati, M. B., and Verstraete, M., 1974, Platelet-aggregating activity in neuraminidase-treated human cryoprecipitates: Its correlation with factor-VIII-related antigen, *Br. J. Haematol.* **26**:645–650.

Vertiev, Yu V., and Ezepchuk, Yu V., 1972, Neuraminidase of *Corynebacterium diphtheriae*, *Fol. Microbiol.* **17**:269–273.

Visser, A., and Emmelot, P., 1973, Studies on plasma membranes. XX. Sialidase in hepatic plasma membranes. *J. Membrane Biol.* **14**:73–84.

Vitti, A., Bocci, V., Russi, M., and Rita, G., 1970, The effect of neuraminidase on the rabbit urinary interferon, *Experientia* **26**:363–364.

von Nicolai, H., and Ziliken, F., 1972, Neuraminidase aus *Lactobacillius bifidius* var *Pennsylvanicus*, *Z. Physiol. Chem.* **353**:1015–1016.

Warren, L., 1959, The thiobarbituric acid assay of sialic acids, *J. Biol. Chem.* **234**:1971–1975.

Warren, L., and Spearing, C. W., 1960, Mammalian sialidase (neuraminidase), *Biochem. Biophys. Res. Commun.* **3**:489–492.

Warren, L., Fuhrer, J. P., Buck, C. A., 1973, Surface glycoproteins of cells before and after transformation by oncogenic viruses, *Fed. Proc.* **32**:80–85.

Watkins, E., Ogata, Y., Anderson, L. L., Watkins, E., III, and Waters, M. F., 1971, Activation of host lymphocytes cultured with cancer cells treated with neuraminidase, *Nat. New Biol.* **231**:83–85.

Webster, R. G., 1972, On the origin of pandemic influenza viruses, *Curr. Top. Microbiol. Immunol.* **59**:75–105.

Weiner, M. S., Bianco, C., and Nussenzweig, V., 1973, Enhanced binding of neuraminidase-treated sheep erythrocytes to human T lymphocytes, *Blood* **42**:939–946.

Weiss, L., and Cudney, T. L., 1971, Some effects of neuraminidase on the *in vitro* interactions between spleen and mastocytoma (P815) cells, *Int. J. Cancer* **7**:187–197.

Weiss, L., Mayhew, E., and Ulrich, K., 1966, The effect of neuraminidase on the phagocytic process in human monocytes, *Lab. Invest.* **15**:1304–1309.

Weiss, L., Fisher, B., and Fisher, E. R., 1974, Effect of neuraminidase on the distribution of intravenously injected walker tumor cells in rats, *Cancer* **34**:680–683.

Wenger, D. A., and Wardell, S., 1973, Action of neuraminidase (EC 3.2.1.18) from *Clostridium perfringens* on brain gangliosides in the presence of bile salts, *J. Neurochem.* **20**:607–612.

Wesemann, W., and Zilliken, F., 1967, Receptors of neurotransmitters. II. Sialic acid metabolism and the serotonin induced contraction of smooth muscle, *Biochem. Pharmacol.* **16**:1773–1779.

White, S., and Mellanby, J., 1969, The separation of neuraminidase from other pathological activities of a culture filtrate of *Clostridium sordellii* CN3903, *J. Gen. Microbiol.* **56**:137–141.

Wildy, P., 1971, Classification and nomenclature of viruses, *in: Monographs in Virology*, Vol. 5 (J. L. Melnick, ed.), pp. 27–75, S. Karger, Basel.

Woodruff, J. J., and Gesner, B. M., 1969, The effect of neuraminidase on the fate of transfused lymphocytes, *J. Exp. Med.* **129**:551–566.

Woolley, D. W., and Gommi, B. W., 1966, Serotonin receptors. VI. Methods for the direct measurement of isolated receptors, *Arch. Int. Pharmacodyn.* **159**:8–17.

Wrigley, N. G., Skehel, J. J., and Charlwood, P. A., 1973, The size and shape of influenza virus neuraminidase, *Virology* **51**:525–529.

Yachnin, S., and Gardner, F. H., 1961, Measurement of human erythrocyte neuraminic acid: Relationship to haemolysis and red blood cell virus interaction, *Br. J. Haematol.* **7**:464–465.

Yaginuma, T., 1972, Uptake of neuraminidase or heat-treated human chorionic gonadotropin by the ovary, *Am. J. Obstet. Gynecol.* **112**:1037–1042.

Yarnell, M. M., and Ambrose, E. J., 1969, Studies of tumour invasion in organ culture. II. Effects of enzyme treatment, *Eur. J. Cancer* **5**:265–269.

Yeh, A. K., Tulsiani, D. R. P., and Carubelli, R., 1971, Neuraminidase activity in human leukocytes, *J. Lab. Clin. Med.* **78**:771–778.

Yogeeswaran, G., Sheinin, R., Wherrett, J. R., and Murray, R. K., 1972, Studies on the glycosphingolipids of normal and virally transformed 3T3 mouse fibroblasts, *J. Biol. Chem.* **247**:5146–5158.

Yokoyama, M., and Trams, E. G., 1962, Effect of enzymes on blood group antigens, *Nature* **194**:1048–1049.

Zhumatov, Kh. Vr., Isaeva, E. S., Chuvakova, Z. K., and Stetsenko, O. G., 1972, Investigation of electrophoretic mobility and immunospecificity of the neuraminidases of influenza virus and host cells, *Byull. Eksp. Biol. Med.* **73**:68–71.

Zvetkova, I. V., 1965, Aldolase of neuraminic acid and neuraminidase in animal tissues, *Biochimica* **30**:407–414.

Index

Acetylcholine, 340
α_1-Acid glycoprotein, 226, 279
N-Acetoglycolyl-4-methyl-4,9-dideoxy-neuraminic acid, see Sialic acid
N-Acetylation of sialic acid, 45
N-Acetyl-β-hexosaminidase, asialoganglioside substrate, 163, 164
N-Acetyl-β-hexosaminidase, ganglioside degradation, 163, 164
N-Acetylneuraminic acid, see Sialic acid
7-O-Acetylneuraminic acid, see Sialic acid
N-Acetyl-4-O-acetylneuraminic acid, see Sialic acid
N-Acetyl-8-O-acetylneuraminic acid, see Sialic acid
N-Acetyl-7,8-di-O-acetylneuraminic acid, see Sialic acid
N-Acetyl-4-O-glycolylneuraminic acid, see Sialic acid
N-Acetylneuraminate: O_2-oxidoreductase, in NGN synthesis, 138
N-Acetylneuraminosyl glycohydrolase, see Sialidase
ACTH, 339
    effect on liver and myocardium sialic acid, 284
    role of sialic acid in, 211
    sialic acid receptor for, 221
Actinomycetales sialidase, 296
N-Acyl derivatives of sialic acid, 28, 29, 38, 39
Acylmannosamine kinase, 132
Acylneuraminate pyruvate lyase, 13
N-Acylneuraminate pyruvate lyase, 45
Adenocarcinoma cells, sialidase effect on, 309
Adhesion, cell, 215-217

Agglutination, cell, 215, 216, 220-223
    erythrocytes, 36
Aggregation, cell, 215-223
Aldol condensation of sialic acid, 10, 36
Aldolase, from *Clostridium perfringens,* 11
Alkaline phosphatase
    chemical and physical properties, 276
    in pregnancy, 278
    sialidase effect on, 174, 278, 307, 308, 310
AL/N cells, 240
Amino acid transport, influence of sialic acid on, 223
Aminodeoxynonulosonic acid (5-amino-3,5-dideoxy-D-glycero-D-galactono-nulosonic acid), 67
Amniocentesis, Tay–Sachs disease, 190, 191
Amylase
    chemical and physical properties, 276
    isoenzymes, 278
    sialidase effect on, 308
Anaphylactic shock, gangliosides in, 224
Antibodies
    sialic acid as receptor for, 211, 212
    to sialidases, 210
Antibody, γ_M-type, 204, 205, 212
Antigens
    A, B, 70, 71
    Allo, H-2, sialidase effect on, 309
    blood groups, 211, 212
    cell surface sialic acid masking of, 205, 206
    H, 70, 71
    heterophile, 71
    histocompatibility, 70, 211
    infectious mononucleosis, 71
    K, 61
    M, 70, 212

Antigens (cont'd)
  N, 70, 212
  O, 61
  of the ox, 70
  P, 70
  θ, 75
  tumor specific, 211
α_1-Antitrypsin
  antigenic determinant, 279
  chemical and physical properties, 276
  in disease, 282
  sialic acid function in, 282
  sialidase effect on, 307
Arylsulfatase, 73, 171
Ascites carcinoma, sialic acid distribution in, 105, 110, 112, 114
Ascites tumor cells, sialidase effect on, 107, 306, 307, 310
Asialoganglioside, G_{A2}, 185
Aspartyl-glycosylamine amidase, in glycoprotein degradation, 166
Asymmetry of sialic acid, 16, 18
ATP-ase, sialidase effect on, 307
Atropinesterase
  antigenic determinant of, 279
  chemical and physical properties, 276
  serum, 73
  sialidase effect on, 278
Autohydrolysis of sialic acid, 22
Autophagy, 170

B-16 cells, sialidase effect on, 312
Baby hamster kidney cells
  hematoside in, 250-252
  sialic acid distribution in, 111
  sialyl transferase in, 251
Bacteriocines, 61
Balb/c cells, 241
Base-cleavage of sialic acid, 11
Base-degradation of sialic acid, 22
N-Benzyloxycarbonyl sialic acid, 39
Bile mucin, 8
Blastogenesis
  galactose oxidase effect on, 224
  periodate-induced, 224
Blood clot formation sialidase effect on, 338, 339
Blood-group antigens, sialidase effect on, 306
Blood group substances, 123
Blood platelets, sialidase effect on, 307

Brain cortex, sialic acid distribution in, 105
Brain excitability, 224
Brain sialidase, 323, 330
  cerebellum, 323
  development, 337
  gray matter, 323
  purification, 326
  substrate specificity, 332
  white matter, 323
Bronchial secretions, 104
Brown fat cells, sialidase effect on, 309

Cancer
  serum sialic acid levels in, 285, 286
  sialidase in management of, 204, 205
Canine distemper, serum sialic acid level, 284
Carbocyanine dye-binding, 285, 286
Carboxypeptidases, acid, and glycoprotein degradation, 166
Cathepsins, 166, 168, 169
  glycoprotein degradation, 166, 168, 169
Cell
  adhesion, sialic acid in, 217
  aggregation
    ionic effect on, 218
    sialidase effect on, 218-220
  cultured, sialidase effect on, 107, 108
  electrophoretic mobility, effect of sialic acid on, 201-203
  lines, 240
    contact-inhibited, 241
    tumorigenic, 241
  proliferation, sialidase effect on, 204, 205
  sialic acid distribution in, 110-113
  surface
    sialic acid in, 201-227
    sialidase effect on, 106-108, 203-207
  surface antigens, masking by sialic acid, 203-207
  surface receptor, sialic acid in, 207-213
Ceramidase, 161
Cerebral cortex cells, sialic acid distribution in, 113
Cerebral tissue, sialidase effect on, 307
Ceruloplasmin, 71, 279, 284
  antigenic determinent, 279
  chemical and physical properties, 276
  sialidase effect on, 172, 173, 308
Cervical glycoproteins, 43
Cervical mucins, 104

Index

Cervix carcinoma cells, sialidase effect on, 308
Chemical carcinogens, effect on gangliosides in cultured cells, 262
Chemical reactions of sialic acid, 22-25
Chemical synthesis of sialic acid, 36-39
Chicken embryo fibroblasts
  gangliosides in, 252
  sialic acid distribution in, 111
Chicken embryo sialidase, 323
Chicken sialidase development in, 336, 337
Chicken wing tumor, sialic acid distribution in, 111
Chloroperoxidase, 73
Cholera toxin, ganglioside receptor for, 268
Cholinesterase, 340
  antigenic determinant, 279
  chemical and physical properties, 276
  serum, 73
  sialidase effect on, 278, 307
Chorionic gonadotropin
  function, 281
  sialic acid content, 281
  sialidase effect on, 279-281, 310
Circular dichroism of sialic acid, 21
Circulating glycoproteins, sialidase effect on, 309
Cirrhosis, 284
Clostridium perfringens
  sialidase, 297
    ionic strength effect, 304
    isoelectric point, 301
    kinetic constants, 302
    pH optima, 302, 303
    purification of, 299, 300
    size of, 301
    substrate specificity, 305
    substrates for, 304, 307-312
Clotting, cell, 216
CMP-sialic acid, 21, 43, 135-138, 141-143, 149
  biosynthesis of, 135
  cellular location of, 115
  feedback inhibitor for ManAc synthesis, 136
  lipid complex of, 61
CMP-sialic acid:ganglioside sialyltransferases, 148, 149
CMP-sialic acid:glycoprotein β-galactosyl sialyltransferases, 143-146
  substrate specificity, 144, 145

CMP-sialic acid:lactose (β-galactosyl) sialyltransferase, 142, 143
CMP-sialic acid:mucin (α-N-acetylgalactosaminyl) sialyltransferase, 146-148
CMP-sialic acid synthetase, 81, 84, 134-136
  nuclear location, 136
Collagenase, glycoprotein degradation, 166
Colominic acid, 5, 43, 61
  biosynthesis of, 141, 142
Colorimetric assay methods for sialic acid, 39-44
Colostrum, 4, 9
Corpus luteum sialidase, 324
Corticosteroids, effect on liver and myocardium sialic acid, 284
Corticosterone, 339
Corynebacterium diphtheriae sialidase, 297
  ionic requirements, 304
  purification of, 299, 300
  size of, 301
  substrates, 307
Crinophagy, 170
Cultured nervous system cells, sialidase effect on, 312
Cyclic AMP, 339

Deacetylation of sialic acid, 46
Decapacitation factor, 338
Dense bodies, 171, 172
Dexamethasone, effect on serum sialic acid, 284
Diabetes
  erythrocyte sialic acid, 283
  insulin effect on sialic acid level, 283
  serum sialic acid level, 283
Diabetes mellitus, 339
Diet, effect on serum sialic acid levels, 286
p-Dimethylaminobenzaldehyde, assay for sialic acid, 40, 44
Diphenylamine assay for sialic acid, 39
Diplococcus pneumoniae sialidase
  ionic requirements, 304
  isoelectric point, 301
  purification of, 299
  size of, 300, 301
  substrates, 309
Disialoganglioside, 16

E. coli U-12, mating inhibition by periodate, 226

Ehrlich's reagent, 2
Electrophoretic mobility of cell, sialic acid role in, 201, 203
Electrostatic charges of sialic acid, role in aggregation, 218
Eledoisin, 341
Endoplasmic reticulum, sialic acid in, 109, 110, 113, 114
Enterokinase, 73
Epiglycanin, 208, 214
Epithelial cells, sialic acid in, 202
Ehrlich ascites tumor cells, sialic acid in, 202
Erysipelothrix insidiosa sialidase, size of, 300
Erythrocytes
 antigens in, *see* Antigens
 sialic acid in, 202
 sialidase, 324, 331
 sialidase effect on, 106-108, 173, 306-308, 311, 312
 sialoglycopeptides in, 70
Erythropoietin, 72
 function, 281
 sialic acid content, 281
 sialidase effect on, 173, 280, 281, 309
Esterification of sialic acid, 23-25
Eubacteriales sialidase, 296
Extracellular secretions
 sialic acid containing macromolecules in, 104
 sialic acid in, 104, 105
Eye sialidase, 324, 331

Factor VIII, 222, 339
 effect of platelet sialyltransferase on, 280
 platelet agglutination by, 222
 sialidase effect on, 280
Fetuin, 71, 320, 329, 331
 chemical and physical properties, 276
 degradation, 166
 injury, role in, 283
 sialic acid, function in, 279
 sialidase effect on, 172
Fibrinogen
 chemical and physical properties, 276
 desialylation of, 280
 sialic acid components in, 280
Fibroblasts
 sialic acid in, 202
 sialidase effect on, 309, 311, 312
Fibrosarcoma, sialidase, effect on, 310, 311

Flat revertants, ganglioside composition of, 257-259
Follicle stimulating hormone, 36, 338
 function, 281
 sialic acid content, 281
 sialidase effect on, 172, 173, 280, 281, 306, 310
Fowl plague virus sialidase, substrate specificity, 319, 320
Friedenreich antigens, 203, 204, 212
 acid treatment of, 212
 sialidase effect on, 204, 212
α-Fucosidase, glycoprotein degradation, 166, 167
Fucosidosis, 167

Galactosaminidase (N-acetyl-α-), glycoprotein degradation, 166, 168
Galactosaminyltransferase, (N-acetyl), 81, 242, 243, 249, 258-260, 262, 265
Galactose oxidase, effect on blastogenesis, 224
α-Galactosidase, 73
 sialidase effect on, 174
β-Galactosidase, 161, 162
 deficiency in G_{M1}-gangliosidosis, 194
 ganglioside degradation, 162, 163, 168
 glycoprotein degradation, 166, 168
 species of, 162, 163
 substrates for, 162, 163, 194
Galactosylgalactosamine, sialic acid linked to, 69
Galactosyltransferase, 242, 243, 249, 259, 262, 265
Gangliosides, 3, 8, 9, 19, 21, 43, 45, 46, 329-332, 340
 in amphibians, 65, 66
 in anaphylactic shock, 224
 biosynthesis of, 148, 149
 biosynthesis in cultured cells, 246-248
 biosynthetic pathway, 242
 in birds, 66, 67
 catabolic disorders, 183-197
 catabolism of, 160-165, 183, 184
 catabolism in cultured cells, 240, 244
 cell—cell contact, effect on, 219, 220
 in cell growth, 75
 in crab, 63
 composition in cultured cells, 241, 245
 cholera toxin receptor, 75, 268
 critical micelle concentration, 329

Index

Gangliosides (cont'd)
  degradative enzymes, 162-165
  degradative pathway, 161, 184
  electro-convulsive shock, effect on, 80
  in extraneural tissues, 81-82
  in fishes, 65
  fucose containing, 82
  glucosamine containing, 82, 162
  in hemolysis, 75
  in intercellular interactions, 75
  isolation from cultured cells, 241
  light effect on, 80
  lysosomal degradation, 189
  in mammals, 75-82
  metabolism
    culture conditions effect on, 248-250
    in cultured cell lines, 244-253
    molecular basis of alteration, 264-267
    RNA virus transformation effect on, 261-264
    in transformed cells, 260-264
    viral transformation effect on, 257-264
  monosialogangliosides, 9
  in molluscs, 63
  in neuronal membranes, 224
  in neurons and glial cells, 80, 81
  nomenclature, 67, 76-79
  in penicillin induced convulsion, 75
  radiochromatograms of, 247
  in reptiles, 66
  sensory and visual stimuli effect on, 80
  serotonin receptors, 75, 211
  sialidase substrate, 160-162
  in Sindbis virus, 60
  source of, 162
  structure of, 67, 76-79, 140
  subcellular distribution, 244-246
  subcellular distribution in brain, 109
  Tay–Sachs, see Ganglioside, G_{M2}
  tetanus toxin receptors, 75, 213
  theta-antigen, 75
  toxin binding, 174
  viral infection, effect on, 257
  viral transformation, role in, 267-269
Ganglioside, G_{M1}, storage, 193, 194
Ganglioside, G_{M2}, 9
  storage, 185, 188, 189, 191
  structure, 185
Gangliosidosis
  animal models, 195, 196

Gangliosidosis (cont'd)
  cell culture models, 196, 197
Gangliosidosis, G_{M1}, 163, 168, 193-195
  clinical aspects, 194, 195
Gangliosidosis, G_{M2}, Type I, see Tay–Sachs Disease
Gangliosidosis, G_{M2}, Type II, 191, 192
  clinical aspects, 191, 192
  pathology, 191
Gangliosidosis, G_{M2}, Type III, 192, 193
Gangliosidosis, G_{M3}, 195
Gas–liquid chromatography, 15
  of sialic acid, 5, 43, 45, 47
Gaucher's Disease, enzyme deficiency in, 240
Generalized gangliosidosis, see gangliosidosis G_{M1}
Globoid cell leukodystrophy, 163
α-Globulins, 284
β-Globulins, 279
γ-Globulins, 277
$γ_M$-Globulins, 204
7 S Globulins, 285
Glucocerebrosidase, 183, 240
D-Glucosaminidase (N-acetyl α-), glycoprotein degradation, 168
Glucosaminidase (N-acetyl β-), glycoprotein degradation, 166, 168
β-Glucosidase, 161, 164
  characteristics, 164, 165
  glycoprotein degradation, 166, 168
  sources, 164
L-Glutamine: D-fructose-6-phosphate, aminotransferase, 126, 127, 136, 137
α-Glutamyltranspeptidase, 73
γ-Glutamyltranspeptidase, sialidase effect on, 307
Glycoamylase, 73
Glycolipid metabolism
  glycosyltransferases, 242-244
N-Glycolyl-4-O-acetyl-neuraminic acid, see Sialic acid
N-Glycolyl-8-O-acetyl-neuraminic acid, see Sialic acid
N-Glycolylneuraminic acid, see sialic acid
Glycolysis, inhibition of, 223
Glycopeptides
  in temperature sensitive mutants, 256
  in transformed cells, 255, 256
Glycoproteins
  changes during cell cycle, 268, 269
  circulating
    sialic acid in protection of, 212, 213

Glycoproteins, circulating *(cont'd)*
  sialic acid receptor for, 212, 213
  composition in membranes, 254, 255
  degradation of, 165-169
  degradative enzymes, 167-169
  degradative pathways, 165-167
  membrane insertion, 150
  mucin type, 144
    structure, 140, 146
  positional specificity, 151
  secretion, 150
  serum type, 144, 146
  sialic acid, incorporation in, 138-140
  structure, 140
α_1-Glycoprotein, 279
α_1-Glycoprotein (3.5S), chemical and physical properties, 276
Glycosidases, lack in genetic disorders, 161, 162
Glycosyltransferases, 138, 139
  acceptor molecules for, 140
  in asialoglycoprotein binding protein, absence of, 71
  assay systems, 243
  biological roles of, 85
  cell density effect on, 249
  cell surface, 151
  in cellular aggregation or adhesion, 216, 219, 222
  galactosyltransferase, N-acetyl, 81
  general properties of, 138-141
  in glycolipid metabolism, 242-244
  host-dependent glycosylation of viral glycoproteins, 60
  intercellular adhesion mediation of, 151
  sialyltransferase, 81, 85
  soluble, 143
Gonadotropins, *see* Hormones
Granulation tissues, sialidase effect on, 312
Granulocytes
  adhesion to vascular endothelium, 222
  sialidase effect on, 311
Gynaminic acid, 1, 67

Hand–Schüller–Christian disease, serum sialic acid level in, 283
Haptoglobins, 71, 284
  chemical and physical properties, 276
Heart-muscle sialidase, 324, 331
  isoelectric point, 327
  purification, 327

HeLa cell
  sialic acid distribution in, 110, 115, 202
  sialidase effect on, 115, 173
Hemagglutination
  sialic acid in, 212, 220, 221
  virus inhibition, 285
Hemagglutinin, 314-317
Hemagglutinin, lobster, sialic acid in, 63
Hemataminic acid, 1, 67
Hematoside, 4, 13, 43
Hematoside gangliosidosis, *see* Gangliosidosis G_{M3}
Hematoside: UDP-GalNAc N-acetylgalactosaminyl-transferase, *see* UDP-GalNAc:G_{M3} N-acetylgalactosaminyltransferase
Hemolysin and O serotype, 61
Hemolysis by *Cryptococcus laurentii,* 75
Hemophilia, 222
Hepatitis, 284
Hepatoglobin, sialidase effect on, 172
Hepatoma cells, sialidase effect on, 106, 107, 306, 308, 310
Hepatoma, mouse, sialic acid distribution in, 112
Heterophagy, 170
Hexosaminidase, (N-acetyl β-), 73, 74, 161, 163
  glycoprotein degradation, 168
Hexosaminidase A, (N-acetyl β-), 163, 164, 188-190
  in deficient adults, 192
Hexosaminidase B, (N-acetyl β-), 163, 164, 188-190
Hf-Sialic acid, 7, 14
Histocompatibility antigens, antisera absorption, 205, 206
Holothuria forskali, 14
Hormone–target cell interaction, sialidase effect on, 338, 339
Hormones
  erythropoietin, 72
  follicle-stimulating sialic acid in biological activities of, 72
  gonadotropin, human chorionic, 72
  gonadotropin, in reptiles, 66
  luteinizing, 72
  sialic acid as receptor for, 211
Horseshoe crab, *Limulus polyphemus,* lectin, 207
Hudson's empirical rotation rules, 16

Hudson's isorotation rules, 20
Hudson's lactone rule, 20
Human chorionic gonadotropin, 338
　sialidase effect on, 172, 173
Hydrocortisone, effect on liver, heart, and serum sialic acid, 284
Hydrolysis of sialic acid, 8
　enzymatic, 9
Hydroxylation, NAN to NGN, 138
Hydroxyl group reactions of sialic acid, 33, 34
Hypercapnia effect on neuronal sialic acid, 224

Immune response
　sialidase effect on, 205-207
Immunoglobulins, 71
Immunoglobulin A
　chemical and physical properties, 277
　sialidase effect on, 307
Immunoglobulin D, chemical and physical properties, 277
Immunoglobulin E, chemical and physical properties, 277
Immunoglobulin G
　chemical and physical properties, 277
　sialidase effect on, 307
Immunoglobulin M, chemical and physical properties, 277
Infectious psychoses, serum sialic acid level, 284
Inflammatory reactions, serum sialic acid level, 283
Influenza virus sialidase
　antigenic properties, 315
　biosynthesis of, 315, 316
　location, 314
　purification of, 316-318
　size, 318, 319
　substrates, 306, 309
Insulin, sialic acid receptor for, 211
Insulin binding, 211, 339
Intercellular interactions, glycosyl-transferases in, 75
Interferon, sialic acid in, 74
　sialidase effect on, 308
Intestinal mucosa, sialidase, 323, 324
Ion transport, sialic acid effect on, 223
Islets of Langerhans, sialidase effect on, 309

Japan *305* virus sialidase, biosynthesis of, 315

2-Keto-3-deoxyaldonic acid, 5, 43, 44
　in plants, 62
Ketoside formation by sialic acid, 23, 25-33
α-Ketosides of sialic acid, 25, 28, 30-33
β-Ketosides of sialic acid, 25-28
Ketosidic configuration of sialic acid, 20, 21
Kidney cells
　hematoside in, 252, 253
　sialyltransferase in, 252, 253
Kidney sialidase, 323, 331
　purification, 326
Kininogen, 72
Klebsiella aerogenes sialidase, 297
Koenigs-Knorr procedure, 20, 25, 28

L cell
　carbohydrate content of, 106
　sialic acid in, 203
　sialic acid distribution in, 105, 110, 114
　sialidase effect on, 107
Lactaminic acid, 1, 67
Lactobacillus bifidus sialidase, purification of, 299
Lactoferrin, 71
　sialidase effect on, 172
Lactosamine, sialic acid linked to, 68
Lactosyl ceramide-β-galactosidase, 183
Lectins, sialic acid specific, 207, 208
Leukemic cells, 285
　sialidase effect on, 173, 307, 308, 311
Leukemic myoblasts, sialidase effect on, 310
Leukocytes
　myelogenous leukemic, glucocerebrosidase activity, 240
　sialidase effect on, 107, 306, 309, 311, 312
Leukocyte sialidase, 324, 331
Lewis blood group, 278
Lewis lung tumor, 285
Lipidoses
　cell culture models, 196, 197
　general characteristics of, 183
Lipofuscin, 171
Liver carcinoma, 284
Liver cells
　sialic acid distribution in, 111, 112, 114
　sialidase effect on, 106, 107, 306
Liver disease, sialic acid changes in, 284, 285
Liver sialidase, 323, 330, 331
　development, 337
Lung sialidase, 323

Lung tumor cells, sialidase effect on, 306
Luteinizing hormone (*see also* Hormones)
    sialidase effect on, 308
Lymph node cells, sialidase effect on, 310
Lymphocytes
    migration, sialidase effect on, 223
    sialic acid in homing, 206, 207
    sialidase effect on, 173, 307-309, 311
    stimulation, sialidase effect on, 224, 225
Lymphocytes, lymphatic leukemia, sialidase effect on, 107
Lymphoid cells, sialidase effect on, 309-311
Lysergic acid diethylamide, 340, 341
Lysosomes, 169
    sialic acid in, 116, 117
    substrate uptake, 169, 170

Macroglobulin, 285
Macroglobulin, α-, 283
Macroglobulin, $\alpha 2$-, 71
    chemical and physical properties, 277
    sialidase effect on, 172
Macrophages, sialidase effect on, 308
Malignant cells, sialic acid content of, 213
Mammalian cells, sialidase effect on, 311
Mammary carcinoma cells, sialidase effect on, 309, 311
Mammary gland sialidase, 323, 331
Mannosamine (N-acetyl)
    biosynthesis, 131, 132
    glucosamine (N-acetyl) interconversion, 131, 132
    mechanism for synthesis, 131
Mannosidase, α-, in glycoprotein degradation, 166-168
Mannosidosis, 167, 168
M blood group antigens, sialic acid in, 212
Meconium, 9
Membranous cytoplasmic bodies, 172, 185, 187
Metachromatic leukodystrophy, 171
8-Methoxy-N-glycolylneuraminic acid, *see* Sialic acid
Methoxyneuraminic acid, 4, 12
Mitochondria, sialic acid in, 114
MN blood group substances, sialidase effect on, 173
Monocytes, sialidase effect on, 307
Monosialoganglioside, 2, 16, 20
    space filling model, 3

Mucins
    biosynthetic pathways, 147
    sialic acid incorporation in, 138-140
Mucopolysaccharides, storage of, 193
Multiglycosyltransferase systems, 150
Mumps virus sialidase, substrate specificity, 320
Muscle, sialic acid distribution in, 111
Muscle cells, effect of sialidase on, 308
Mycoplasma, sialic acid as receptor for, 210, 211
Mycoplasmatales sialidase, 296
Myeloma cell sialic acid, 202
Myxovirus sialidase, purification of, 318

NAN-aldolase, 133-135
N blood group antigens, sialic acid in, 212
Neuraminic acid
    structure, 2
    terminology, 1-3
Neuraminic acid methylketoside, methyl ester, 13
Neuraminidase, *see* Sialidase
Neuraminlactose, *see* Sialyllactose
Neurotransmission, sialidase effect on, 338, 341
Newcastle disease virus sialidase
    biosynthesis of, 315, 316
    purification of, 317
    size, 319
    substrate specificity, 319, 320
Niemann–Pick's disease, 3
Nomenclature of sialic acids, 1, 2, 6, 7
Nuclei, sialic acid in, 114, 115
Nucleotide sugars, synthesis of, 130, 131

Oncogenicity, role of mammalian sialidase in, 341
Oral bacteria, sialidase of, 313
Orcinol determination of sialic acid, 2, 39-41
Orosomucoid, 71, 277, 320
    degradation, 166
    in disease, 279
    sialidase effect on, 172, 306
Orthomyxovirus, sialidase in, 296
Ovary cells, sialidase effect on, 307
Oxytocinase
    chemical and physical properties, 277
    sialic acid function in, 278, 279
    sialidase effect on, 306

Index

Pancreatic islets, sialidase effect on, 312
Parainfluenza virus, adsorption and infectivity, 210
Paramyxovirus sialidase, 296
Pasteurella multocida sialidase, 297
  size of, 300
  substrates, 309
Periodate oxidation
  cellular receptor destruction by, 201
  effect on blastogenesis, 224
  of sialic acid, 12-14, 34-36, 42
  site of action, 202, 209
Periodate-resorcinol assay for sialic acid, 41
Peritoneal cells, sialidase effect on, 311
Phagocytosis, sialic acid in, 223, 224
Phagocytosis inhibition, 223
1,10-Phenanthroline assay for sialic acid, 44
Phosphatase, acid, 72
Phosphatase, alkaline, 72
Phosphoanhydrides, unsymmetrical, 129, 130
Phosphoglucose isomerase, 126, 127
Pituitary gonadotropin, sialidase effect on, 279, 280
Pituitary hormone, function, 281
Pituitary luteinizing hormone, sialic acid content, 281
pK_a of sialic acid, 22
Plasma membrane, sialic acid in, 105-113
Plasma membrane glycoproteins, sialidase effect on, 309
Plasma sialoglycoproteins, 275-288
Plasminogen, 72
Platelet agglutination, role of sialic acid in, 221, 222
Platelet aggregation, 221, 222
Platelet sialidase, 324
Platelet sialic acid, 202
Pneumococcus sialidase, 297, 298
  substrates, 307
Polymorphonuclear leukocytes, sialidase effect on, 310
Polyoma virus, in cell transformation, 240
Prednisolone, effect on brain, liver, and serum sialic acid, 284
Pregnancy, sialidase effect on, 308
Protagon, sialidase effect on, 310
Protein transport, influence of sialic acid, 223
Prothrombin, 72
  sialidase effect on, 309
Pseudocholinesterase, sialidase effect on, 306

Pseudomonadales sialidase, 296
Py cells, sialidase effect on, 308
Pyrophosphatase, nucleotide, 73
Pyrophosphorylase, 129, 135
Pyruvate lyase, N-acylneuraminate, 15, 45

Receptor destroying enzyme (RDE), 123, 124, 201, 208, 280 (*see also* sialidase)
Reproduction, sialidase effect on, 338
Residual bodies, 171, 172
Resorcinol determination of sialic acid, 40, 41, 44
Retinal cells, sialidase effect on, 311
Reverse aldolization of sialic acid
  chemical, 10-12
  enzymatic, 11-13
Ribosome-free membranes, sialidase effect on, 311
Rickets, serum sialic acid level in, 283
RNA viruses 261, 262

Saliva sialidase, 324
Sandhoff's Disease, *see* Gangliosidosis, G_{M2}-Type II
S- and N-ketosyl derivatives of sialic acid, 29, 33
Saponification of sialic acid esters, 24, 25
Sarcoma cells
  sialic acid in, 202
  sialidase effect on, 106, 107, 307
Sea urchin eggs, 14
Sea urchin jelly coat, 104
Sendai virus sialidase
  purification of, 316, 317
  size, 319
  substrate specificity, 320
Sero-lactaminic acid, 67
Seromucoid, 331
  recovery of, 282
Serotonin, 340, 341
  sialic acid receptor for, 75, 109, 211
Serum alkaline phosphatase, sialidase effect on, 306
Serum factor, cytotoxic, 204, 205
Serum sialoglobulins, 279
Sialic acid
  O-acetylation, 128, 138
  O-acetylneuraminic acid, 203
  acid hydrolysis of, 22, 23, 43
  acidity of, 22
  activation of, 135

Sialic acid (*cont'd*)
  amount at cell surface, 202, 214
  in amphibians
    frogs, 65
    toads, 66
    triton, 65
  anabolic pathway outline, 125, 126
  anabolic reactions of, 123-152
  antibody receptor, 211, 212
  antigens, *see* Antigens
  in arylsulfatase, 73
  in atropinesterase, 73
  in bacteria
    bacteriocines, 61
    colominic acid, 61
    endotoxin, 61
    gram-negative, 60, 61
    gram-positive, 60, 61
    hemolysin, 61
    in capsular material, 61
    K serotype, 61
    lipid–CMP–NANA complex, 61
    O-antigen, 61
    pathogenicity, 60, 61, 84, 85
  in bile, 74
  biosynthesis of, 125-135
    acetylation in, 128
    N-acetylglucosamine phosphates interconversion of, 129
    N-acylmannosamine in, 131-135
    N-acylmannosamine-6-PO_4 in, 134
    N-acylneuraminic acid-9-PO_4 in, 134
    fructose-6-PO_4 as precursor for, 126-128
    glucosamine formation in, 126-128
    glucose as precursor, 125
    nucleotide sugars in, 129-131
    pathway, 137
    phosphoenol pyruvate in, 134, 135
    pyrophosphorylase in, 129-131
    pyruvate in, 133
    regulation of, 136, 137
  in birds
    brain gangliosides, 66, 67
    collocalia mucoid, 66
    egg white, 66
    egg yolk gangliosides, 67
    salt gland and osmoregulation, 67
    serum, 66
    tyrosine *O*-glycoside, 67
  in brain, 66, 74
  in bulbourethral gland, 75

Sialic acid (*cont'd*)
  cell aggregation, role in, 217-223
  in cell–cell interaction, 215-223
  cell membrane location, 215
  cell quantity, 202, 203
  in cell surface, 201-227
    antigen masking by, 203-207
    changes in cancer cells, 202, 203
    regeneration, 205
  cellular adhesion, role in, 217
  in cephalic glands, 66
  in cephalochordates, 64
  in cervix, 75
  chemical reactions and derivatives, 22-36
  chemistry of, 10-36
  chain termination by, 147, 150
  in chloroperoxidase, 73
  in cholinesterase, 73
  cleavage, 8, 9
  colorimetric and fluorometric assays of, 39-44
  composition, 2, 4, 36
  configuration, 16-20
  conformation, 2, 16, 18
  control of subcellular distribution, 116, 117
  coulombic effect, 205, 215, 218, 219
  in Crustacea, 63
  crystallization of, 9, 10, 201
  cyanosis, 224
  definition of, 67
  derivative biosynthesis of, 137, 138
  detection in cell surface, 202
  determination by fluorescent assay, 44
  determination by thiobarbituric acid assay, 278
  diastereoisomers of, 36
  diphenylamine reagent for, 39
  direct Ehrlich reaction for, 43-45
  distribution in cell, 105-116
  distribution in eukaryotic cells, 103-117
  in echinoderms, 63, 64
  effect upon circulating glycoproteins, 206
  endoplasmic reticulum, 109, 110, 113, 114
  enterokinase, 73
  enzymatic assay for, 45
  enzymatic hydrolysis of, 43
  enzymatic synthesis of, 36
  evolution of, 84-86
  extracellular, 104, 105
  feedback regulation in synthesis of, 136

Sialic acid (cont'd)
  fish
    cyclostomes, 65
    elasmobranchs, 65
    lampreys, 65
    teleosts, 65
  fluorescent assay for, 44
  forms of
    N-acetoglycolyl-4-methyl-4, 9-dideoxy-neuraminic acid, 6, 7, 13, 14, 43, 63
    N-acetylneuraminic acid, 60, 61, 64-69, 83
    N-acetyl-4-O-acetylneuraminic acid, 4, 6, 12, 66, 69, 83
    N-acetyl-8-O-acetylneuraminic acid, 6, 13, 65, 83
    N-acetyl-7,8-di-O-acetylneuraminic acid, 83
    N-acetyl-4,9-di-O-acetylneuraminic acid, 6, 13
    N-acetyl-3-fluoroneuraminic acid, 38
    N-acetyl-4-O-glycolylneuraminic acid, 84
    N-acetyl-7 or 8-O-glycolylneuraminic acid, 84
    N-acetyl-tri-O-acetylneuraminic acid, 83
    N-glycolyl-4-O-acetylneuraminic acid, 84
    N-glycolyl-8-O-acetylneuraminic acid, 84
    N-glycolylneuraminic aicd, 63-69, 83, 203
    8-methoxyl-N-glycolylneuraminic acid, 63
    7-O-acetylneuraminic acid, 83
  function of, 150-152, 172-174
  functional groups of, 10, 22
  in α-galactosidase, 73
  in gangliosides, see Gangliosides
  in α-glutamyltranspeptidase, 73
  in glycoamylase, 73
  N-glycolation, 128
  glycoprotein receptor, 212, 213
  in glycoprotein turnover, 151
  in gonadotropins, 66
  hemagglutination, role in, 220, 221
  hematoside, 60
  hemichordate, 64
  Holothuria forskali, 64
  homopolymer of, 5, 21
  hormones, receptor for, 211 (see also Hormones)
  host-dependent glycosylation of viral glycoproteins, 60, 84, 85

Sialic acid (cont'd)
  in humans
    N-acetylneuraminic acid, 83
    N,O-diacetylneuraminic acid, 84
    N-glycolylneuraminic acid, 83, 84
  hydrochloric acid reagent, 39
  in hypercapnia, 224
  infectivity, 60
  infrared spectra, 18, 36
  isolation of, 4, 8, 10, 201
  isotopically labelled, 39
  2-keto-3-deoxyaldonic acid relationship, 62
  ketoside derivatives, 25-33
  ketoside linkages, 19
  lactones, 16
  lactonization of, 24
  lipid-bound at cell surface, 108, 109
  in lysosomes, 116, 117
  malignant cell culture, loss in, 214
  in mammals, 67-84
    free, 67, 68
    in brain, 68
    in cerebrospinal fluid, 67
    in gastric tissue, 67
    in seminal vesicles, 68
    in sialuria, 68
    in urine, 68
    in glycoproteins, 69-75
      N-acetyl-β-D-hexosaminidases, 73, 74
      acid phosphatase, 72
      alkaline phosphatase, 72
      immunoglobulins, 71
      interferon, 74
      kininogen, 72
      nucleotide pyrophosphatase, 73
      phosphatase, 72
      plasminogen, 72
      prothrombin, 72
      semen, 75
      serum glycoproteins, 71, 72
      spermatozoa, 75
      thrombin, 72, 73
      urine, 74, 75
    in heterosaccharides, 68, 69
      galactosylgalctosamine containing, 69
      lactosamine containing, 68, 69
      lactose containing, 68, 69
      UDP-containing, 69
  mass spectrometry of, 12, 15, 16
  mechanism of acid degradation, 23

Sialic acid (cont'd)
  methanolysis of, 8, 9, 45, 46
  methyl ester formation, 23
  in mitochondria, 114
  model of, 2, 3
  in molluscs, 63
  mutarotation, 18
  mycoplasma receptor, 210, 211
  natural occurrence, 59-86
  in normal cells, 213, 215
  nuclear magnetic resonance of, 18
  in nuclei, 114, 115
  occurrence in vertebrates and invertebrates, 5, 8
  optical rotation of, 18, 19, 36
  oxidation and reduction, 34-36
  paper chromatography of, 5
  periodate oxidation of, 34, 35
  periodate-resorcinol method, 41
  pK_a values, 22
  in phagocytosis, 223, 224
  in plants, 61, 62
  in plasma membrane, 105-113
  in platyhelminths, 63
  polyhydroxy side chain interactions, 226, 227
  protein-bound at cell surface, 108, 109
  in protozoa, 62, 63
  purification of, 9, 10
  pyrophosphorylase activation of, 135, 136
  quantification of, 39-47
  reactions with acid and alkali, 22, 23
  reactions of alcoholic hydroxy groups, 33, 34
  reactions of the carboxyl groups, 23-25
  receptor for antibodies, 211
  receptor at cell surfaces, 207-213
  receptor for lectins, 207
  receptor for mycoplasma, 210
  receptor for tetanus toxin, 213
  receptor for viruses, 208-210
  reducing properties, 34
  regeneration at cell surface, 205
  regulation of metabolism, 136
  renal threshold value, 104
  in reptiles, 66
  resorcinol method for, 41
  role of, 223-227
  role in cerebral tissue, 224
  in Rous sarcoma, 60
  S- and N-ketosyl derivatives of, 29

Sialic acid (cont'd)
  in sialoglycolipids, 60
  in sindbis virus, 59, 60, 85
  in skin, 66
  sources of, 4, 5, 9, 13-15
  space filling model of, 3
  species, 2, 4, 6-8, 12, 13, 15
    at cell surface, 203
  sperm capacitation, role in, 225
  staining at cell surface, 202
  stereochemistry of, 16-22
  structure of, 38, 39, 124, 133
  subcellular distribution, 104
  9-substituted, 34
  sulfuric-acetic acid reagent, 40
  synthesis, 36-39
  synthesizing enzymes, 132
  tetanus toxin, receptor, 213
  thin-layer chromatography of, 10
  thiobarbituric acid procedure
    molar extinction coefficients, 42
    β-formylpyruvate, chromogen for, 35
  in transformed cells, 213-215
  in transport, 223
  N-, O-trifluoroacetyl derivative of, 46
  trimethylsilylether derivative of, 13, 15, 16, 45
  trypsin release of, 108
  tryptophan-perchloric acid reagent, 40
  in ungulic acids, 82
  unsaturated derivative of, 14
  in vesicular stomatitis, 60
  in viruses, 59, 60
  virus receptor, 208-210
  X-ray diffraction of, 11, 18, 36
  O-Sialic acid, 1
Sialic acid aldolase, see NAN-aldolase
Sialic acid monophosphate, in sialyl transfer to platelets, 221
Sialic acid-9-phosphatase, 134, 135
Sialic acid-9-phosphate synthetase, 134, 135
Sialidase
  agglutination, effect on, 220, 221
  aggregation, effect on, 217-220
  amino acid transport, effect on, 223
  bacterial, 296-313
    biological roles of, 305-313
    cellular location, 297, 298
    characteristics, 297, 298
    induction, 297, 298
    inhibitors, 302, 304, 305

Index

Sialidase, bacterial (cont'd)
  ionic strength effects, 302-304
  isoenzymes, 301
  properties of, 301
  purification of, 298-300
  secretion of, 297, 298
  size of, 300, 301
  subunit structure, 301
 cell proliferation, effect on, 204, 205
 cell surface, effect on, 203-207
 cell transplantability, effect on, 204-207
 cellular adhesion, effect on, 217
 chicken development in, 336, 337
 competitive inhibitors, 33
 discovery of, 295
 function of, 295
 in ganglioside degradation, 160-162
 ganglioside as substrate, 244
 in glycoprotein degradation, 166, 167
 in Golgi, 325
 granulocyte adhesion, effect on, 222
 immune response, effect on, 205-207
 inhibition of, 13, 14
 ion transport, effect on, 223
 lectin binding, effect on, 207, 208, 225
 lymphocyte migration, effect on, 206, 223
 lysosomal, 324, 325
 mammalian, 322-341
  activators, 333-335
  assay of, 327, 328
  biological roles of, 337-341
  development of, 336, 337
  endogenous activity of, 332, 333
  inhibitors of, 333-335
  ionic effects on, 333, 334
  ionic requirements of, 333-335
  kinetic constants for, 329
  organ distribution of, 323, 324
  pH optima of, 328, 330, 331
  physical properties, 328-336
  purification, 326, 327
  role in oncogenicity, 341
  subcellular distribution, 324, 325
  substrates for 330-333
  substrate specificity, 328, 329, 332, 333
  temperature stability of, 335, 336
 microbial
  experimental uses of, 306-312, 322
  kinetic constants for, 302
  pH optima, 302, 303

Sialidase (cont'd)
  substrates for, 302, 304-312
  in nerve endings, 325
  occurrence in microbes, 296-299
  phagocytosis, effect on, 223
  in plasma membrane, 324, 325
    function in, 288
  platelet aggregation, effect on, 221, 222
  in small intestine, 323
  source of, 188
  specificity of, 9, 19-21, 28, 29, 209, 215
  sperm capacitation, effect on, 225
  substrates of, 28, 29
  terminology, 295, 296
  in tissue culture cells, 341
  viral, 313-322
    antigenic properties of, 314-316
    biological roles for, 320-322
    genetics of, 313-316
    inhibitors of, 320
    ionic requirements, 320
    kinetic constants, 319, 320
    morphology, 313-316
    occurrence, 313
    pH optima, 319
    physical properties of, 210
    properties of, 319, 320
    purification of, 316-318
    role of, 209, 210
    size, 318, 319
    substrates, 319, 320
    in viral infectivity, 210
Sialocompounds
  sialidase effect on half life, 338
  role of sialic acid in, 286-288
Sialoenzymes, circulating, 278, 279
Sialoglobulins, serum, 279
Sialoglycolipids, see Gangliosides
  cellular mechanism of degradation, 169-172
  in echinoderms, 64
  in molluscs, 63
Sialoglycopeptides, 329, 332
  cell surface release by trypsin, 108
Sialoglycoproteins, 329-332
  in abnormal physiological states, 282-286
  cellular mechanism of degradation, 169-172
  degradation of, 165-167
  hormone, 279-282
  plasma, 275-288
  in transformed cells, 253-257

Sialohormones, sialidase effect on, 280, 281
Sialo-macromolecules, biosynthesis of, 138
Sialopolyanion, 285
Sialuria, 14, 33
Sialyllactose, 19, 20, 22, 68, 69, 320, 329-331
  biosynthesis of, 142-146
  isomers of, 22
  structure of, 140, 143
  sulfate, 329
Sialyltransferases, 81, 85, 139-149, 221, 339
  acceptor molecules, 140
  acceptor specificity, 139, 142
  for colominic acid synthesis, 141, 142
  effect of cell transformation on, 214
  exogenous acceptors, 140, 141
  glycolipid specific, 148
  linkages synthesized by, 139
  particulate, 142
  reaction, 139
  soluble, 145
  substrates for, 253
  in transformed cells, 255, 256
Sialyltransferase I, 242, 243, 249, 259, 262
Sialyltransferase II, 242, 243, 249, 259, 262, 265
Sialyltrisaccharides, 331
Simian virus 5 sialidase, location, 314
Simian virus 40
  in cell transformation, 240, 241
Skin graft, sialidase effect on, 311
Spermatozoa, sialidase effect on, 311
Sperm capacitation
  role of sialic acid in, 225
  sialidase effect on, 225
Spirillaceae sialidase, 296
Sphingolipid composition in tumors, 240
Spleen cells, sialidase effect on, 310
Spleen sialidase, 323
Steroid hormones, effect on blood sialic acid, 284
Streptococcus sialidase, 297
  ionic requirement, 304
  purification of, 299
Streptomyces sialidase
  purification of, 299, 300
  size of, 301
Structure determination of sialic acid, 10-16
Structure of sialic acid, 2
Submandibular glands, 15
Submaxillary glycoproteins, 43, 329

Submaxillary mucin, 2, 4, 8, 9, 12, 13, 19, 124, 166, 320, 329
  degradation, 166
  pentasaccharide portion, 148
Sulfophosphovanillin assay for sialic acid, 44
Sulfuric-acetic acid determination of sialic acid, 40
Synaptic membranes, sialidase effect on, 307
Synaptic structure, sialidase effect on, 312

3T3 Cells, 241
  sialic acid distribution in, 111
TA3 murine cancer cell
  sialic acid in, 203
  sialidase effect on, 311
Tay–Sachs disease, 172, 184, 191
  clinical aspects, 184
  enzymology of, 189, 190
  metabolic defect in, 185, 188, 189
  pathology of, 184-187
  prenatal diagnosis, 190, 191
Tay–Sachs ganglioside (G_{M2}), 20, 21, 43
Testes, sialidase in, 323
Tetanus toxin
  gangliosides as receptors for, 213
  role of sialic acid in fixation of, 213
  sialic acid, receptor for, 213
  sialidase effect on, 312
Thiobarbituric acid assay, 327, 328
Thomsen effect, 203, 204, 212
  role of sialidase in, 204
Thrombin, 72, 73
Thrombocytes, effect of sialidase on, 309
Thrombocytopenia, 222
Thymocytes, sialic acid distribution in, 110
Thyroglobulin, 71
  function of, 281
  sialic acid content of, 281
  sialidase effect on, 172, 281
  sialylation, 280
Thyroid-stimulating hormone, 280
Thyroxine-binding globulin, 284
Tissue culture cells, sialidase effect on, 308, 312
Toxins
  cholerae, receptor for, 75
  tetanus, receptor for, 75
Transferrin, 279, 329
  chemical and physical properties, 277
  sialidase effect on, 306, 307, 309

Index

Transformed cells
  glycopeptides in, 255, 256
  sialic acid content of, 213
  sialoglycoproteins in, 253-257
  sialyltransferase in, 253, 254
Transport, sialic acid in, 223
Trichomonas foetus sialidase, 296
Thiobarbituric acid assay, 5, 22, 35, 40, 42-44
Trophosblastic cells, sialidase effect on, 310, 311
Tryptophan-perchloric acid determination of sialic acid, 40
Tubercular meningitis
  serum sialic acid level in, 283

UDP-GalNAc:G_{M3} N-acetylgalactosaminyltransferase in cultured cells, 247, 248, 250, 257, 266
UDP-N-acetylglucosamine, metabolic conversion, 130
Ungulic acid, 82
Uridine diphosphate, sialoheterosaccharide-containing, 69
Urinary glycoprotein
  sialidase effect on, 307
  urinary tract epithelium sialidase, 324, 331

Vibrio cholerae sialidase, 296-298
  ionic requirements, 304

Vibrio cholerae sialidase (*cont'd*)
  ionic strength effect, 304
  isoelectric point, 301
  kinetic constants, 302
  pH optima, 302, 303
  purification of, 298, 299
  size of, 300, 301
  substrate specificity, 305
  substrates, 304, 306, 312
Vibrio comma, see Vibrio cholerae
Viral transformation, role of gangliosides in, 267-269
Viruses
  evolution of sialic acid in, 60, 84, 85
  infectivity and sialic acids in, 60, 84, 85
  myxo, 201, 208-210
  paramyxo, 208-210
  sialic acid as receptor for, 208-210
  sialoglycolipids in, 60
  sialoglycoproteins in, 60
von Willebrand disease, 222

Walker tumor cells, sialidase effect on, 312
Wheat germ, *Triticum vulgaris,* lectin, 207

X-Irradiation, effect on gangliosides of cultured cells, 262
β-Xylosidase, glycoprotein degradation by, 166

Yucaiba virus sialidase, pH optimum, 319